Hammer & Silicon: The Soviet Diaspora in the US Innovation Economy

Immigration, Innovation, Institutions, Imprinting, and Identity

This deeply personal book tells the untold story of the significant contributions of technical professionals from the former Soviet Union to the US innovation economy, particularly in the sectors of software, social media, biotechnology, and medicine. Drawing upon in-depth interviews, it channels the voices and stories of more than 150 professionals who emigrated from 11 of the 15 former Soviet republics between the 1970s and 2015, and who currently work in the innovation hubs of Silicon Valley and Boston-Cambridge. Using the social science theories of institutions, imprinting, and identity, the authors analyze the political, social, economic, and educational forces that have characterized Soviet immigration over the past 40 years, showing how the particularities of the Soviet context may have benefited or challenged interviewees' work and social lives. The resulting mosaic of perspectives provides valuable insight into the impact of immigration on US economic development, specifically in high technology and innovation.

Sheila M. Puffer is University Distinguished Professor and Professor of International Business and Strategy at the D'Amore-McKim School of Business, Northeastern University, Boston, USA. She served as Program Director of the Gorbachev Foundation of North America, and is an Associate at the Davis Center for Russian and Eurasian Studies at Harvard University. Business and management in the former USSR are a major focus of her 160 publications, including *Behind the Factory Walls: Decision Making in Soviet and US Enterprises*.

Daniel J. McCarthy is University Distinguished Professor and the Alan S. McKim and Richard A. D'Amore Distinguished Professor of Global Management and Innovation at the D'Amore-McKim School of Business, Northeastern University, Boston, USA. He is also an Associate at the Davis

Center for Russian and Eurasian Studies at Harvard University. He has over 110 publications, including four editions of *Business Policy and Strategy*, as well as *Business and Management in Russia*, *The Russian Capitalist Experiment*, and *Corporate Governance in Russia*.

Daniel M. Satinsky is an attorney, business consultant, and independent scholar, and an Associate at the Davis Center for Russian and Eurasian Studies at Harvard University. He served as Board President of the US-Russia Chamber of Commerce of New England, Inc., from 2001 to 2016. He is editor of the *Buyer's Guide to the Russian IT Outsourcing Industry* and author of *Industrial Giants, Entrepreneurs, and Regional Government: The Changing Business Environment in Yaroslavl' Oblast, 1990–1999*, amongst other publications.

Hammer & Silicon: The Soviet Diaspora in the US Innovation Economy

Immigration, Innovation, Institutions, Imprinting, and Identity

SHEILA M. PUFFER
D'Amore-McKim School of Business, Northeastern University

DANIEL J. MCCARTHY
D'Amore-McKim School of Business, Northeastern University

DANIEL M. SATINSKY
Business Consultant and Independent Scholar

CAMBRIDGE
UNIVERSITY PRESS

CAMBRIDGE
UNIVERSITY PRESS

University Printing House, Cambridge CB2 8BS, United Kingdom

One Liberty Plaza, 20th Floor, New York, NY 10006, USA

477 Williamstown Road, Port Melbourne, VIC 3207, Australia

314–321, 3rd Floor, Plot 3, Splendor Forum, Jasola District Centre,
New Delhi – 110025, India

79 Anson Road, #06–04/06, Singapore 079906

Cambridge University Press is part of the University of Cambridge.

It furthers the University's mission by disseminating knowledge in the pursuit of
education, learning, and research at the highest international levels of excellence.

www.cambridge.org
Information on this title: www.cambridge.org/9781107190856
DOI: 10.1017/9781108120302

First published 2018

Printed in the United States of America by Sheridan Books, Inc.

A catalogue record for this publication is available from the British Library.

Library of Congress Cataloging-in-Publication Data
Names: Puffer, Sheila M., author. | McCarthy, Daniel J., author. |
Satinsky, Daniel M., author.
Title: Hammer and silicon : the Soviet diaspora in the US innovation,
economy : immigration, innovation, institutions, imprinting, and
identity / Sheila M. Puffer, Northeastern University, Boston,
Daniel J. McCarthy, Northeastern University, Boston, Daniel M. Satinsky,
Foresight Science & Technology, Inc.
Description: New York : Cambridge University Press, 2018.
Identifiers: LCCN 2017055513| ISBN 9781107190856 (hardback) |
ISBN 9781316641262 (paperback)
Subjects: LCSH: Technological innovations – United States. | Former
Soviet republics – Emigration and immigration. | United States – Emigration
and immigration. | BISAC: BUSINESS & ECONOMICS / Entrepreneurship.
Classification: LCC HC110.T4 P84 2018 | DDC 338/.0640973–dc23
LC record available at https://lccn.loc.gov/2017055513

ISBN 978-1-107-19085-6 Hardback
ISBN 978-1-316-64126-2 Paperback

Contents

Foreword

Hammer & Silicon is a model of social science research, but its subject matter would also make for a great novel. The book juxtaposes two places that are literally "worlds apart:" the Soviet communist regime during its decline, collapse, and subsequent disintegration (the Hammer) and the dynamic regions of entrepreneurship and innovation that emerged at roughly the same time in the US (the Silicon). The protagonists of this unlikely collision – and the focus of this fascinating book – are the highly educated scientists and engineers who left the Soviet Union and settled in Silicon Valley and the Boston-Cambridge areas in the late twentieth and early twenty-first centuries.

The book's authors, all established scholars of Russian studies, interviewed 157 members of the Soviet diaspora. The interview results provide a rich tapestry of individual trajectories that differ due to ethnic, cultural, and family circumstances, but nevertheless accumulate to illuminate strong cross-cutting themes at the core of the book. We learn that the earliest Soviet immigrants to the US beginning in the 1970s were refugees escaping virulent anti-Semitism, or, in later years, the economic dislocations following the collapse of the Soviet Union. Only more recently have Russian-speaking immigrants come to the US seeking additional education and/or economic opportunity. This latter wave has more in common with the Asian immigrants who typically come to the US for higher education, and then stay on to work in fast-growing technology regions.

Some of the most engaging parts of the book are the first-hand accounts, mostly in the words of immigrants themselves, of the experience of being raised in the former Soviet Union (with its authoritarian and bureaucratic institutions, pervasive dissembling and cynicism, distrust of business, and highly personalized trust) and adapting to the US and to technology centers (where entrepreneurship is a social good and work is organized around teams, collaboration,

open exchange, customer service, and generalized trust). *Hammer &*
Silicon details the challenges these immigrants face adapting to a new
language and unfamiliar institutions while also redefining their own
identities. The book's theoretical contributions lie in a systematic
analysis of the role of institutions, imprinting, and identity formation
in the immigration process. This is the most sophisticated work I've
seen on the experience of highly educated immigrants making the
transition between such different worlds.

The authors argue convincingly that while the Russian-speaking
community in the US is smaller than the more visible Chinese and
Indian diasporas, their impact has been disproportionate because
they represent the "best and brightest" mathematicians, physicists,
chemists, biologists, and engineers from the former Soviet Union.
Their evidence makes it clear that Russian-speaking immigrants, like
their Asian counterparts, have been a source of considerable talent,
creativity, and entrepreneurial capability for the US economy. They
have started profitable businesses and they work in leading American
universities, medical centers, and multinational companies.

In my work I refer to the highly educated immigrants to Silicon
Valley as the "New Argonauts"–an allusion to the ancient Greek
myth of Jason and the Argonauts, who sailed in search of the Golden
Fleece, testing their mythic heroism while seeking earthly riches and
glory. These journeys, like those of their Russian-speaking
counterparts, were only possible because of the opening of national
borders to increased global migration in the post World War II era. The
strong lesson of this book is that closing US borders to highly educated
immigrants is short-sighted and likely self-defeating. Highly skilled
immigrants are essential contributors to the entrepreneurial and
technological dynamism that distinguish the US economy today.

Another important policy lesson from this book is for places aspiring
to participate in global technology networks. The new Argonauts–
whether from Russia, China, India, Israel, or Taiwan– succeed in
large part because of investments by their home countries in world-
class science and engineering institutions. Elite higher education may
be available only to a small segment of the youth in these countries, but
those who have access to it, and are willing to take the risks of
migration, can ultimately benefit not only themselves and the US, but
also their home countries. Building a high-quality educational

infrastructure takes decades, but the alternative for any country is to fall further behind in the global economy.

Hammer & Silicon doesn't dwell on this, but the book provides ample evidence that the Russian-speaking Argonauts have become part of an international technical community that circulates among dynamic regions in the US, Asia, and Europe. Soviet diaspora members have seeded technology activity in Ukraine, Russia, Lithuania, Romania, Armenia, Estonia, and others, providing employment, technical know-how, advice, funding, and other opportunities for their home-country counterparts. In short, the mobility of highly educated workers–which depends on keeping national borders open–provides benefits to regions around the world. Even places that seem as unlikely as the former Soviet Union.

AnnaLee Saxenian
Dean and Professor, School of Information
Professor, Department of City and Regional Planning
University of California, Berkeley

Acknowledgments

We begin our acknowledgments by recognizing that we owe a huge debt to the eminent scholars whose work has informed our own over the past several decades. Among them are the legendary Sovietologists Joseph Berliner, Abram Bergson, and Marshall Goldman, who were regular attendees at the Economics Luncheon Seminar organized at Harvard University's Davis Center for Russian and Eurasian Studies where we three authors continue to be Center Associates. We have learned a great deal from them as well as from the dozens of scholars who presented their work there over the past three decades. We also owe a tremendous debt to Loren Graham of MIT and Irina Dezhina of Skolkovo Institute of Science and Technology for their work on Soviet science and the 1990s brain drain from that country.

As with any successful book project, beyond the authors and, in this case, even beyond the interviewees, there were numerous contributors who in various ways played key roles in the successful completion of this book. We acknowledge here the many individuals, groups, and institutions that played important roles for which we authors are extremely grateful while recognizing that the responsibility for the book's content remains with us.

We first thank the team of transcribers of the 157 interviews, the major transcribers being Northeastern University research assistants Ryan Donohue and Jacklyn Gronau, as well as professional transcriber Daina Krumins. Other Northeastern students who provided transcription services were Veronique Falkovich, Lily Gacicia, Ruth Leifer, Alina Samarova, and Rachael Volpert. Ryan Donohue also provided the majority of research assistance involved with analyzing transcripts, while other Northeastern students, Jacklyn Gronau and Rohit Kogta, provided additional assistance. D'Amore-McKim School of Business staff members who provided administrative support were Jenny (Evgeniia) Bagnyuk, Magda Drici, Michael Marafitte, Grace

Oliveira, and Oxana Tkachenko. Other sources of support important to completing this book are D'Amore-McKim colleagues former Dean Hugh Courtney, Senior Associate Dean Emery Trahan, and Group Coordinators Christopher Robertson of the International Business and Strategy Group and Marc Meyer of the Entrepreneurship and Innovation Group. Colleagues at Stanford University Graduate School of Business were also instrumental in sponsoring Sheila Puffer's sabbatical there from January to September 2015, including Senior Associate Dean for Academic Affairs, Madhav Rajan, and especially Professor Charles O'Reilly. We also acknowledge University of California, Berkeley, Dean AnnaLee Saxenian for her encouragement to us in adding to the literature on immigrant technical professionals in the United States.

In addition to the many interviewees who provided referrals to other interviewees, we would like to especially thank Anna Dvornikova, Maria Eliseeva, and Evgeny Zaytsev, who provided referrals to numerous interviewees, as well as Kate Carleton, Walter Chick, Ivan Correia, Douglas Fraser, Ingrid Larsson, Peter Larsson, Bob Nelson, Lindsey Sudbury, Martina Werner, and Maury Wood. We also thank others who provided background insights based on their association with technical professionals both in the United States and in the former USSR. They are Dmitry Dakhnovsky, Ekaterina Evstrateyva, Tatiana Fedorova, Richard Golob, Alexander Ivanov, Julia Ivy, Anna Lamin, Peter Loukianoff, Tatiana Lysenko, Mykola Lysetskiy, Katia Epshteyn Ostrovsky, Olga Rodstein, Maxim Russkikh, Paul Santinelli, Amir Sharif, Joel Schwartz, and Vera Shokina. We are also indebted to those who provided ideas and insights for the book title and cover design: Ralph Dinneen, Annika Fraser, Douglas Fraser, Liane Middleton, Marlene Puffer, and Maury Wood, and we gratefully acknowledge Carol Fraser and Dorian Scheidt for creating the map of interviewees' birthplaces.

We would also like to acknowledge that our work was facilitated by attending and networking at events and conferences sponsored by various organizations and associations, including the American Business Association of Russian-Speaking Professionals (AmBAR), the Global Technology Symposium, the Davis Center for Russian and Eurasian Studies at Harvard University, New England Russian-Speaking Entrepreneurs (NERSE), Silicon Valley Open Doors, the US–Russia Business Council, and the US–Russia Chamber of Commerce of New England.

We extend our deepest and most heartfelt thanks to the 157 people who devoted their valuable time to be interviewed, usually for one to two hours and sometimes longer. We owe a great debt to them for their willingness to trust us, to answer our specific questions, and to share their stories candidly and willingly. Additionally, we thank them for having done so in English rather than in their native languages. We hope the interview experience was valuable for our interviewees, perhaps giving them an opportunity to reflect on their lives and gain insights and perspectives about themselves. As interviewee Alexei Masterov said: "I feel like I'm learning something about myself in the process of this conversation because I never spoke about it this way, especially in English. So it's interesting." We also hope that the interviewees gain insights not only from reflecting on their own experiences, but also from the shared experiences of all 157 interviewees included in this book, and that they can appreciate the commonalities and differences among their compatriots who shared having been born in the former Soviet Union before becoming contributors to the US innovation economy.

We would, of course, be remiss without thanking the Cambridge University Press team that shepherded our manuscript to successful completion, including Valerie Appleby, Commissioning Editor for Business and Management, Assistant Commissioning Editor Stephen Acerra, Editorial Assistants Kristina Deusch and Toby Ginsberg, Marketing Executive Ellena Moriarty, Project Manager Sunantha Ramamoorthy, and Content Manager Bronte Rawlings, as well as others who reviewed and approved our book proposal and worked on various aspects of the production and marketing phases to ensure the quality of our product and its dissemination to institutions and individuals.

A major project of this type and proportions obviously requires substantial financial resources to support the numerous activities and individuals in the many phases of its development and completion. The Alan S. McKim and Richard A. D'Amore Distinguished Professorship of Global Management and Innovation provided major funding for the book. Other sources of funds from Northeastern University's D'Amore-McKim School of Business included the International Business and Strategy Group research fund, the Dean's Faculty Travel Fund, and the Center for Emerging Markets. We acknowledge and thank them all for their generous help in accomplishing various aspects of this book.

About the Authors

Sheila M. Puffer is University Distinguished Professor and Professor of International Business and Strategy at the D'Amore-McKim School of Business, Northeastern University, Boston. She is also an Associate at the Davis Center for Russian and Eurasian Studies at Harvard University. In 2015, she was a visiting research professor at the Graduate School of Business at Stanford University where she interviewed Silicon Valley entrepreneurs and other professionals from the former Soviet Union about their contributions to the US innovation economy. Dr. Puffer has more than 160 publications, including eighty refereed articles and eleven books. She has been recognized as the leading scholar internationally in business and management in Russia, the former Soviet Union, and Eastern Europe according to a 2005 *Journal of International Business Studies* article. She also ranks as the most published author (tied with coauthor D. McCarthy) in the *Journal of World Business* from 1993 to 2003. She has been ranked in the top 5 percent of authors worldwide who published in the leading international business journals from 1996 to 2005, according to a Michigan State University study. She is fluent in French and Russian. She earned a diploma from the executive management program at the Plekhanov Institute of the National Economy in Moscow, and she holds a BA (Slavic Studies) and an MBA from the University of Ottawa, Canada, and a PhD in business administration from the University of California, Berkeley.

Daniel J. McCarthy is University Distinguished Professor and the Alan S. McKim and Richard A. D'Amore Distinguished Professor of Global Management and Innovation at the D'Amore-McKim School of Business, Northeastern University, Boston, and is also an Associate at the Davis Center for Russian and Eurasian Studies at Harvard University. He is cofounder, codirector, and chair of the strategy advisory council of Northeastern's Center for Entrepreneurship Education. Additionally, he is cofounder of the Northeastern University Venture Mentoring Network and a member of the steering committee, as well as a board member for IDEA, the Northeastern University Venture Accelerator. Dr. McCarthy has more than 110 publications, including four editions of *Business Policy and Strategy*, as well as *Business and Management in Russia*, *The Russian Capitalist Experiment*, and *Corporate Governance in Russia*. He served as the lead director of Clean Harbors, Inc., a multibillion dollar NYSE-listed company, and has consulted in North America and Europe for more than forty companies. Early in his career, he was cofounder and president of a public company, Computer Environments Corporation, and served as a director on its board and also on the board of its sister public company, Time Share Corporation, as well as on a number of private company and nonprofit boards. Dr. McCarthy ranks as the most published author (tied with coauthor S. Puffer) in the *Journal of World Business* from 1993 to 2003, and he has been ranked in the top 5 percent of all authors worldwide who published in the leading international business journals from 1996 to 2005, according to a Michigan State University study. He is also one of the top three scholars internationally in business and management in Russia and Central and Eastern Europe, based on a *Journal of International Business Studies* article analyzing publications in thirteen leading journals from 1986 to 2003. Professor McCarthy holds AB and MBA degrees from Dartmouth College and the Tuck School of Business, and a DBA from Harvard University.

Daniel M. Satinsky is a business consultant and independent scholar. For more than twenty years, he has provided market entry and commercialization services to Russian and US technology companies. In this capacity, he has traveled extensively throughout Russia and the former Soviet Union. He has also written and spoken on topics related to business, innovation, and technology. Selected publications include *Industrial Giants, Entrepreneurs and Regional Government – The Changing Business Environment in the Yaroslavl' Oblast 1991–98*; he is coauthor of a New York Academy of Sciences study of worldwide innovation best practices and their application to Russia, *Yaroslavl Roadmap 10–15-20*, and editor of *Buyer's Guide to the Russian IT Outsourcing Industry*. He served as President of the Board of the US–Russia Chamber of

Commerce of New England for more than fifteen years, and is an Associate at the Davis Center for Russian and Eurasian Studies at Harvard University. He holds a Master of Law and Diplomacy degree from the Fletcher School of Law & Diplomacy at Tufts University, a JD from Northeastern University Law School, and a BA from James Madison College of Michigan State University.

Map 1 Birthplaces of Interviewees in the Former USSR
Map developed by Carol Fraser and Dorian Scheidt using Tableau Maps,
© 2018. www.tableau.com. Used with permission. Birthplaces added by the
authors.

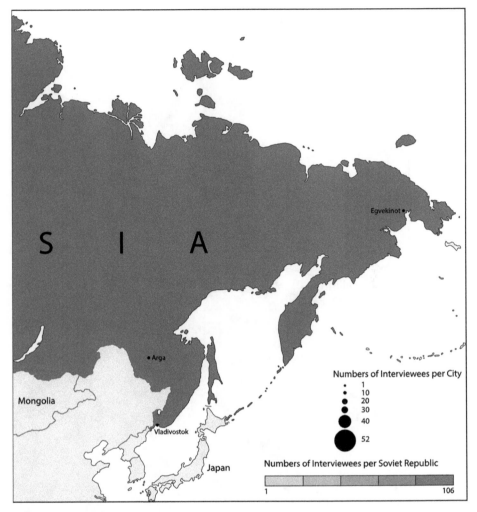

Map 1 (cont.)

Map 2 Birthplaces of Interviewees: Detail of the Western Region of the Former USSR

Map developed by Carol Fraser and Dorian Scheidt using Tableau Maps, © 2018. www.tableau.com. Used with permission. Birthplaces added by the authors.

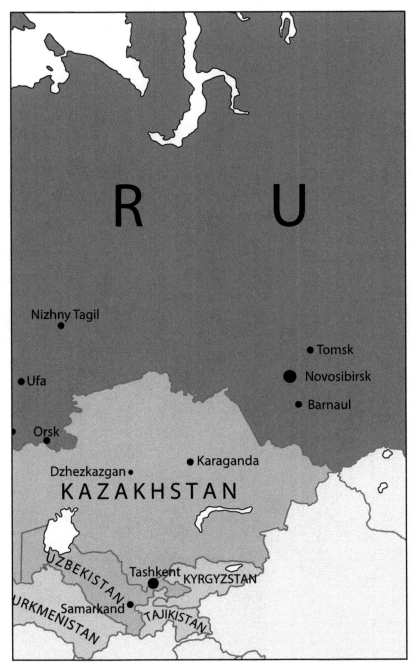

Map 2 (cont.)

Introduction

"Immigrants. We get the job done."[1]

This quote from the hit Broadway show, *Hamilton: The Musical*, epitomizes one of the core American portrayals of immigrants as hardworking people who come to the United States seeking economic, social, and personal opportunities and who are willing to work hard to achieve their goals. In the musical, the characters depicting Alexander Hamilton and Marquis de Lafayette were referring to immigrants at the time of the American Revolution nearly 250 years ago. Today, immigrants are currently the topic of intense controversy, with competing political narratives regarding their role in the US economy. One often neglected element in this controversy is the tremendous contribution to the US technology sector that has come from immigrants, particularly those who came after the Immigration Reform Act of 1965. These contributions have been well recognized in industry, as evidenced by the recent establishment of two organizations founded by prominent leaders of US technology companies advocating for immigration policies that would continue attracting and retaining scientific and technical professionals from abroad. One group, FWD.us, was founded by high-profile technology executives including Mark Zuckerberg of Facebook, Bill Gates of Microsoft, and Marissa Mayer of Yahoo. Another is Partnership for a New American Economy whose founders include former Microsoft CEO Steve Ballmer and former New York City mayor Michael Bloomberg who heads the huge diversified business communications company, Bloomberg L.P.

This book, *Hammer & Silicon: The Soviet Diaspora in the US Innovation Economy*, presents the story of one specific group of immigrants who were part of the global migration of talent attracted to the United States in the later decades of the twentieth century and early part

---

[1] Lin-Manuel Miranda, *Yorktown (The World Turned Upside Down)*. From *Hamilton: The Musical* (New York: Grand Central Publishing, 2015).

of the twenty-first century after US immigration reform. This talent pool helped fuel US economic growth and world leadership in high technology. Presented in this book are immigrants from the former Soviet Union who tell their stories of their generally unrecognized role in this era of US technology leadership.

About the Title

The title of this book incorporates two of the most powerful symbols of the twentieth century, one associated with the Soviet Union and the other with the United States. The hammer and sickle is well-known as the symbol of the USSR, and the silicon chip is widely used as a symbol of the high-technology industry in the United States. The full title of the book points to the emigration and brain drain before, and particularly during, the collapse of the Soviet Union and its impact on US science and technology. This process is illustrated through interviews with a remarkable and talented group of individuals who came to the United States and who ended up becoming significant contributors to the US technology sector.

The Soviet hammer of industry and sickle of agriculture were the symbolic representation of the forced transformation that turned the Russian Empire of the nineteenth-century czars into a world power. That symbol was intended to represent the worker–peasant alliance of social forces in an idealized depiction of the Soviet Union. As such, that symbol represented the command or centrally planned economy of the Soviet state and was ubiquitous in the country's flag, documents, uniforms, and government buildings. In the post–World War II world, the Soviet Union began to emerge as a technological rival to the United States, particularly with the launch of the Sputnik space satellite in 1957. It was a rival with enormous human capital in science, engineering, and mathematics, one that was institutionalized in a network of state-financed universities, research institutes, laboratories, and specialized enterprises.

A major US response to this technological threat was the silicon chip, which became the building block of a transformational technological era. In the 1950s, the Santa Clara Valley on the San Francisco peninsula was the birthplace of the silicon chip, or silicon-based integrated circuit, and was the home of the world's first silicon chip producer, Fairchild Semiconductor. The Soviet satellite launch was the impetus for creating the National Aeronautics and Space Administration (NASA) in 1958 and also for increasing government spending to accelerate technology that depended on semiconductors. As a result, a large number of silicon chip or semiconductor companies sprang up in the San Francisco Bay Area, giving rise to the widely recognized designation "Silicon Valley" to much of the San Francisco peninsula. The technology developed there paved the way for the rapid development of the microprocessors that undergird the computer hardware, software, communications, and social media sectors, and also for a much broader blossoming of US innovation in biotech, medical instruments, robotics, and artificial intelligence. This technology explosion enabled the United States to leapfrog over the Soviet Union and, in many respects, contributed to the USSR's economic demise.

Genesis of the Book

The authors have devoted the past thirty years to research, publication, presentations, and consulting on a wide variety of business and management issues in the Soviet Union and its successor countries. In doing so, it has been obvious to us that a significant number of technical professionals from that part of the world are now working in the US innovation economy, but this awareness is absent from most business and technology research, as well as in the popular press. Over the past decade, both immigration and innovation in general have become increasingly high-profile topics in the media, society, and the corporate world, as well as in academia. Recent studies by the Kauffman Foundation show conclusively that immigrants are twice as likely to found new businesses than are native-born Americans.[2] While this is a general measure of entrepreneurship, further work indicates the critical role of immigrants in the innovation process.

   One of the seminal works on immigration and innovation was *The New Argonauts: Regional Advantage in a Global Economy* by AnnaLee Saxenian, published in 2006. That book focused on the critical role of immigrants from Taiwan, China, India, and Israel in Silicon Valley and their subsequent contributions back to their home

[2] "Kauffman Compilation: Research on Immigration and Entrepreneurship," Ewing Marion Kauffman Foundation (October 2016), 3. http://www.kauffman .org/what-we-do/resources/kauffman-compilation-research-on-immigration-and -entrepreneurship.

countries. As explained in this book, the movement of these immigrants to Silicon Valley and the United States generally was not possible until after the US Immigration Reform Act of 1965. The Act's legal framework allowed for immigrants of all nationalities to enter the United States, including the influx resulting from the decline and dissolution of the Soviet Union. Missing from Professor Saxenian's insightful analysis and that of others was any mention of the ex-Soviet immigrants that we knew to also be part of this immigration–technology nexus. We know of no research that has been published on the impact of the Soviet diaspora on the US innovation economy. This book is the story of this exodus and its impact on the US technology sector in Silicon Valley and the Boston-Cambridge area.

We tell this story from both personal and professional points of view. Two authors who do not have direct family ties to Russia or the Soviet Union conceived this book. Professors Sheila Puffer and Daniel McCarthy have no family genealogy that can be traced to that part of the world. In writing the book, they were joined by Daniel Satinsky who has attenuated family ties to the Russian Empire through his grandparents on his father's side, who emigrated to the United States from Ukraine in 1911. All three authors' professional lives are entwined in the stories presented in this book.

Sheila Puffer, herself an immigrant, serendipitously began learning Russian as an undergraduate at Laurentian University in her native Canada on the recommendation of a professor to study that challenging language. That background led her to spend a year in the Soviet Union after completing her MBA, earning a diploma in management from the Plekhanov Institute of the National Economy in Moscow. Nearly a decade later, it was her good fortune to have worked with Professor Paul Lawrence of Harvard Business School in the late 1980s as an author of the book, *Behind the Factory Walls: Decision Making in Soviet and US Enterprises,*[3] an opportunity that launched her research focus on the Soviet Union.

Daniel McCarthy became interested in business and management in the Soviet Union in the late 1980s, when Mikhail Gorbachev's *perestroika* and *glasnost* policies opened that country for research

---

[3] Paul R. Lawrence, Charalambos A. Vlachoutsicos, Igor Faminsky, Eugene Brakov, Sheila Puffer, Alexander Naumov, Elise Walton, and Vitaly Ozira, *Behind the Factory Walls: Decision Making in Soviet and US Enterprises* (Boston: Harvard Business School Press, 1990).

opportunities, and Daniel and Sheila began doing research together. He later formed a connection at his Harvard alma mater to faculty members who had been involved in the famous Harvard Soviet Interview Project of the 1950s, including Abram Bergson and Joseph Berliner.[4]

Daniel Satinsky made his first trip to the Soviet Union on a study and travel trip for lawyers in 1984 and began serious involvement with joint venture businesses in the region after completing a mid-career master's degree at Tufts University's Fletcher School of Law and Diplomacy in 1991. In the two decades that followed, he traveled to the former Soviet Union more than 100 times, visiting many different locations in that enormous and varied country.

Together, the three authors have accumulated decades of experience engaged in study, publishing, business, and general interaction with people of the former Soviet Union. In doing so, they personally witnessed the dramatic historic events that took place there at the end of the twentieth and beginning of the twenty-first centuries and that span the major time period in this book.

Talent Flow from the USSR

In 2014, the authors conceived of a book to focus on the topic of the human impact on the US technology sector of the decline and collapse of the Soviet Union. The distinctive foundation of this book is the 157 in-depth interviews we conducted in the leading innovation hubs of Silicon Valley and the Boston-Cambridge area from January 2015 through March 2016. We developed a semistructured interview protocol of topics for interviews of an hour or more with each person, with most interviews running about an hour and a half and some as long as three hours. With a goal of 150 interviews, the authors interviewed seventy-nine entrepreneurs and other technical professionals in Silicon Valley and conducted an additional seventy-eight interviews with a comparable group in the Boston-Cambridge area. The 157 interviewees came to the United States over roughly the past forty years from eleven of the fifteen republics of the former Soviet Union. Of those, forty-three, or 27 percent, were women. These individuals arrived in the United States in what we have designated as three waves from the 1970s through 2015.

---

[4] Harvard Project on the Soviet Social System. http://hcl.harvard.edu/collections/hpsss/about.html

In analyzing the information collected through the interviews, we have utilized established social science conceptual tools. Specifically, we apply institutional theory, imprinting theory, and identity theory to explore the complexities underlying the significant role played by the Soviet diaspora in the technology sectors of the US economy. The book probes how Soviet institutions and institutional voids, as well as imprinting from Soviet times and its aftermath, were instrumental in shaping the identities of our interviewees. We have also placed this process in the historical context of the times and the consequent impact on who emigrated, why they emigrated, and how they ended up in the United States. Based on this analysis, the book explores the rich interview narratives regarding the immigration experience and work-related experiences such as entrepreneurial activities, mentoring, and teamwork, as well as how interviewees overcame obstacles and sought opportunities to utilize their technological and scientific expertise. Some interviewees were sponsored by US employers while others held refugee, student, investor, or exceptional talent visas. We have utilized the term "immigrant" and the related terms "immigration" and "emigration" rather loosely since not all interviewees held the status of immigrant, with some not choosing to do so and others unable to do so at the time of the interview. Like French General Lafayette, mentioned at the outset of this Introduction, these individuals did not have US citizenship, yet they contributed significantly to the country. Others were like Alexander Hamilton, the first secretary of the US treasury, who emigrated from the British West Indies and became a citizen.

In presenting the various chapters, we have excerpted sections of the interviews that highlight interesting and important aspects of individuals' experiences in the former Soviet Union and in the United States, as well as during their immigration experience. Many interviewees have quotes in multiple chapters. While we would have liked to feature each person on every topic, we had to make difficult decisions about which portions of the interviews to include. We also emphasize that the interviewees constitute a convenience sample rather than a statistically based sample of the population.

Commonality of Interviewees

The individuals included in this book represent many different nationalities, religions, and cultural backgrounds, as well as different histories and circumstances, reflecting the vast expanse of the Soviet Union. In addition, there has been growing differentiation of the independent countries' paths of development after the dissolution of the Soviet Union. This is particularly acute in the differences between Ukraine and the Russian Federation even before the 2013 Maidan protests and change of government in Ukraine. Amid this complexity, we have also observed strong elements of commonality that come from the interviewees' shared Soviet history, culture, and educational system, as well as the Russian language.

We acknowledge a bias in our presentation of an emphasis on the Russian Federation that reflects the numbers and the power concentrated there as the center of the former Soviet government and its continuing dominating influence as the largest, by geography and population, of the independent countries formed after the dissolution of the USSR in 1991. However, our interviewees come from throughout the former Soviet Union, and we have included overviews of relevant developments in other countries, primarily in Ukraine, Belarus, Georgia, Armenia, and Kazakhstan as the main centers of post-Soviet emigration. The Baltic countries of Latvia, Lithuania, and Estonia have a separate historical trajectory and less connection to Soviet culture and institutions and thus are not covered with the same level of detail.

For those who came to the United States in this time period, the influence of the Soviet past remains very strong, although producing differing effects and reactions. In order to describe the commonalities of this group while respecting the individual differences of nationality, culture, and individual personality, we have focused on the common elements that were forged through the overlay of the Soviet system. This overlay consists of common institutional experiences of the highly centralized and standardized Soviet Union, a common educational system, and common cultural reference points that come from the centralized Soviet political and administrative systems. The other element that binds them together is the Russian language as the state language and lingua franca of the diverse nationalities and religions that were contained within the Soviet Union. Regardless of their attitudes toward the Soviet system from a political point of view, these individuals share a common background that influenced both their opportunities in the US technology sector and their adaptation to the country's innovative and entrepreneurial business culture.

Changing geographical and statistical methods during this period complicate the task of presenting an accurate numerical picture for all three waves. Prior to 2000, US immigration statistics counted all immigrants from the Soviet Union as having that country of origin despite the fact that the Soviet Union dissolved in 1989. Thus, it is very difficult to construct an accurate picture for the 1990s. After 2000, statistics have been recorded on the basis of the fifteen new countries that emerged. For our purposes, only the total numbers of immigrants for the three waves are presented in order to give an idea of magnitude, if not a precise statistic. We have used indicative numbers where possible to give this overall picture.

We use the terms "Soviet Union" and "USSR" interchangeably, as well as the terms the "former Soviet Union" and the "former USSR." We also use the terms "Russian Revolution" and "Bolshevik Revolution" interchangeably to refer to the 1917 revolution that overthrew the czars. In presenting names of cities and regions, we use the Russian spelling and the name of that city or region contemporaneous with the speaker. We recognize that the names used for cities can have strong political undertones. A number of cities in the former Russian Empire had their names changed by the Soviets to honor Bolshevik or revolutionary figures. For instance, St. Petersburg became Leningrad and Nizhny Novgorod became Gorky. Most such names were restored to their prerevolutionary status after the dissolution of the USSR. Our interviewees sometimes use the old name and sometimes the Soviet name. We have left the name as it was at the time they were referring to it, but use the contemporary name when speaking about the present. We are also aware of the sensitivity of spelling city and other place names in national languages other than Russian. We have chosen to use the transliteration of the Russian spelling since it is the most recognizable to Western readers. To give readers an appreciation for the wide variety of locations in which our interviewees were born, we have created a map[5] of the former Soviet Union indicating their birthplaces, and also include an enlarged view of the highly populated western region. Interviewees' birthplaces spanned the entire length and breadth of that vast country: from Murmansk, Russia, the world's largest city north of the Arctic circle; to Meghri, Armenia, in the south near the

[5] Map developed by Carol Fraser and Dorian Scheidt using Tableau Maps, © 2018. www.tableau.com. Used with permission. Birthplaces added by the authors.

border with Iran; and from Kaliningrad, the far western territory separated from the rest of Russia by Belarus; to the remote settlement of Egvekinot, the most easterly outpost of Russia located on the Bering Sea. And while recognizing that Siberia is part of Russia, we specify Siberia when appropriate in recognition of that vast geographical territory and its distinct culture within Russia. The 157 interviewees were born in eleven of the fifteen Soviet republics: Russia (106); Ukraine (23); Belarus (6); Armenia (5); Uzbekistan (4); Moldova and Georgia (3 each); Kazakhstan, Estonia, and Latvia (2 each); and Azerbaijan (1). No interviewees were born in Lithuania, Kyrgyzstan, Tajikistan, or Turkmenistan. As for cities, 44 percent of interviewees were born in Moscow (52) or St. Petersburg (18). The Ukrainian cities of Kharkov (7) and Kiev (5) were next, followed by Novosibirsk, Tashkent, and Yerevan with four each. The remaining cities had three or fewer interviewees born in that location.

We also note that we have highlighted in bold the names of interviewees when introducing their quotes for ease of locating them in the text. When they are referred to in contexts where they are not quoted, we have kept their names in regular typeface. In each chapter, the type of descriptive information included about the interviewees, such as age at the time of coming to the United States and current position title, varies according to its usefulness in putting their comments in context.

Technology Sector Impact

This book is not meant to be a systematic, statistical study of technology companies started by immigrants from the former Soviet Union nor of other companies to which they have made major contributions. However, we will illustrate their contributions anecdotally through interview excerpts. Immigrants from the former Soviet Union historically have had a significant impact on company formation in the United States, with a 2011 study by the Partnership for a New Economy reporting that they or their children founded twenty-eight Fortune 500 companies, including Google, Oracle, United Technologies, Occidental Petroleum, Qwest Communications, Omnicom Group, Avnet, Viacom, Home Depot, CBS, and Polo Ralph Lauren.[6] Other

---

[6] Steven A. Ballmer et al., Co-chairs of the Partnership for a New American Economy, *The "New American" Fortune 500*, Partnership for a New Economy

technical and scientific contributions of major importance to the innovation economy include those of Soviet immigrants Alexander Poniatoff, the 1940s founder of the audiotape manufacturing firm Ampex, and Yakov Rekhter, the 1980s codesigner of BGP, the core routing protocol of the Internet.

Our interviewees' contributions range from professional positions in iconic companies like Google, Facebook, Apple, and Microsoft; to founding industry leaders like IPG Photonics and PTC; to initiating numerous startups and conducting groundbreaking research for the products of the future. As a result of their strong educational backgrounds in mathematics and basic science, the majority of them are clustered in biotech, pharma, and medical products, as well as software, IT, communications, and the Internet. Also notable is the significant number who became serial entrepreneurs, angel investors, and venture capitalists engaged in fueling next-generation innovation.

Themes of the Book: The Five I's

The themes of the book are centered on five i's: immigration, innovation, institutions, imprinting, and identity. We discuss each one below.

Immigration

This book focuses on immigration from the former Soviet Union to the United States that took place beginning in the early 1970s through 2015, and, within that immigration, on those who contributed to the high-technology sector. To put this story in context, we review the overall patterns of emigration from the USSR during this period and the corresponding patterns of immigration to the United States. The research and interviews made it clear that within this period of time there were actually three separate waves that corresponded to different geopolitical conditions affecting the Soviet Union and the successor states that emerged after the country's dissolution.

The First Wave, from 1972 to 1986, was primarily Soviet Jews. During this period, the number of Soviet Jews allowed to emigrate fluctuated depending on the state of relations between the USSR and

(no location, June 2011). http://www.nyc.gov/html/om/pdf/2011/partnership_for_a_new_american_economy_fortune_500.pdf

the West. When the Soviet Union was seeking closer trade relations in the early to mid-1970s, more Jews were allowed to emigrate. At the time of the USSR's involvement in the conflict in Afghanistan and Western nations' boycott of the 1980 Moscow Olympics, the numbers were restricted again. When President Gorbachev began to decrease tension with the West in the mid-1980s, the number of Jews allowed to emigrate went up again. In that period, comparatively few immigrants came to the United States, and almost all of those who came were Jewish. Some joined the tech sector, as we shall see in the interview excerpts.

The Second Wave began in 1987, during the *perestroika* period, with a dramatic expansion of permissions to emigrate, initially mainly for Jews, ethnic Germans, and Armenians. When the Soviet state collapsed in December 1991, a torrent of people began to emigrate as a result of the economic and political turmoil and the establishment by the governments of the newly independent nations of the legal right to leave that had not existed under the Soviet Union. This is the major period of the brain drain that brought a broad range of the scientific intelligentsia to the United States, not only Jews.

The Third Wave began at the turn of the twenty-first century with the ascension to power of Vladimir Putin and an upturn in world oil prices that allowed Russia, especially, to strengthen its economy. In that period, institutions were stabilized, accompanied by an emphasis on integration of Russia into the world economy, including into the increasingly important knowledge economy. The result was a change in the motivation for coming to the United States, one moving away from necessity and more toward opportunity-seeking. Ukraine, as a successor state and one of the main sources of emigration, is not an energy exporter and therefore did not share in Russia's bounty of increased oil prices. Thus, a push to emigrate due to economic reasons continued from Ukraine as well as from other former republics.

This book examines these three waves of immigration in terms of the political and economic institutional contexts of each and the nuances that emerged from our interviews in terms of motivation, education, and culture in each wave. The book provides an in-depth look at the political, social, economic, and educational forces that coincided to allow and motivate these people to emigrate. It also describes the window of opportunity to do so and the ability of individuals to transform their Soviet-acquired education, training, and capabilities into contributions to the US innovation economy.

Innovation

We analyze the paradox of immigrants contributing to innovation in the United States while coming from a country that had made private business illegal and, consequently, that had great difficulty designing systems to commercialize the world-class scientific knowledge that these talented citizens possessed. The key to understanding this paradox lies in the nature of Soviet science and mathematics education that emphasized problem-solving without providing a practical outlet for these capabilities. It is an established fact that the Soviet Union produced some of the world's leading scientists, especially in the field of mathematics. This was reflected in massive engineering projects like the electrification of the country in the 1930s, construction of advanced weapons manufacturing facilities, initiation of the space program, and development of atomic energy industries. Many of the people interviewed initially entered math and science programs to avoid the politically charged social science disciplines. Once transplanted into the new societal and institutional circumstances of the United States, many were transformed into innovators as well as contributors to development and other phases of commercialization by actively utilizing their advanced education in basic research and its resulting application.

Institutions

Institutional theory provides an important part of the analytical framework by which to examine Soviet immigration and subsequent contribution to innovation in the United States. We examine the impact of Soviet institutions that were intended to develop rules, laws, informal influences, and cultural norms for a complete socialist and secular transformation of society after the 1917 Bolshevik Revolution. In many cases, formal Soviet institutional norms conflicted with deep familial and cultural traditions. We discuss how the inability of Soviet institutions to maintain their legitimacy during the later stages of the Soviet Union, particularly among the intelligentsia and even more particularly among Jews, was a critical influence promoting emigration in the First Wave. In the Second Wave, the collapse of Soviet educational and scientific institutions allowed for broader emigration and simultaneously pushed many dedicated scientific researchers out of the country due to lack of salary and support compounded by questions about their future. The stabilization of new or reformed institutions, beginning around 2000, changed the equation again, leading to an era of choice of domestic or foreign locations for those who wanted to pursue innovation and business.

We also look at how new institutional norms influenced the behavior of these immigrants once they entered the United States. In particular, we examine how immersion in the institutional innovation environment provided a common experience with other technical professionals, including entrepreneurs and researchers in the hotbeds of technology innovation in Silicon Valley and Boston-Cambridge. These institutions introduced a new set of values, rules, and cultural norms that became layered on top of Soviet institutional influences or that, in some cases, completely replaced those previous institutional influences.

Imprinting

A second theoretical foundation of the book is imprinting, a well-known social science theory that has been applied to many disciplines including migration studies, multilingualism and multiculturalism, and work adjustment in business organizations. Imprinting refers to the process that takes place during a formative period of life in which individuals learn and incorporate important features of their environment into their conception of the world that tend to persist even through subsequent changes in their environment. Usually this takes place during youth and early adulthood, but imprinting can also take place during a period of great change in an individual's life, like immigration to a new country. Imprinting is a means of examining the impact of societal institutions on individual personality and value development.

We use imprinting as a tool to analyze how the interviewees' life experiences and their reactions to Soviet institutional influences facilitated or hindered their ability to contribute to the US innovation economy. While exploring the impact of the communist political ideology and of the centrally planned economy, we also looked at the extent of absorption or rejection of those values among our interviewees. We also explore the extent to which these professionals have accepted organizational imprinting from the US firms where they worked and from the US culture of innovation. Of particular interest are the different roles of scientists and researchers in the Soviet system and in the United States, including the critical link between science and commercial activity that differentiates the US system.

The book also examines factors that have influenced how interviewees have resolved potential conflicts between societal imprinting from the Soviet Union and the general societal imprinting in the United States, including organizational imprinting within companies. Ultimately, the way a person resolves the potential conflict of diverse imprinting influences shapes his or her current social identity.

Identity

Identity refers to how people view themselves or consider who they are. Research has found people to self-identify with multiple and overlapping identities depending on their past and current social circumstances. In the USSR, each person had an official identity both as a citizen of the Soviet Union and as a member of a specific nationality that was indicated in their internal passport and other official documents. This system arose out of the Bolsheviks' political conceptions about how to sort out and organize the multinational, multiethnic Russian Empire that they supplanted.

Formally, each person was designated as being a member of a certain nationality as part of a promised system of equal treatment of all and recognition of semi–self-government in traditional homeland geographic areas within the broader Soviet state system. While these concepts and their implementation created many issues arising from the gap between theory and practice, the most critical for the purposes of this book concerns the treatment of Jews. While this is a complex topic that cannot be explored completely within the bounds of this work, the important fact is that Jews were considered to be a nationality, the same as Armenians, Ukrainians, Russians, Tatars, Uzbeks, Kazakhs, and the like, within the Soviet system. Despite a formal commitment to equality for all nationalities, the Soviet state was unable or unwilling to root out anti-Semitic bias. This failure had a decisive impact on the First Wave of immigrants and an important impact on the Second Wave.

Beyond the issues of nationality and religion are critical identity issues having to do with professional, cultural, and social values that are also revealed in the interviews. As a general matter, the book explores the interaction of the identities formed in the former Soviet Union with the newly encountered values, expectations, and work routines of US society and its innovation economy. This inquiry looks at the extent to which interviewees cast aside, retained, or modified attitudes, values, and behaviors imprinted in Soviet or post-Soviet society. They revealed to us many different new identities formed through the immigration experience, including notions of nationality and religion, as well as scientific, business, and professional concepts.

A Brief Look at the Chapters

The book consists of three parts. Part I: Analytical Framework, which includes Chapters 1 through 4, sets the analytical framework for understanding interviewees' backgrounds, the immigration process, and their subsequent adaptation to the US technology sector. Part II: Immigrants' Experiences, Integration, and Contributions, consisting of Chapters 5 through 9, draws even more extensively from our interviews to provide an in-depth look at the contributions the interviewees have made and continue to make to the US technology sector. Part III: Conclusion, consisting of Chapter 10, presents a summary and our conclusions.

Chapter 1 presents the theoretical foundations for the book grounded in institutional theory, imprinting theory, and identity theory. These social science tools are essential for organizing and evaluating the interview material and showing its relevance to the US innovation economy.

Chapter 2 applies institutional theory to the Soviet political, economic, and social systems as the key overall institutional influences on interviewees. This chapter presents an overview of the fundamental components of the Soviet system and its institutions and explains how they were catalysts for emigration through the inequities and other issues of discontent experienced by many talented professionals and intellectuals. We examine how interviewees evolved from being objects acted upon by the Soviet system to becoming persons of action based on their own fundamental beliefs or values that were transmitted through family, traditions, opportunity-seeking, or pure youthful rebellion.

Chapter 3 presents an institutional analysis of the Soviet educational system as one of the principal institutional influences forming the capabilities and outlooks of our interviewees. This chapter provides an overview of the emphasis on mathematics, science, and engineering from elementary grades through graduate education that aimed to advance that country's military and industrial objectives. The expertise and skills learned in this system were critical to our interviewees' later capability for making contributions to the US technology sector.

Chapter 4 presents a historical synopsis of emigration from the Soviet Union after the early 1970s and the process of entry into the United States. The chapter presents a political and legal overview of both the exit from the Soviet Union and later from its successor countries to entry into the United States. It also introduces the differentiation of the waves of immigrants to the United States into three separate periods, each with its own significant characteristics. Wave One took place between 1972 and 1986 and was primarily a period of Jewish emigration. Wave Two, spanning 1987 through 1999, was a period of increased exit due to relaxed emigration controls, economic chaos, and the ensuing brain drain. Wave Three, from 2000 through 2015, was a period of economic stabilization, internationalization of technology, and emigration precipitated more by choice than necessity. The framework of these three waves is critical to the analysis presented in the book.

Chapter 5 discusses the activities of our immigrant innovators as startup entrepreneurs in Silicon Valley and Boston-Cambridge. It presents interview materials by immigration wave and by industry sector. The industry sectors are divided into three categories: (1) biotech, pharma, and medical products; (2) software, Internet, communications, and IT; and (3) other industry sectors. The chapter looks at aspects of high-tech startups that are common to entrepreneurs, as well as the particular strengths or weaknesses of our immigrant entrepreneurs. The experience of these immigrants illustrates the limits of Soviet institutional imprinting, with its antagonism toward private business, on this group of immigrants.

Chapter 6 presents another aspect of contribution to the US innovation economy, in this case, by academic and industry scientists, researchers, and managers. This chapter presents interview material from each of the three wave periods and from the three sector categories used in Chapter 5 to illustrate significant activities directly impacting technology innovation. Again, this is a critical area of the innovation economy to which our interviewees bring special talents and experience. Chapter 5 and Chapter 6 illustrate the profound institutional impact of the Soviet educational system on the ability of these immigrants to contribute in fundamental ways to US high technology, and the interview excerpts are testimony to the personal tenacity and fortitude of these talented individuals.

Chapter 7 examines the impact of new institutions on the process of imprinting and identity as the interviewees adapted to the US cultural environment. The process of adaptation involved facing challenges and securing sources of support in that process. The chapter reviews challenges in adapting to a new language and culture, as well as challenges with new living conditions in daily life. Interviewees describe their experiences and the support they received from mentors, role models, and networks.

Chapter 8 looks at issues in workplace adaptation. The chapter briefly presents experiences interviewees had in the workplace in the former USSR, as well as challenges they faced in skill areas necessary to succeed in the United States. These skills include areas common to anyone in the high-tech workplace but have specific resonance among people with different institutional contexts, imprinting, and identities. The chapter looks at teamwork, managerial and leadership styles, communication, and attitudes toward trust.

Chapter 9 continues the application of institutional theory and imprinting on the formation of new aspects of identity given the dramatically changed circumstances of our interviewees. This complex and disruptive process could well facilitate their adaptation and contribution to the US innovation economy. In describing their identities, many responded that "American" was their primary identity, while many others noted that they were "American-Russians," as did others from other former republics like Ukraine and Armenia. Some interviewees described their identities by referring to their homelands first, such as "Russian-American," as did some from other former republics in referring to their homelands. Another large group described their identities as being Russian, while others responded that they were Jewish. In contrast to ethnicity, some identified with their professions or interests, while a fairly large group emphasized the complexity of their ethnicities and experiences. Finally, a rather large group dealt with complexity by considering themselves to be global citizens, internationalists, or cosmopolitans.

Chapter 10 summarizes the five i's of our analytical framework. Each topic is addressed separately and synthesized toward the end of the discussion. We highlight similarities and differences in these five dimensions across the three waves of immigration. We then discuss the results of our interviews in the context of the roles these interviewees are playing in the US innovation economy as entrepreneurs, executives, managers, scientists, venture capitalists, and other high-tech professionals. We offer our conclusions about the characteristics of this unique set of immigrants and their relevance for US immigration policy and assimilation practice.

PART I

Analytical Framework

1 | *Theoretical Foundations*
Institutions, Imprinting, and Identity

The Union of Soviet Socialist Republics (USSR) was created as the result of the 1917 Bolshevik Revolution in Russia whose goal was to implement radically new institutions and to instill in the population a radically new worldview that corresponded to the broad-ranging Bolshevik social and political ideology. As part of this massive transformative experiment in human engineering, the Bolsheviks, led by Vladimir Lenin, pursued a forced modernization that transformed the predominantly peasant agricultural economy of the Russian Empire into the industrial and scientific world power, the USSR. A component of this process included changing a mostly illiterate population into one of the most literate in the world, creating a massive base of highly educated professionals in science and technology by the early 1960s. This transition took place at tremendous human cost, and, by the end of the 1980s, the Soviet Union was stagnating and ultimately collapsed. The consequences of this collapse are well understood on a political level but poorly understood in terms of the impact on science and innovation. Our goal in this book is to evaluate the impact of this historic process on the Soviet scientific and technical intelligentsia and, in turn, on the US innovation economy.

We develop a theoretical framework to analyze the two i's of immigration and innovation by drawing on three prominent social science theories: institutions, imprinting, and identity. All three have been applied to many disciplines including migration studies, multilingualism and multiculturalism, economic development, and work adjustment. We use these theories as an organizing framework for our in-depth interviews to provide insights into the contributions, attitudes, and behaviors of entrepreneurs and other technical professionals from the former USSR. The two i's of immigration and innovation, as well as theories of institutions, imprinting, and identity, comprise the five i's of the book's subtitle. We discuss each theory below.

Institutional Theory

Institutional theory examines the processes by which social structures, including rules, norms, and routines, become accepted as guides for people to engage in social behavior.[1] Institutions have been defined as "the humanly devised constraints that structure human interaction."[2] Formal, regulative institutions include branches of government, including their departments that create and enforce procedures and regulations, as well as the judicial system and the law enforcement apparatus. Other important formal institutions are educational systems, work organizations, and religions.[3] In contrast, informal, cultural-cognitive institutions stem from cultural traditions affecting beliefs, values, and behaviors over the course of centuries[4] and can be transmitted through family members, partners, and friends in various settings, including leisure and volunteer activities.[5]

The main institutions of the Soviet Union were the governmental and administrative apparatus that had shadow oversight of government, educational, and social organizations through the parallel units of the Communist Party and the KGB (Committee for State Security), the comprehensive network of communist youth organizations, and the politicized university and scientific research institutions. We also note the continuing influence of informal, cultural-cognitive institutions surviving from the prerevolutionary period that continued as traditions in the new environment.

While the main role of institutions is to create stability and predictability in a society, dissatisfaction or disillusionment with institutions can create instability and result in beliefs and practices being replaced by new ones. This process, known as *deinstitutionalization*, typically

---

[1] Sheila M. Puffer and Daniel J. McCarthy, "Institutional Theory," in *Wiley Encyclopedia of Management*, ed. Cary L. Cooper, 3rd edition, Volume 6 (New York: Wiley, 2015), 1–5.

[2] Douglas C. North, *Institutions, Institutional Change and Economic Performance* (Cambridge, UK: Cambridge University Press, 1990), 3.

[3] Paul J. DiMaggio and Walter W. Powell, "The Iron Cage Revisited: Institutional Isomorphism and Collective Rationality in Organizational Fields," *American Sociological Review* Volume 48, Number 2 (April 1983): 147–160.

[4] W. Richard Scott, *Institutions and Organizations: Ideas and Interests*, 3rd edition (Los Angeles: Sage, 2008).

[5] Patricia H. Thornton, William Ocasio, and Michael Lounsbury, *The Institutional Logics Perspective: A New Approach to Culture, Structure, and Process* (New York: Oxford, 2012).

occurs incrementally.[6] In the case of the Soviet Union, there was a growing disillusionment among the intelligentsia, particularly among the Jewish intelligentsia, beginning in the 1960s. During the period of stagnation in the late 1970s and early 1980s under Leonid Brezhnev and his successors, there was growing cynicism among the population, as reflected in the oft-repeated saying, "we pretend to work and they pretend to pay us." The growing deinstitutionalization accelerated in the *perestroika* period initiated by Mikhail Gorbachev in the mid-1980s, culminating in a monumental institutional collapse at the end of 1991.

The institutions supporting the communist regime and the centrally planned economy were abruptly dismantled, with no new institutions ready to take their place. The ensuing chaos resulted in *institutional voids*,[7] referring to the relative absence or illegitimacy of formal institutions such as governments, courts, and law enforcement agencies.[8] In conditions of weak formal institutions, organizations and individuals often rely heavily on informal, cultural-cognitive institutions such as using trusted personal networks, exchanging favors, and other cultural traditions.[9]

Institutional theory can thus be a framework for understanding the dramatically changing institutional landscape that individuals from the former USSR experienced in that country and its successor states. They experienced to varying degrees the institutions supporting communist

[6] Douglas C. North, *Institutions, Institutional Change and Economic Performance* (Cambridge, UK: Cambridge University Press, 1990).

[7] Tarun Khanna and Krishna G. Palepu, "Why Focused Strategies May Be Wrong for Emerging Markets," *Harvard Business Review* Volume 75, Number 4 (July–August, 1997): 41–54.

[8] Sheila M. Puffer, Daniel J. McCarthy, and Max Boisot, "Entrepreneurship in Russia and China: The Impact of Formal Institutional Voids," *Entrepreneurship: Theory and Practice* Volume 34, Number 3 (May 2010): 441–467.

[9] Ruth C. May, Sheila M. Puffer, and Daniel J. McCarthy, "Transferring Management Knowledge to Russia: A Culturally Based Approach," *Academy of Management Executive* Volume 19 (May 2005): 24–35; Daniel J. McCarthy and Sheila M. Puffer, "Interpreting the Ethicality of Corporate Governance Decisions in Russia: Utilizing Integrative Social Contracts Theory to Evaluate the Relevance of Agency Theory Norms," *Academy of Management Review* Volume 33 (2008): 11–31; Snejina Michailova and Kate Hutchings, "National Cultural Influences on Knowledge Sharing: A Comparison of China and Russia," *Journal of Management Studies* Volume 43, Number 3 (May 2006): 383–405; Mike W. Peng, "Institutional Transitions and Strategic Choices," *Academy of Management Review* Volume 28, Number 2 (April 2003): 275–296.

ideology and the apparatus of the centrally planned economy, followed by the chaos of the institutional voids created by the collapse of the USSR and the fragile and uncertain new formal – and questionably more democratic and market-oriented – institutions that emerged from 1992 to 2000. Many also experienced the various economic cycles in their homelands after 2000. Interviewees reveal the impact of these institutional upheavals through stories about their lives during these dramatically different eras, including everyday life, their educational experiences, and their careers.

After enduring the roller coaster of tumultuous change, these entrepreneurs and technical professionals then had to find ways of adapting to the new institutional landscape of the United States. In this instance, another construct of institutional theory, *institutional distance*, is useful. In contrast to institutional voids that occur within a country, institutional distance refers to the extent of similarity or dissimilarity between institutions in different countries.[10] When the institutional distance is high, building trust and positive relationships can be challenging.[11] Regarding informal, cultural-cognitive institutions, Americans tend to be relatively trusting of strangers and have a broad view of friendship, whereas people raised in the former USSR tend to be distrustful of those outside their close personal networks and reserve friendship for a small circle of people with whom they share deep, mutual dependence. Individuals trying to adapt to new work situations in new countries with high institutional distance from their country of origin may suffer from the liability of foreignness, which can add to the social and monetary costs of doing business.[12] For technical professionals from the former USSR, that liability of foreignness stemming from a limited understanding of US formal institutions and business practices might hinder cultural adaptation and result in some degree of social isolation and even in earning salaries below their actual capabilities.

---

[10] Dean Xu and Oded Shenkar, "Institutional Distance and the Multinational Enterprise," *Academy of Management Review* Volume 27, Number 4 (October 2002): 608–618.

[11] Tatiana Kostova, "Transnational Transfer of Strategic Organizational Practices: A Contextual Perspective," *Academy of Management Review* Volume 24, Number 2 (1999): 308–324.

[12] Lorraine Eden and Stewart R. Miller, "Distance Matters: Liability of Foreignness, Institutional Distance and Ownership Strategy," *Advances in International Management* Volume 16 (2004): 187–221.

In summary, institutional theory can help show how emigrants from the former USSR faced the challenges of learning to understand the new institutions they encountered in the United States and adapting their attitudes and behaviors to become effective contributors to the US innovation economy. The Soviet political, economic, and social institutions that served as catalysts for migration are discussed in Chapter 2, and the Soviet educational institutions that provided the capability for eventually contributing to the US innovation economy are discussed in Chapter 3.

Imprinting Theory

Institutions are a primary source of imprinting and identity creation among individuals through the process of institutional logics that encompass "beliefs, practices, values, assumptions, and rules that shape cognition and guide decision-making in a given social context."[13] *Imprinting* refers to "a process whereby during a brief period of susceptibility, a focal entity develops characteristics that reflect prominent features of the environment, and these characteristics continue to persist despite significant environmental changes in subsequent periods."[14] A *period of susceptibility* denotes a time during which an individual is particularly susceptible to external environmental influences. The individual "comes to reflect elements of the environment at that time," and the characteristics developed during the sensitive period persist "even in the face of subsequent environmental changes."[15] One major period of susceptibility is an individual's formative years that occur from ages six to twenty-five, a life stage that has been documented in research, including communist imprinting.[16] However, new conditions can lead

---

13 Patricia H. Thornton and William Ocasio, "Institutional Logics and the Historical Contingency of Power in Organizations: Executive Succession in the Higher Education Publishing Industry, 1958–1990," *American Journal of Sociology* Volume 105, Number 3 (1999): 801.

14 Christopher Marquis and Andras Tilcsik, "Imprinting: Toward a Multilevel Theory," *Academy of Management Annals* Volume 7, Number 1 (March 2013): 199.

15 Ibid.

16 Carrie M. Brown and Wells Ling, "Ethnic-Racial Socialization Has an Indirect Effect on Self-Esteem for Asian-American Emerging Adults," *Psychology* Volume 3, Number 1 (2012): 78–81; Grigore Pop-Eleches and Joshua Tucker, "Communist Socialization and Post-Communist Economic and Political Attitudes," *Electoral Studies* Volume 33 (March 2014): 77–89.

individuals to replace some or all earlier imprinting. For example, in the workplace, "during periods of role transition – including periods of organizational and professional socialization – individuals are particularly susceptible to influence ... because of the great uncertainty regarding role requirements."[17] Emigration and subsequent experiences in American society and the US workplace reflect such role transition under conditions of great uncertainty. Imprinting theory also recognizes that individuals' cognitive models "can be challenged and replaced with scripts and schema that are more congruent with the new environment."[18] We believe that this replacement process occurred to varying degrees with our interviewees as they transitioned to new roles in the United States. Their identities were not necessarily replaced, but could have built upon previous imprinting in their homelands.

Imprinting is also a form of social learning in that it is the "time-sensitive (i.e., occurs at sensitive stages of life) learning process (i.e., a stamping process whereby the focal entity reflects elements of its environment) that initiates a development trajectory (i.e., produces persistent outcomes)."[19] There also may be a "sedimentation of imprints, whereby new imprints are layered upon earlier imprints."[20] In this book, interviewees recount incidents of the learning process in specific environments at key stages of their lives, giving insights into how that imprinting affected their subsequent development. Their stories depict the potential replacement or repurposing of older imprinting acquired in the former USSR or its layering with newer imprinting acquired from their transition as they acclimated to the US environment and, particularly, to the roles they came to fulfill in the innovation economy. Research on how communist institutions might have imprinted

[17] Blake K. Ashforth and Alan M. Saks, "Socialization Tactics: Longitudinal Effects on Newcomer Adjustment," *Academy of Management Journal* Volume 39, Number 1 (1996): 149.

[18] Gina Dokko, Steffanie L. Wilk, and Nancy P. Rothbard, "Unpacking Prior Experience: How Career History Affects Job Performance," *Organization Science* Volume 20, Number 1 (2009): 55.

[19] Blake D. Mathias, David W. Williams, and Adam R. Smith, "Entrepreneurial Inception: The Role of Imprinting on Entrepreneurial Action," *Journal of Business Venturing* Volume 30, Number 1 (2015): 12.

[20] Zeki Simsek, Brian Curtis Fox, and Ciaran Heavey, "What's Past Is Prologue: A Framework, Review, and Future Directions for Organizational Research on Imprinting," *Journal of Management* Volume 41, Number 1 (2015): 303.

individuals[21] has found that a communist "attitudinal legacy"[22] persists over time and that communism affects an individual's values and behaviors, including the propensity to become an entrepreneur.[23] Additionally, different forms of communism[24] during the various waves of migration, and the subsequent conversion of post-Soviet successor states to more market-oriented economies, would likely have created different imprints. These interviewees, for better or for worse, would likely have carried elements of this early imprinting into their new lives in the United States. Their new experiences in a capitalist system would clearly have challenged some of those imprints while potentially reinforcing other positive ones. Since communist ideology castigated market capitalism, we would expect such individuals to undergo replacement imprinting during their role transitions in this strikingly new environment.[25]

The replacement and sedimentation processes could well have differed among interviewees depending on the wave of their migration and their age at the time. We would expect that replacement imprints would be most challenging for those who migrated as more mature adults since they would have experienced a longer period of imprinting in their home countries. And the strong desire of many to leave the former USSR might indicate that they had already been imprinted in a way that reflected little influence from that environment. For such people, migrating to the United States might not have resulted in a major new imprint since they might already have undergone such a process to some extent. This possibility might well prevail for many

---

[21] Pop-Eleches and Tucker, "Communist Socialization and Post-Communist Economic and Political Attitudes," 77–89; Michael Wyrwich, "Can Socioeconomic Heritage Produce a Lost Generation with Regard to Entrepreneurship?" *Journal of Business Venturing* Volume 28 (2013): 667–682.

[22] David Blanchflower and Richard Freeman, "The Attitudinal Legacy of Communist Labor Relations," *Industrial and Labor Relations Review* Volume 50, Number 3 (1997): 438.

[23] Wyrwich, "Can Socioeconomic Heritage Produce a Lost Generation with Regard to Entrepreneurship?" 667–682.

[24] Pop-Eleches and Tucker, "Communist Socialization and Post-Communist Economic and Political Attitudes," 77–89.

[25] Elitsa R. Banalieva, Sheila M. Puffer, Daniel J. McCarthy, and Vlad Vaiman, "The Impact of Communist Imprint Prevalence on the Risk-Taking Propensity of Successful Russian Entrepreneurs," *European Journal of International Management* Volume 12, Number 1/2 (2018): 158–190.

interviewees since they were not typical Soviet citizens, but instead individuals who had decided to leave a system and environment that they did not identify with and in which they may have felt marginalized.

Many chose the route of math and science in the vaunted Soviet educational system, subjects among the few fields that rewarded performance over political loyalty, helping to shield people from the worst of the negative political imprinting. Still, their subsequent imprinting in the United States could have involved replacement or sedimentation processes. Regardless, we expect that most interviewees gained at least partial new imprints and identities after migrating to the United States, while understanding that some impacts of Soviet imprinting likely remained. Chapters 2 and 3 examine how institutional imprinting from political, economic, social, and educational institutions in the former USSR might have affected the people we studied, both negatively and positively. Chapter 4 examines the migration experience and potentially new imprinting, while Chapters 7 and 8 explore the influence of subsequent imprinting as these individuals dealt with cultural adaptation to US society and work organizations.

Identity Theory

The imprinting process shapes an individual's identity. *Identity* refers to "how individuals view themselves, and might self-identify if asked, 'Who are you?'"[26] Identity theory explains the relationship between self and society and the consequences for individuals from identity-related processes.[27] Specifically, identity theory "is a social psychological theory in the field of sociology that attempts to understand identities, their sources of interaction in society, their processes of operation, and their consequences for interaction and society. The theory brings together in a single framework the central roles of both meaning and resources in human interaction and

---

[26] Alyson Meister, Karen A. Jehn, and Sherry M.B. Thatcher, "Feeling Misidentified: The Consequences of Internal Identity Asymmetries for Individuals at Work," *Academy of Management Review* Volume 39, Number 4 (October 2014): 490.

[27] Michael Hogg, Deborah J. Terry, and Katherine M. White, "A Tale of Two Theories: A Critical Comparison of Identity Theory with Social Identity Theory," *Social Psychology Quarterly* Volume 58, Number 4 (1995): 255–269.

purpose."[28] An identity can be viewed as "a system that controls its inputs of self-relevant meanings: the basic process of identity operation is known as identity verification."[29] People try to ensure that the meanings about themselves in a situation correspond to the meanings in their identity standards. Identities operate at both conscious and unconscious levels, and an identity is seen as consisting of four basic components: identity standard, an input, a comparator, and an output. An *identity standard* is a set of meanings, and individuals compare their perceptions of meanings to that standard.[30]

As with imprinting, an individual's identity is formed primarily during the preadult years that generally last from ages six to twenty-five.[31] As children reach adolescence, they begin forming more complete ideological commitments regarding their place in the world, and their core identity becomes largely fixed.[32] Identity typically remains stable in adulthood, resulting in "habits of mind" being created and repeated in response to environmental stimuli.[33] An individual's identity is acquired by imprinting from the environment that is fundamental to shaping it. While a particular imprinting–identity relationship can be relatively permanent, it can also be altered as a result of a major change in an individual's environment.

Identity is an inherent component of the self-concept, and the self is seen as fundamental to understanding human thought, affect, and behavior.[34] In contemporary social science, "selves" are now referred to as "identities." Two basic ways of examining multiple identities within individuals are from an internal or external framework. The internal focus is on how an individual's multiple identities function together and within the overall identity-verification process, including how multiple identities are switched on and off. The external focus examines how an individual's multiple identities are anchored in the social structure, including the degree of commitment to multiple identities and the way each identity is connected to the social structure.

---

[28] Peter J. Burke and Jan E. Stets, *Identity Theory* (New York: Oxford University Press), frontispiece.

[29] Ibid., 68.   [30] Ibid., 62–68.

[31] Erik Erikson, *Identity, Youth, and Crisis* (New York: Norton, 1968), 161.

[32] Ibid.

[33] Katsuhiko Shimizu and Michael A. Hitt, "What Constrains or Facilitates Divestitures of Formerly Acquired Firms? The Effects of Organizational Inertia," *Journal of Management* Volume 31, Number 1 (2005): 52.

[34] William James, *The Principles of Psychology* (New York: Holt, 1890).

These include multiple role identities within a single group, the same role identities in different groups, and different role identities within intersecting groups.[35] Some individuals have been found to self-identify with multiple and overlapping identities.[36] Thus, although a person's identity may have been heavily influenced during earlier imprinting in one environment, identity is susceptible to change when engaging with a very different environment. In the workplace, self-awareness is considered a career meta-competency that facilitates the learning of other competencies,[37] and a clear sense of self and identity is essential for making effective career decisions and achieving success.[38] This insight is critical since the process of imprinting and potential identity change could affect the ability of immigrants to contribute to the US innovation economy.

Four sources of identity change have been found.[39] First, changes in the situation disrupt meanings and can cause distress and uncertainty. When it is not possible to change the situation to match one's identity, the discrepancy is reduced by changing the identity standard to match the situation. Second, identity conflicts occur when multiple identities conflict with each other and are activated simultaneously. Role conflict and status inconsistency can be viewed as identity conflicts. The degree of commitment to and salience of each identity, as well as the connectedness of identities to a person's other identities, can affect the adjustment of identities. Third, identity–behavior conflicts refer to engaging in a behavior that is inconsistent with one's identity standard, such as an ethical standard. This situation could arise when people do not foresee the unintended consequences of a behavior or decision. Over time, such small changes can lead a person to deviate from an identity standard and result in an identity change. Fourth, taking the role of another is a way of adapting to social situations when interacting with others. By doing so, people come to internalize the other person's expectations, and, in the process, their identity comes to include these expectations as standards.

---

[35] Burke and Stets, *Identity Theory*.   [36] Ibid.
[37] Douglas T. Hall, *Careers In and Out of Organizations* (London: Sage Publications, 2002).
[38] Brad Harrington and Douglas T. Hall, *Career Management and Work–Life Integration* (Thousand Oaks, CA: Sage Publications, 2007).
[39] Burke and Stets, *Identity Theory*, 180–184.

Foregone alternative identities may not, or even need not, be suppressed: "[i]nstead, some of them become internalized in the self-concept and continue to exert influence."[40] This can happen as a result of "events that remind the person about the road not taken and critical moments can make alternative selves salient."[41] Reminders would include meeting people from one's home country, while critical moments would include achievements and failures at work or personal milestones such as the birth of a child. Such influences can be positive or negative. Specifically, comparing one's current self with a better or worse alternative self can create positive or negative emotions or affect; can generate knowledge of one's strengths or weaknesses; influence satisfaction with one's job, career, work–life balance, and life in general; and increase motivation to change or maintain the status quo. In short, "it is the verification of identities that makes people feel good in general and feel good about themselves especially."[42] This expanded view of multiple identities opens up the possibility that some prior identities may not be abandoned, but remain part of the self-concept and continue to influence one's self-concept and professional life.[43]

Identity formation and identity change have been extensively studied from the perspective of migration and immigration. The early years after immigration are very important in an immigrant's integration and ability to understand the new market and business environment of the host country.[44] National origin is crucial in immigrant entrepreneurs' self-definition, being a foundation of constructing and reconstructing one's identity. The elements of national identity take different forms, evolve over time, and even serve different purposes. Immigrants to the United States typically have a dual sense of national identity, wishing to maintain important characteristics and values of their homelands while also being "good" Americans. Furthermore,

---

[40] Otilia Obodaru, "The Self Not Taken: How Alternative Selves Develop and How They Influence Our Professional Lives," *Academy of Management Review* Volume 37, Number 1 (January 2012): 47.

[41] Ibid., 42–43.

[42] Alicia D. Cast and Peter J. Burke, "A Theory of Self-Esteem," *Social Forces* Volume 80, Number 3 (March 2002): 1047.

[43] Obodaru, "The Self Not Taken," 34–57.

[44] Nona Kushnirovich, "Economic Integration of Immigrant Entrepreneurs," *Entrepreneurial Business and Economic Review* Volume 3, Number 3 (January 2015): 9–27.

entrepreneurship has been found to be the most stable dimension of immigrant entrepreneurs' self-definitions, providing a self-assuring element while other dimensions of identity, such as nationality and immigrant status, can cause instability and stress.[45] Additionally, economic success has been found to influence new identity development, and immigrant entrepreneurs tend to feel more at home in their new countries when they feel that native citizens appreciate their success.[46] Chapter 9, "Identity: A Constellation of Influences," is replete with interviewees' comments that illustrate influences from institutions and imprinting in both the former USSR and the United States that have shaped their identities. The comments demonstrate the positive and negative impacts of prior identities, as well as sedimentation or replacement imprints on work satisfaction, performance, and adaptation to the US environment. These topics are explored in Chapter 7, "Cultural Adaptation: Challenges and Sources of Support," and in Chapter 8, "Workplace Adaptation: Developing Soft Skills." Other sources of influence on identity, such as imprinting from the institutions of the former Soviet Union are explored in Chapter 2, "Soviet Political, Economic, and Social Institutions: Catalysts for Migration," while the influence of the educational system is discussed in Chapter 3, "Soviet Educational Institutions: Capability for Contribution." The findings from our 157 interviews appear consistent with those in other recent research. Ludmilla Isurin's book on Russian immigrants to the United States, Germany, and Israel discussed changes and transformations in cultural perception, self-identification, and attitudes toward first-language maintenance.[47] Katerina Bodovski took a personal approach to these same questions in her research-based memoir about growing up in the Soviet Union and emigrating at a young age to the United States.[48] Others who have taken an oral history approach to understand the Soviet Union and its collapse include a major work by Donald Raleigh on Soviet baby boomers who grew up

---

[45] Beata Glinka and Agnieszka Brzozowska, "Immigrant Entrepreneurs: In Search of Identity," *Entrepreneurial Business and Economic Review* Volume 3, Number 3 (December 2015): 55–71.

[46] Ibid.

[47] Ludmilla Isurin, *Russian Diaspora: Culture, Identity, and Language Change* (New York: de Gruyter Mouton, 2011).

[48] Katerina Bodovski, *Across Three Continents: Reflections on Immigration, Education, and Personal Survival* (New York: Peter Lang Publishing, 2015).

during the Cold War.[49] And the Belarusian author, Svetlana Alexievich, was the 2015 Nobel Prize recipient in literature for her book based on interviews of people of the same generations as our interviewees but who remained in their homelands. Her work provides a rich and complex record of the experiences and worldviews of people from all walks of life from the various republics of the former USSR.[50]

A series of complex questions of identity will be explored in this book that are not obvious to the general public outside the former USSR, who typically refer to the people from that vastly diverse region as "Russians." A narrow definition of Russian ethnicity would classify Russians as being culturally connected to the Russian Orthodox religion and being of Slavic origin.[51] However, the Russian Empire and later the Soviet Union contained many different nationalities that were brought together by policies of first Russification and later Sovietization, such that they came to share a common cultural touchpoint with the Russian language and Russian and Soviet culture.[52] The vast majority of those who immigrated to the United States in the time period covered in this book were initially Jews but later included significant numbers of Ukrainians, Belarusians, ethnic Russians, Armenians, Georgians, and Kazakhs. Smaller numbers also came from other former Soviet republics.

Most studies of emigration from the Soviet Union and post-Soviet states focus on Jewish emigration, such as the recent book by Zvi Gitelman that, like Isurin's book noted earlier, analyzed emigration to the United States, Israel, and Germany.[53] In contrast, this book focuses on immigrants to the United States from a wide variety of ethnicities and from most of the republics of the former Soviet Union.

---

[49] Donald J. Raleigh, *Soviet Baby Boomers: An Oral History of Russia's Cold War Generation (Oxford Oral History Series)* (New York: Oxford University Press, 2012).

[50] Svetlana Alexievich, *Secondhand Time: The Last of the Soviets – An Oral History* (New York: Random House, 2016).

[51] Stephen J. Gold, "Russia," in *The New Americans: A Guide to Immigration Since 1965*, eds. Mary C. Waters and Reed Ueda (Boston: Harvard University Press, 2007): 580.

[52] Ibid.

[53] Zvi Gitelman. *The New Jewish Diaspora: Russian-Speaking Immigrants in the United States, Israel, and Germany* (Rutgers, NJ: Rutgers University Press, 2017).

This diverse group shares a common Soviet political, economic, social, educational, and cultural background. In summary, we apply theories of institutions, imprinting, and identity to help explain how immigrants from the former Soviet Union have made contributions to the US innovation economy. This perspective brings together the five i's that constitute the subtitle of this book.

2 | Soviet Political, Economic, and Social Institutions

Catalysts for Migration

Our analysis of key institutions examines those that to a large extent catalyzed the exodus of the Soviet technical intelligentsia that migrated to the United States. The formal institutions that are relevant to our discussion are the Soviet government and its communist ideology, the centralized command economy, national republics, youth organizations, the Communist Party, and the Soviet educational system and research institutions. We also consider the impact of the informal institutions of family and inherited culture and of the unwritten but societally understood policies of anti-Semitism. We will describe how the impact of each of these institutions varies during the three separate waves of immigration that we have identified.

Institutional theory thus can be a framework for understanding the dramatically changing institutional landscape that people experienced throughout their lives in the Soviet Union and its successor states. The individuals featured in this book reveal the impact of these institutional upheavals through stories about their lives during these dramatically different eras, including everyday life, their educational experiences, and their careers. As we stated in Chapter 1, the main role of institutions in any society is to create stability and predictability. When there is dissatisfaction or disillusionment with those institutions, there is an incremental process of deinstitutionalization. The personal stories recounted in our interviews illustrate the progression of deinstitutionalization among the educated technical elite. In the case of the Soviet Union, there was a growing disillusionment among a section of the educated elite during the late Soviet period and into the *perestroika* period from 1972 to 1986, which is covered by Wave One interviews. The late *perestroika* period and then the first decade following the collapse of the Soviet Union, from 1987 to 1999, was a time of institutional collapse and then rebuilding, which is covered by Wave Two interviews. The Wave Three period beginning with the turn of this century is characterized by much more ambiguity about the institutions in interviewees' home countries, and, as such, their choices of country of residence are more about personal opportunity rather than only hardship caused by institutional influences.

It is critically important to place the experience of our interviewees within the historical context of the chaos of the World War II period and the failed and repressive institutions of that time, the consequences of which remain in interviewees' family memories. We relate one such a story that illustrates the extraordinary hardships that countless people faced as they made their way to freedom through emigration necessitated by the extraordinary institutional upheaval of the wartime years. **George Gamota**, CEO of Science Applications International, fled Ukraine when he was a mere ten years old. He and his family journeyed by horse and wagon through war-torn Europe and eventually emigrated to the United States as refugees in 1949. Gamota's story provides insight into the historical trauma of experiences that most interviewees in all waves would have experienced or heard about in regard to their parents and/or grandparents. Gamota, who later earned a PhD in engineering at the University of Michigan and enjoyed a highly successful career in business and government, was a migrant virtually from birth. He recalled: "I was born in May 1939 and by August we were fleeing because Stalin and Hitler had signed a pact that allowed Stalin to take over eastern Poland, Galicia, which is now in western Ukraine. My grandfather was the justice minister and my father, who was twenty-two years old, was a conscript, and both of them were arrested by the Polish authorities in Poland. My other grandfather in southeastern Galicia was a colonel in the Austrian army, which then became the Ukrainian army. He marched with his column to help the Ukrainian government there fight the Russians. He got captured and was sent to Kazakhstan and spent four years or so in jail and we never saw him again. Stalin probably got rid of him in the 1930s. My grandfather paid off the jailers and got out after a relatively short time, while my father was in jail for a year and then was finally let go."

Gamota continued the story of his family's flight through Europe: "By August 1939, the Soviet Army started coming and so we fled. We fled to Poland, to Krakow, and we stayed there for two years. Then the Germans, the Nazis, pushed everybody back and so my family was pushed back to Galicia. My father was a forest engineer, and I was about five years old when we were visited, I don't know how often, by the Nazis to kind of check on us. The Ukrainian underground army that was fighting the Russians and the Nazis also visited us. But in 1944, it was clear that we had to leave. And so in the middle of that summer, the four of us went on about a six-week journey by horse and buggy to Vienna. My brother was four years older than I was. He was on crutches at that time. He had some kind of a disease that was debilitating. But with God's help, we made it to the outskirts of Vienna, and we saw Vienna up in flames and burning because it was being bombed by the Allies. And so we couldn't go to Vienna, so I guess my father knew the territory, and, being a forest engineer, he went west of Vienna to Beethoven's Woods. That's kind of romantic. During the raids we had to hide. The women hid in nearby cheese caves while the men and older boys hid in the forest. That's where the war ended, and so we thought that we could start our lives again. But then the Soviets decided to take over Austria and they were supposed to stop, and they never stopped. By that time, it was nine months or a year later, we only had one horse and the wagon was still there, and so with one horse and a wagon, we started fleeing the Soviet tanks. We had to cross the Danube to get to the Allies' side. As we were going by that side of the river, the Soviet tanks were behind us and almost all the bridges were blown up. Finally we saw one bridge that wasn't completely blown up, and my parents decided that my mother and I would walk across and my brother would stay in the wagon. My father took the wagon across and finally we made it over, and within a few minutes that bridge got blown up. But we were safe."

Yet more trials lay ahead. Gamota continued: "Well, there were terrible things. On the road through the mountains there were all kinds of bandits and deserting soldiers who were attacking fleeing civilians and robbing them of food and valuables. One time, in Slovakia probably, when we heard shots up ahead, we stopped and waited, and then dusk came and we decided to go. And this was the scene: I was a kid then, but I remember horses dead, a man and his wife dead, and two kids who were shot. Splattered across the road, and you just sort of . . .," his voice trailed off. After a moment, he continued: "Once we crossed the Danube, we were finally safe, but the feeling of real safety did not occur until we were on a ship sailing to America. Once I got to America, all of those memories sort of disappeared." Gamota then recalled how those images that he had put aside came back to him decades later, memories triggered by the terrorist attacks in the United States on September 11, 2001, when his son was working in Manhattan and was briefly unaccounted for. Gamota's sudden flashback to a long-suppressed traumatic childhood attests to the power of imprinting, especially when such events occur at an impressionable young age.

Three Waves of Immigration and Different Institutional Experiences

In the Introduction, we presented an outline of the three waves of immigration, which will be explained in greater detail in Chapter 4. However, it is worthwhile to repeat the summary here as we examine the varying institutional impacts in each of the three waves.

The First Wave took place from 1972 to 1986. During that period, those who were allowed to emigrate from the Soviet Union were primarily Jews. This was a peculiarity of geopolitics that will be explained in Chapter 4. We will explore why Jews wanted to leave the Soviet Union in such large numbers when the opportunity arose. Furthermore, we will describe the institutional impact on them of communist youth organizations, their educational experiences, and Soviet cultural institutions.

The Second Wave began in 1987, in the *perestroika* period, and is associated with the rapid decline and sudden collapse of Soviet institutions. This period saw a dramatic expansion of emigration, particularly among the intelligentsia, regardless of ethnicity or nationality. This period was punctuated by the dissolution of the Soviet Union into separate new countries in 1991, and continued through the end of the 1990s.

The Third Wave, which began in 2000, was a period of reestablishment of institutional life in all of the fifteen newly independent countries. This growing stability of institutions was constructed upon the foundation and institutional memory of the Soviet Union, but now includes specific national and cultural differences in each independent country.

In examining the institutional contexts for the three waves, it is perhaps stating the obvious that those who emigrated had both motivation and opportunity to do so. Particularly in the early waves, many interviewees rejected the prevailing institutional values and the imprinting and identity formation associated with them. These institutional influences are critical to understanding the ability of these individuals to contribute to the US technology sector and the new identities forged under the influence of US institutions, as will be explored in the later chapters. We also note that the story of immigration is often fraught with personal hardship and can be accompanied by demonstrations of personal strength. We will highlight important categories of institutions and note that, while presented separately here, in real life they overlap and condition the processes of imprinting and identity. We will focus on the formal and informal institutions that served as catalysts for emigration.

Formal Institutions: Soviet Government and Ideology

The majority of the First Wave of immigrants were independent-thinking intellectuals who rejected the ideological worldview presented by Soviet institutions. **Isaac Fram**, Principal of Biomedical Imaging Solutions, was part of this First Wave, initially emigrating to Israel from Riga, Latvia, and coming to the United States in 1969 at age twenty-one: "Basically, I was always interested in history and knowing what was going on. As ironic as it may seem today, I was also responsible for informing the rest of the class about political events that were taking place. The problem was that the sources of information that we chose to call newspapers came from the major newspapers, *Pravda*, which means 'truth,' and *Izvestiia*, which means 'news.' The well-known joke was that in *Pravda* there was no '*izvestiia*,' and in *Izvestiia* there was no '*pravda*.' It was a joke at the time, but anyway, the other sources we used to listen to were the Voice of America and Radio Free Europe, which kind of opened up to us how it was in different places. So the moment we left the Soviet Union, my thought was how great it is to be like a bird. We were flying from Moscow to Vienna, and I felt, 'I'm free as a bird.' I could fly anywhere and I would never see that place again because I built up antipathy toward the regime there."

Many of those interviewed were strong believers in socialist ideology in their youth and then began to question and change those beliefs, illustrating the process of deinstitutionalization among the intelligentsia. **Slava Epstein**, Distinguished Professor of Biology at Northeastern University, who came from Moscow in 1988 at age twenty-nine, described his personal metamorphosis from being a fervent communist in younger years to seeing its weaknesses and irrationality during university years through discussions with his future wife and their friends: "Well, until I was seventeen or eighteen years old, I was a devout communist. Absolutely. The only thing I regretted in my life at the time was that I couldn't become a member of the Communist Party before I turned eighteen because that was a rule. I never became a member of the Party because by the time I became eligible for it, the nonsense had disappeared. Much of the transition I described I owe to Moscow State University. However, the number one influence, by far, was my future wife who came from a very different background, the intelligentsia. We were dating then, and she and her friends introduced me to literature, actually school-banned books, those you would go to prison for having. But it was all unfocused, and people felt it was OK to share them even with people you don't know. So I started reading, and then it took half a year to fully realize the stupidity of my prior beliefs. Pieces fell into place as the picture of the world changed 180 degrees for me and then became reasonable, as opposed to being totally irrational as it was before that."

Gene Shkolnik, Cofounder and CTO of CoachUp, who came from Moscow in 1992 at age seventeen, noted: "I definitely remember a lot of propaganda. I was a very impressionable kid. Not that I believed in all that stuff, but there was a lot of talk about Communist ideals, what's right and what's not right. I read a lot as a kid, and a lot of the books available to us at that time contained some form of propaganda. There was certainly a good amount of brainwashing going on in my school, too. I was a Pioneer after being an Oktiabrionok. For better or worse, I didn't get to become a Komsomolets. By then, *perestroika* had started, and I no longer had to join Komsomol, which otherwise would have been required for me to get into college."

Another prevalent form of passive resistance in the intelligentsia came through avoiding the more politicized professions and choosing math- and science-related pursuits. **Eugene Shablygin**, Founder and CEO of WWPass, who came from Moscow in 1992 at age thirty-two, described his choice of a career in science as a conscious avoidance of politics: "I certainly believe that communism is a very wrong theory, but everything was all 'blessed' by communism, so you could not be an honest scientist unless you did some science that was not politically engaged. That's why physics and math and chemistry can be the only possible occupations for an honest scientist. This gives you opportunities beyond just politically oriented activities. Because, you know, I hate lying. I always have, and, by the way, that's why I cannot be a politician. I call things the way I think they are."

Formal Institutions: Centralized Command Economy

The collapse of the political institutions at the end of Wave One and beginning of Wave Two also meant the collapse of the centralized command economy, which in turn brought shortages and uncertainty that provided an added push to emigrate. **Ksenia Samokhvalova,** Senior User Experience Specialist at MathWorks, who came in 2003 at age twenty-three, grew up in the Russian city of Gorky, subsequently renamed Nizhny Novgorod. For her, the shortages of the time were just something to be endured: "I think the '80s were pretty peaceful. I think I was in kindergarten first, and then when I was seven I went to school. I remember that rationing coupons were needed to buy things. We were stockpiling vodka because it was part of that ration, and nobody was drinking it because it was poor quality. But it was a good currency to pay people, a bottle of vodka or two, depending on the job. I also remember having ration coupons for macaroni and butter. The other thing was that sometimes I remember these things, and I'm like, 'Wow, I totally forgot.' And then it floods back into my memory of how my parents would bring a hunk of kielbasa from work. We would eat it for a while and it would be wrapped in this gray paper and it would have our last name on it. And butter was not easy to come by. But we were not starving. In the '90s, they wouldn't pay people sometimes for a while, but my dad's job would always pay, I don't know, maybe because it was defense. I feel like I grew up really happy, it was life. We weren't starving or really suffering much. Still, we didn't grow up with a lot of pocket money. If I needed to buy ice cream, I would ask my parents, since it wasn't available all the time."

The uncertainty of those times left a deep impression on **Mark Kofman,** CEO and Cofounder of Import2, who came from Tallinn, Estonia, in 2011 at age thirty. He described the protests he witnessed as a child in the late 1980s: "We were living downtown, in a house which was between the TV station and the radio station, so it was actually pretty scary because that was the place where at the time all the action was happening. We were expecting that, because if something was about to happen, then it would be in this place because that's where you want to keep control over the television and things like that. I remember that our parents were pretty scared, and there were a lot of preparations and different barricades around this place. So those kinds of things you remember, but I think we were too small to understand what exactly was going on, so it was more like a visual impression."

Other interviewees had family experiences in which the new conditions of the Russian version of the free market brought stress and physical danger. **Maxim Matuzov,** Search Program Manager at Apple, came in 2009 at age twenty-one, from Murmansk, a Russian port city near Finland: "The 1990s were a pretty chaotic time with restrictions, but mostly lots of questions for my parents. I just remember that my dad had been the second in command on a fishing trawler. However, during that time in the 1990s, he started doing business, and he actually got into some serious stuff. I remember we had to leave the city for about two weeks because, you know, it seemed there was some vengeance involved. I guess, because he was doing some 'gray' business. And there was probably a criminal element involved. So we had to leave for two weeks for Belarus because of this. We didn't really see it, and we didn't want to, so I remember my parents took me there with my brother. I think it was during the mid-1990s, I can't remember, but I was in school. I can't remember that much, actually, but our lives might have been in danger."

Andrey Doronichev, Senior Product Manager at Google, came from Moscow in 2012 at age thirty. He recounted the dangers of business life in the early 1990s with a personal story from around 1993: "Well, when I was a teenager the political situation was already calming down. The harshest moments of my life were when I was twelve. That was probably the worst. There was a moment when we found my uncle, who was my dad's business partner, in the entry hall of our building, with a dumbbell in his skull. There was blood everywhere. He survived but that was scary. There were really bad attacks, and we had to hide."

The institutional collapse of the Soviet Union brought a corresponding crisis of beliefs, ideals, and worldview that had a dramatic impact on Second Wave immigrants. **Umida Gaimova Stelovska,** Founder and CEO of ParWinr, who came in 2009 at age thirty, was raised on a cotton-growing collective farm near Samarkand, Uzbekistan. She recalled her unusual childhood in an agricultural region far from the center of Soviet power: "I was twelve years old when the Soviet Union collapsed and Uzbekistan declared independence. I remember how from a very young age we participated shoulder-to-shoulder with the grownups planting and picking cotton in the fields. We didn't see the world behind those vast farm fields." In a revealing comment that was likely true for many citizens during that period, she stated: "I believe I am the in-between generation that grew up during both the Soviet regime and Uzbekistan's independence, the generation that witnessed the great change from communism to capitalism. I remember that after the Soviet Union collapsed, we didn't have textbooks. Because all of our books were written from the Soviet Union's point of view, everything had to be thrown away. We now had our own nation, history, and culture we could be proud of. I also remember that, before our independence, we were told that Lenin was our godfather, while Timurlan[1] was portrayed as the torturer and killer ruler, and we were made to feel ashamed of his actions. After independence, we had to learn that history in reverse, and Timurlan, the Great Amir Timur, was our hero. For the next ten years, I feel like our generation was somehow lost.

"Especially observing my parents' heartbroken moments was not easy for me as a kid. I clearly remember the frustrations of my parents and their hopes for a better future. During Soviet times, my father couldn't even buy a car because it was not allowed for workers in the cotton industry, and for that he would have gone to prison, as did so many people in the 1980s. I also remember the night when communism was over, and my mother asked my dad: 'What are we going to do now?' I think she didn't know whether she should be happy or sad, because she was one of the big believers in communism and socialism. That night, there was the first currency change and many became poor, because whatever you had as savings now had zero value and was lost. Independence made living harder, especially for the people who lived in the rural countryside. Almost everyone started losing jobs. Suddenly, most of our Russian and other nationality neighbors started leaving the country. Later, while we expected we would live free and happy

---

[1] Amir Timur, or Tamerlane (1336–1405), conquered an empire from Russia to India, and from the Mediterranean Sea to Mongolia, with Samarkand as its capital.

forever, it was impossible to have a better lifestyle with the corruption going on. It was very hard to see my father trying to make a living for the family, and seeing my mom going from office work to farm fields because it was hard for her to find any job. We worked hard as kids in the fields to help our family. Even after independence, we kids also became free labor for our government."

Formal Institutions: National Republics

The Soviet Union was based on a formal system of political institutions of national republics and subregional nationality districts that is quite different from Western political institutions. When the Bolsheviks overthrew the Russian Empire, they inherited a multinational territory with many different open or incipient national movements that sought independence. The ideology of the Bolshevik Party was to unite the working class of all countries, regardless of nationality. However, they had to also come to terms with the realities of national cultures, histories, and political yearnings. So, alongside the system of a centralized government and economy, they developed administrative governmental units that corresponded roughly to the geographic homelands of different nationalities.

This political approach to uniting different nationalities in one political state resulted in the structure of the Union of Soviet Socialist Republics, containing fifteen separate republics and numerous nationality-based subunits within each. The setting of borders of these republics and subunits is a matter of some controversy and a source of ongoing conflicts to the present day. What is important for our inquiry is to establish this dual institution of identity as a fact of Soviet life. Each Soviet citizen's passport and other identification documents contained a notation of nationality, regardless of his or her actual place of residence. People had a social, cultural, or ethnic identity linked to their nationality and background and an overall political identity as Soviet citizens.

The political institutions of the Soviet Union based on these national republics came under increasing pressure from nationalist forces during *perestroika* and ultimately brought forth fifteen new countries out of the disintegration of Soviet institutions. The institutional crisis of the Soviet Union was perhaps most acute in the national republics in which nationalism and independence became the popular mood of the time.

As part of the deinstitutionalization of the late Soviet period, increasing conflicts and demonstrations erupted in the Baltic republics of Estonia, Latvia, and Lithuania, and in the Caucasus republics of Armenia, Georgia, and Azerbaijan. This period of deinstitutionalization was closely followed by a period of institutional collapse, with the dissolution of the Soviet Union and a long period of reconstituting new institutions with varying degrees of success.

As the Soviet Union dissolved, all of the former constituent countries suffered a period of economic decline as they sought to transition to a new economic model. The Baltic republics quickly moved into the European economic sphere, enabling them to recover more quickly. Russia and Kazakhstan were eventually able to stabilize themselves primarily as producers of oil, gas, and other natural resources sold in international markets. Ukraine, Belarus, Georgia, and Armenia are not major energy producers, and all depended upon Russia as the traditional market for their manufacturing and agricultural production. None of them had much success in international markets, and each has had various political problems with Russia, except for perhaps Belarus and Armenia. As a result, Russia and Kazakhstan were able to revive spending on science, research, and education in the early part of this century as world energy prices rose, thus reducing the economic pressure for emigration. Meanwhile, Ukraine, Armenia, and Georgia, in particular, have experienced ongoing economic pressure for the educated technical elite to emigrate, as will be explained in Chapters 3 and 4. This process of conflict and confusion was part of the motivation for emigration for a number of interviewees from various Soviet republics that became independent countries.

The Baltic countries saw the earliest and most active independence activities. For many who became adults during that time, opposition to the institutions of the Communist Party and the KGB had an immediate impact on their imprinting and sense of identity. **Sten Tamkivi,** Cofounder and CEO of Teleport and the first General Manager of Skype, first came from Tartu, Estonia, in 1994 as a sixteen-year-old exchange student. He described the influence of the Party and the KGB on his family and his own perception of the world: "I was born in 1978 and entered the first grade in 1985. Independence came in 1991 when I was in the sixth grade, when you kind of understand what's going on around you. So the events started that led to independence. I think the Estonian national colors were brought out in public in 1988. They were prohibited during the Soviet occupation. But, at that time, the changes were still in the making, and the KGB was everywhere, so everyone was very careful. But, at the same time, I still remember this dichotomy that on one side we had an understanding of the world beyond Estonia. We studied the English language, had some insight into what was happening in the UK, studied some subjects in English, and maybe even were studying some international geography, maybe different than at some other schools. At the same time, we had to memorize poems about Lenin. And that contrast was made even harder since I and others had to speak proper Russian. So I was in constant conflict with my Russian teachers because the material they were teaching was so politically in contrast with what I was getting from outside. During the Soviet occupation, when entrepreneurship was a criminal offense, I looked at my dad and my dad's friends, and people he went to school with, and saw what these people ended up doing later. One route was to join the Party and become an official and work the system to your benefit. But there were also people who didn't want to go that route and didn't join the Communist Party. Very often they ended up in academia. And academia became a sort of catch-all of those who, in a free society, might really want to become academicians, and others might actually want to become entrepreneurs or something else. I remember in 1983, when I was still very young, my dad got to do research in the Netherlands for half a year. I was the most popular kid on the block because I had Legos that he brought back for me. But in order to work abroad, it was still necessary for him to travel via Moscow because everything was tightly controlled. During that six months, my mom was always under KGB surveillance. My dad was heavily pressured to join the Party, just to get permission to go abroad. He didn't join, and then he faced KGB stuff continuously after that. When you interfaced with the free world, it made you not want to join the Party."

Armenia also underwent trying times in the struggle for independence. **Nerses Ohanyan**, Director of Growth Analytics at Viki, who came from Yerevan in 1998 at age fourteen, remembered the conflicts of that time: "In 1991, that's when we saw a sort of revolution happening. I remember the days of Soviet occupation and Russian soldiers around our house. I remember them coming into our house and getting food. They were freezing outside. They were there, and they didn't want to be there, and we didn't want them to be there."

Georgia went through an analogous process, as described by **Shalva Kashmadze**, Product Manager at Pocket Gems, who came from Tbilisi in 2011 at age twenty-five: "It happened really suddenly. People would probably have known in the late '80s that something was happening because there were huge lines, and you couldn't buy some basic goods. So people knew that something was amiss. But then it dissolved into a civil war in Georgia in 1992. That was totally unpredictable, and no one knew that it would happen. I think of things, but probably don't remember what happened in 1990 or 1991. But I do remember 1992, because that year I went to elementary school, and that's when we had the civil war going on. So, obviously, the memories are kind of fragmented, but whatever memories are there, it's like they're really kind of drastic. So I do remember inflation, and I do remember huge lines to buy bread. I remember a central administrative system of distributing food because there was simply not enough for everybody. We had tickets for bread, and you had one per day. You would go to a bakery. Then you would wait in a huge line for a delivery car with bread to arrive, and it wasn't always on time. And you also had a certain amount you could buy that day. You could not buy more than that."

Ukraine, too, has struggled to establish a stable economy and legitimate political institutions. As such, it has been a source of significant numbers of immigrants to the United States. Even those who immigrated as part of Wave Three bear the imprint of the hardship and disorder in their home country. **Iryna Everson**, Manager of Procurement at Pattern Energy Group, came from Kiev in 2008 at age twenty-six: "I remember that it was not very good times because Ukraine had to figure out how to live on its own, without any other countries. So it was definitely challenging times, and as a little kid what I remember is that there was a limited supply of clothes and foods in the stores. So when I would come to my class in school, I would see probably three or four people wearing the same sweater. People had similar things because we had limited choices of things to buy. During those times, there was an increased level of crime, too, and my parents would always warn me, 'Don't go with others. If somebody approaches you, don't go with them.' This was because I was going to school myself. I was exposed to all sorts of bad scenarios at that very early age. So I was a little scared child, I must admit. I think my parents wanted to protect me from the crime that was around. So I guess they made me kind of a scared child."

Another Ukrainian, **Nick Bilogorskiy,** Director of Security Research at Cyphort, was from Kharkov and came in 2006 at age twenty-five after spending a several years in Canada: "I do remember how the economy started changing and the geopolitical situation started changing and how it affected me. It was difficult. There were bread lines and inflation, and they kept changing the currency in Ukraine: You had karbovanets, then finally khryvnia. But it was still some of the happiest times." His last comment is indicative of the mixed feelings that many experienced during that dramatic transition period.

Formal Institutions: Young Pioneers, Komsomol, and Communist Party Membership

A substantial number of interviewees had been members of the Young Pioneers, the Komsomol, or both. There was wider variation among interviewees as to their parents' Party membership, with one, both, or neither belonging. The widely different experiences of our interviewees go beyond membership itself to include various motivations to join or not and experiences as members ranging from very positive to unhappy ones. Beyond politics, the benefits of membership could be many, while resisting membership could bring severe restrictions on activities and could thwart educational opportunities and careers. Even as youths, the benefits of membership, including admission to schools, special educational programs, and travel, were incentives to join. By and large, these organizations had more influence on imprinting and identity formation while being less of a catalyst for emigration.

Some interviewees gave fairly extensive views of their experiences in communist youth organizations. One was **Sergei Kovalenko,** Software Engineer at Verizon, originally from Gomel, Belarus, who came in 1995 at age twenty-three: "I was a Pioneer, which is like a children's communist organization. All children go through it. It is like Scouts and stuff, with communist ideology. In school, I was actually engaged in a lot of this communist activity like the Komsomol. I actually saw myself, my future, somewhere with the Party because, you know, it was where you wanted to be. It was visible, and it gave you opportunities. So, you know, you do this work and you are respected. You go places to different events. We had a lot of competitions in singing and dancing and all sorts of theatrical performances. I loved that, and then we would have to organize all sorts of work. And I quite early understood that sometimes organizing work is a lot more fun than actually doing it. I remember that one of my jobs was to be in charge of the labor education sector. About once a year, we had to clean up the schoolyard. Sometimes we had to go and collect scrap metal and give it to the State. It wasn't paid, but it was part of you, like what American students do, volunteering. There you were required to volunteer for the State. I would be in charge of actually organizing these groups, deciding who goes where, and creating a list of people, who brought what, and so on. You know, keeping the books."

Serial entrepreneur and investor, **Max Skibinsky,** who came from Moscow in 1996, remembered his time in communist youth organizations and the benefits they brought: "We had circles, modeled after Vienna Circles, in every school. And then there was this system that was like Soviet Boy Scouts, the Young Pioneers. As a member, you could go to the Palace of Pioneers that covered the district of Moscow. I remember we had rocket modeling, we had astrophysics, and we had math. And it was free, and you could go there and say, 'I want to enlist in this circle.' And you had an instructor who would give you classes after school, and it would always be practical. With one circle you would build rockets. With another circle, again in this whole fifth- and sixth-grade environment, you could go to this huge radio observatory and stay overnight with your instructor and you would do some sort of astronomical observations. If you weren't participating in those circles, you weren't a cool kid. Membership had two parts: first you had some minimal inclination to do it personally, but the second part is that you have all these opportunities at your fingertips. Literally, all you needed to do was write your name and persuade your parents, 'Hey I'm going to this circle, and I'm going to build rockets.'"

Other interviewees had significant experiences that deeply affected their overall attitude to Soviet life and that eventually led to emigration. **Vladimir Torchilin,** Distinguished Professor of Pharmaceutical Sciences at Northeastern University, came from Moscow in 1991 at age forty-five. He emphatically stated: "I hated being a Young Pioneer. My mother always told me how afraid she was. Once I came back after the Pioneers meeting. I was ten or eleven, and told my mother: 'I hate it so much, I just cannot stand it there.' And she explained to me that I had to do it if I didn't want to cause harm to her and the family, even if I didn't like it. I just had to shut up and quietly sit in the corner.

Probably, because I'm very individualistic, I just cannot stand that we all were taught the same thing, were supposed to think the same way. I didn't like it. Maybe at that time I had some other reasons, but it was most probably because of that."

Michael Barenboym, President of Baren-Boym Company, who came from Moscow in 1990 at age twenty-four, described a formative experience in the Komsomol: "I was part of the Komsomol, but the first bell that rang inside my head when thinking about it was when I was in eighth grade. One of the guys in my group was also Jewish. His father was dying, and the only procedure that could have helped him was in Israel, so they decided to emigrate. The KGB harassed them. It was probably in the late 1970s or early 1980s. At that time, the director of the school came to me and said, 'Because you're Jewish, you have to stand up at the Komsomol meeting and just be a part of us and say how bad those people are.' And that's when I realized that life is not that simple. And I refused to do that. They threatened to take my Komsomol membership away. And I said, 'That's fine. I will not punish another person for believing that they want to be free.' So that's when I started growing up, I think."

Some interviewees described a pragmatic attitude toward membership that illustrates the deinstitutionalization that accelerated during the Gorbachev years. **Marat Alimzhanov**, Lead Staff Scientist at Acceleron Pharma, described the transition from pragmatism to collapse of these organizations. Although born in Russia, he grew up primarily in Tselinograd (now Astana), Kazakhstan: "Yeah, I was in the Komsomol. I had to go through the Komsomol organization because I was told that if you don't join, you can forget about going to university. So I went, like pretty much everybody in my class, no exceptions really. And then when I got into Moscow State University, they had a Komsomol organization. But that was the time when things started to change. It was Gorbachev's time, and all of a sudden nobody really cared about it anymore. It was supposed to have formal meetings, but nothing was happening, so it was like, 'Wow, OK.' Somebody was still collecting the membership dues, but that was it." Alimzhanov first came to the United States in 1994, on a research travel fellowship, and later immigrated in 2002 at age thirty-one.

Many interviewees, in addition to their own experiences with communist youth organizations, described their parents' involvement or noninvolvement with the Communist Party. These descriptions range from explaining their parents' disdain for the Party and refusal to join, to a few who seemed to embrace the Party's ideology, to many who realized that Party membership was virtually required to progress in one's career, with the occasional exception of a scientific route. The Communist Party's institutional role in Soviet life, one way or another, had a great deal of influence on the imprinting and identity of interviewees. The period of immigration for Wave One and particularly for Wave Two interviewees was one of growing deinstitutionalization of the Communist Party, along with many other Soviet institutions.

A financial executive who came from Moscow in her twenties in the 1990s described the deep hold the Party had on her parents and, by implication, others in that generation: "Both of my parents were Party members. Somehow they both believed in the idea of socialism and communism. They truly believed this on the one hand, and on the other hand, I also think it was common knowledge that without Party membership you could not really build a career at that time in Russia. Actually, all of my grandmothers, grandfathers, and both of my parents were Party members by belief. But again, I don't know what was heavier on their minds. They believed in the ideals, but also thought it was important to make a career in the country at that time."

Anya Kogan, User Experience Lead and Manager for AdWords Display at Google, who came in 1990 from St. Petersburg at age eight, described a similar family dynamic: "My mother's family actually resisted emigrating for a long time. My mother's father was a devout communist in the truest, purest sense of the form. He really believed that communism was the way things needed to be. He was extremely pro-Russian, extremely pro-communism. He had a fairly important role in the Party."

A pragmatic approach analogous to what we heard about communist youth organizations was described by **Alex Petetsky**, Director of Software Development at PatientKeeper. Originally from Berdychiv, Ukraine, he came in 1991 at age twenty-four: "My father was a Party member. There was this very interesting structure that in order to get promoted you pretty much had to be a Party member. My mom, however, never took an interest. To my mom, with the background of a large family that was really struggling financially, she was like, 'I don't want to be a communist, not because of some ideological reason, but because I don't want to pay dues every month, every couple of months.' So that was my mom. But really, neither she nor my father were political in any way." A similar pragmatic reason was provided by **Alexander Vybornov**, Product Line Manager of Medical Products at IPG Photonics who came in 1993, from St. Petersburg. He described the situation for his parents as being a way to move ahead while being Jews. "Absolutely they were Party members. It was the only way to get advanced. For Jews in Russia, the bar was set even higher. So it was almost inevitable: If you wanted to make it, most Jews went into an academic career or you had to be a member of the Party. Otherwise you were automatically treated as an outsider, and the typical route for them was emigration."

Some parents managed to build careers without joining the Party. **Alexey Bulychev**, Principal Scientist at Moderna Therapeutics, who came from Moscow in 1992 at age twenty-four, discussed the case of his father: "Being a professor at the university, he never joined the Communist Party. That pretty much meant there was a glass ceiling for his career path. He still managed to grow well in the organization due to his talents and scientific achievements, but not to the levels that were reserved for those who were members of the Communist Party."

Formal Institutions: Soviet Educational and Research Institutions

The critical role of educational and research institutions will be explored in detail in Chapter 3. In terms of acting as a catalyst for emigration, the key aspect of these institutions was the unstated but clearly understood anti-Semitism expressed as quotas for Jews, particularly in the First Wave of emigration after 1965. This exodus was accelerated by the collapse of support of universities and research institutions at the beginning of the Second Wave.

Dmitri Petrov, the Michelle and Kevin Douglas Professor of Humanities and Sciences at Stanford University who came in 1990 from Moscow at age twenty-one, described his experience with entrance exams at the Moscow Institute of Physics and Technology (PhysTech), where he applied at the precocious age of fifteen: "I mentioned to my parents that it was strange that everybody looked Jewish in the group that I was taking the oral exam with. And even though my last name is Petrov, my mother's maiden name is Jewish, and you were required to write your mother's maiden name on the papers. And so I just looked at the situation and thought it was funny: all the blond Russians were at the other end of the hallway, and everybody who looked dark and had Jewish names was right there with me. I told my father, and he mentioned it to a friend whose wife's father was President of the Academy of Sciences and who was, in many ways, a very pure man. And he really hated anti-Semitism in particular. When he heard about this, he called the rector, the head of PhysTech, and said, 'I heard some rumors that there is some anti-Semitism in your Institute. Can you please check?' So the rector wrote down my name, and they called my father. He was surprised to get a phone call and was asked to come to the Institute. He didn't know what it was about. Then they said, 'Okay let's look at your son's exams. I hope you didn't bother the President for nothing.' They looked at my exams, and they noticed that a lot of problems I had done correctly were marked as incorrect. The guy was a mathematician, and he said, 'OK, fine. He should be admitted.' My father came home and said, 'You're starting tomorrow.' The academic year had already started, and students had already been in class for two weeks. And when I asked, 'What happened?,' my father said, 'Well, I went there and looked at your exam and it was not graded correctly and so they reversed the decision.' So I got in. Later, I learned that basically there were no Jews at PhysTech at the time – like zero." When asked whether more Jews were admitted after his father had intervened, Petrov said that was not the case: "No. Anti-Semitism really was institutionalized completely. They were not going to change the institutional way of doing it. I don't know where the policies were coming from. All I know is that some institutes were known to be completely anti-Semitic, and some were not. And PhysTech was not the worst, but one of the bad ones at that time. Because, as I said, half-Jews were allowed to get in if they were exceptionally good."

Another Jewish PhysTech graduate was Epiphan Systems Founder and CEO, **Mike Sandler**, who came from Moscow to Canada in 1991, and later to Silicon Valley. He agreed with Petrov: "I believe that PhysTech evidently had quotas for Jews, and all the faculties had one Jewish student per about 100. That's what I've seen there. And it was much better than Moscow State University, MIFI [Moscow Engineering and Physics Institute], or the Bauman Institute. Those universities rejected Jews completely until 1983, when Andropov came to power."

The push to emigrate in the 1990s was primarily the result of institutional collapse in the former USSR. **Boris Berdnikov**, Staff Software

Engineer at ITA Software who came from St. Petersburg in 1995 at age nineteen, described those times and the pressures that they put on university professionals: "The times were chaotic. In my case, I understood I wasn't responsible for the family. I didn't have to feed the family. From where I was, I really couldn't see the positive changes, or I could see very little of it. I remember, though, that 1991 and 1992 were tough times. I remember my parents were not paid regularly on salary days. My mom was teaching in a community college. My dad was teaching in university and doing his programming assignments every now and then on the side. So he was being paid irregularly. I remember this one particular time where we literally didn't have money to buy bread. It seemed like my dad did get his salary that day, but he was still at work. He would only come back in the evening, and the store would be closed by then. It's not like we were going to be really hungry or anything, but it felt uncomfortable."

Informal Institutions: Family and Culture

As we indicated in Chapter 1, formal institutions give shape to the rules and stated objectives of a society, particularly in the institutions of a centrally controlled society like that of the Soviet Union. Rejection of those institutional rules or personal hardships were principal catalysts for emigration. It is also clear that the chaos brought on by the collapse of those formal institutions increased even further the pressure to emigrate. The existence of informal sources of imprinting and identity, like those from family and societal culture, provided alternative views of the world for our interviewees and helped them develop coping mechanisms.

　　Timur Shtatland, Development Scientist II at New England Biolabs, one of the early Wave Two immigrants, came from Kiev, Ukraine, in 1988 at age twenty. He spoke about the private life of his family, as opposed to the face they showed to Soviet society: "We were *refuseniks*[2] until 1988. So, during the time that I grew up, it was clear that, sooner or later, we would emigrate. Just keep your mouth shut. Keep your mouth shut about all the anticommunist thoughts of my parents and my grandfather." Like others, he was aware that saying

---

[2] "Soviet citizen and especially a Jew refused permission to emigrate." www.merriam-webster.com/dictionary/refusenik

anything against the communist regime could be used against him and his family and potentially interfere with their plans to emigrate.

Another Wave Two immigrant, **Stas Gayshan**, Managing Director and Founder of the Cambridge Innovation Center, came with his family in 1992 at age ten. He spoke about the centrality of family in defining his early life in Moscow: "I think you could say easily that we were poor. But we didn't know it, because you don't really know what you're missing. I think that's probably the best way to describe our family. We had all the things you needed to survive. Maybe not all the things that you needed to live well, and just defining what those things are could be interesting. We had a color TV and a washing machine, but I also remember the rationing and shortages growing up."

Coping in a society with continuous shortages was a fact of life for most families and an important imprinting influence for interviewees in Wave One and Wave Two. **Anna Scherer**, Science Writer at Custom Learning Designs, came from Sevastopol in the Ukrainian region of Crimea in 1994 at age eleven. She recalled the influence of those days on her early worldview: "Well, my first word as a baby was '*defitsit.*' It means shortage. So that's a pretty good sign of the times. You know, if you ask any kid, I think they'll tell you they had a happy childhood. That's what I would say about my childhood. But thinking about it as an adult, and as a parent myself, living in completely different conditions, it was normal, right? The world I lived in was the same world that all the other people I knew lived in, so I didn't know that it could be different. So it was normal, and therefore, it didn't make me unhappy in general. But the environment in which we grew up in did change over time. I felt like there was always a shortage of everything. It was so pervasive in my life, but probably wasn't the case in, like, Moscow or Kiev all the time. But in Sevastopol, there was always a shortage of food, you know, a shortage of basics like clothing and things like that."

Parents and other family members provided for the protection and cultivation of their children. Another early Wave Two immigrant, **Valerie Gordeski**, Senior Engineer II at Raytheon, came in 1995 at age twenty-two. Born in Vladimir, Russia, she recalled the protective mechanisms of her family life: "My mom and my grandmother did a really good job of shielding me from a lot of it each day, even though we had food shortages and we had two-hour long lines we had to stand in, and we had rationing and all that stuff. I thought that it was just normal." The protective instincts of her mother and grandmother allowed her to see her life as normal, rather than imprinting her with negative memories and experiences.

Informal Institutions: Anti-Semitism

In the early days of the Soviet Union, the Bolsheviks had great difficulty in dealing with the large Jewish minority inherited from the Russian Empire. Within their nationality framework, they decided to designate Jews as a separate nationality rather than a religious denomination. In line with the policy of geographic homelands for nationalities, a Jewish Autonomous Region was created in the Russian Far East on the border with China. This arrangement ignored the historical fact that the center of Eastern European Jewry was in Eastern Ukraine and Galicia. Furthermore, it did not deal with deep-rooted anti-Semitism that had been historically exploited for political purposes and that continued to infect the postrevolutionary Soviet Union.

Despite an official proclamation of equality of all nationalities, the Soviet government did not actually seek to end anti-Semitism. Instead anti-Semitism became an undeclared but well-understood institutional element of Soviet policy, particularly with regard to education, as will be addressed in Chapter 3. As noted earlier, Wave One emigration was restricted to Jews, Armenians, and Germans, with Jews constituting the majority of the immigrants to the United States. It is important to recognize anti-Semitism as an informal institution that stimulated Jewish emigration when the door was cracked open to leave during the First Wave of the 1970s and then much more broadly opened in the Second Wave of the late 1980s and 1990s.

Several early Wave One interviewees were victims of the purge of members of the dissident scientific intelligentsia in the early 1970s. **Alexander Gorlov**, Professor Emeritus of Engineering at Northeastern University, was exiled as a political dissident and left Moscow in 1976 at age forty-five. His application to study physics at Moscow State University had not been accepted because he was Jewish. Instead, he attended the Moscow Institute of Transportation Engineers, graduating with a PhD specializing in tunnels and power stations: "I worked in the Central Institute of Automation and Design and became chairman.

Andrei Sakharov[3] was my boss then." Around that same time, Gorlov's friendship with Aleksandr Solzhenitsyn[4] resulted in life-changing repercussions for Gorlov: "Solzhenitsyn came to my family with his wife, and we became great friends. I very much sympathized with what he was doing at the time and tried to help him, so that is how our friendship happened." The traumatic incident at a dacha, where he was assaulted by KGB agents and that he describes in his autobiography,[5] along with his friendship with Solzhenitsyn, brought many other problems. Gorlov and his entire research staff ended up being dead-ended in their careers, even though Gorlov said, "I was an engineer and project manager and not a Communist Party member, nor a dissident, and my association with Solzhenitsyn was purely a personal friendship between our two families." Finally, he was forced to emigrate under a direct order by Soviet President Yury Andropov.

Another member of Wave One, **Michail Pankratov**, President and CEO of MMP Medical Associates, came in 1974 at age twenty-six, emigrating as a Jewish dissident. He was born in a prison camp in Arga, in the Russian Far East near the Chinese border, where his mother was serving a sentence for anti-Soviet activities: "I was returned to my mother when she was released in 1954, when I was six years old. My mother was then a free person in exile for life, but still a free person in Siberia. But I was very lucky. I had a good upbringing with people who took care of me while my mother was finishing her sentence. My parents were sent to permanent lifetime exile in Siberia and were settled in a very small village with no civil rights. They couldn't vote. They didn't have passports. They couldn't travel. I spent four years there in Siberia. Then, in 1956, my father was rehabilitated with his civil rights restored, and we were allowed to leave the exile but not go to any big city. We couldn't go back to Moscow. We couldn't go back to St. Petersburg, where my father lived. We couldn't go to Samara or Gorky. We could only go to some small city. At that time, in 1958, we

---

[3] Sakharov, a nuclear physicist and dissident, was awarded the 1975 Nobel Peace Prize for his human rights activism. www.nobelprize.org/nobel_prizes/peace/la ureates/1975/sakharov-bio.html

[4] Solzhenitsyn, trained in mathematics and physics, was a dissident who was awarded the 1970 Nobel Prize in Literature "for the ethical force with which he has pursued the indispensable traditions of Russian literature." www.nobelprize .org/nobel_prizes/literature/laureates/1970/solzhenitsyn-facts.html

[5] A. M. Gorlov, *Incident at a Summerhouse: A Chance Encounter at Solzhenitsyn's Dacha* (Boston, 1990).

moved to a small town, Zhygulevsk, on the Volga River, where my mother's older brother lived, and we moved in with them for a few months. And that's where I lived until I graduated from high school. Then I went to medical school in Gorky and had no trouble getting in because I was a silver medalist from my high school. I always was very good at learning and science. My curiosity was always unbounded. For as long as I can remember, I always wanted to be a doctor."

When asked about highlights of his experiences at the university, Pankratov replied: "A lot of rebellion, a lot of problems with the KGB. Things were not very good, and I was arrested and released for anti-Soviet activity several times. I was a sort of student organizer. If the administration didn't do something right, I would organize students, and it was not taken lightly. The main accusation by the administration was distribution of underground publications. The materials included information about the Brodsky trial and the 1967 Trial of the Four.[6] As a result of all that, I was never allowed to really pursue my career, so after graduating I worked as a laborer loading wet laundry into large washing machines. I had to move from city to city so the KGB would lose track of me. Luckily, they were not very efficient, so every time I would move from Gorky back to the town where my parents lived, it would take them a year or two to find me. So, that was my life. Thank God, in those days, the KGB was very inefficient. But when the KGB really started to make a criminal case against me, my wife said, 'Look, you have no future here, so we have to leave.' And she convinced me to apply for emigration, and we emigrated in 1974. I had equal chances, in my opinion, of being sent back to Siberia because there were criminal cases open against me, or they would let me go. I don't think anyone knew what they would do."

The interviewees from the end of Wave One and early in Wave Two explained their reasons for emigrating as being the result of finding their life prospects limited by anti-Semitism and of living against a background of fear of physical violence to themselves and their families.

Alex Polinsky, CEO of OncoTartis and CEO of Everon Biosciences, who came from Moscow in 1988 at age thirty-two, described how his

---

[6] Underground publications or *samizdat* were prohibited by the authorities and were passed secretly from friend to friend. The Brodsky trial and the 1967 Trial of the Four refers to the Soviet cultural figures and human rights dissidents Yuri Galanskov, Alexander Ginzburg, Alexey Dobrovolsky, and Vera Lashkova.

Jewish nationality identity influenced his life and his decision to emigrate: "My grandmother on my father's side would celebrate holidays and invite all of us, so from the very beginning I was exposed to Passover and other Jewish holidays. But my parents were not religious, although they were very proud of being Jewish, and they definitely thought that Jews were very special, mentioning great scientists who were Jewish, and all that. This is how I grew up, although it felt inconvenient at first. My parents said that in the first grade I came to them after my first experience with people in school. I told them, 'I don't know about you, you can do what you want, but I don't want to be a Jew. It's just not good, not comfortable.' That was in first grade. After that, they started helping me to become comfortable with my Jewish identity. Fortunately, I was very big: by the fourth grade, I was as big as if I were in the eighth grade. So, in addition to being proud of being Jewish, I was able to physically punish anybody who would say something anti-Semitic. Not often, maybe three or four times a year, but it was there."

Polinsky next described his university admission experience: "It just so happened that at the school that I was studying in, I was very well prepared. So I wanted to enter the Moscow University chemistry department, which was the top. We knew that at Moscow University the Dean at the time was a person of great integrity and not anti-Semitic, and the chemistry department was among very few that would allow many Jews. Still, my parents taught me that I had to get A++ where non-Jews could get As or Bs. There were also two more big chemistry-related institutes in Moscow that taught applied and industrial chemistry. I entered an evening program Young Chemists School at one of those industrial schools during my high school years. By the way, in my class there was the grandson of Leonid Brezhnev. He was also named Leonid, and he liked chemistry, and then he went to Moscow University like me. After two years at the Young Chemists School, I was admitted to the industrial institute without taking exams, so at least I had a backup in case there was a problem at Moscow University. Then, when I took exams for admission there, I was more relaxed because if I failed I would go to the industrial school, not straight to the army.

"I was admitted to Moscow University in 1973, and was there for five years. Then I went on for my PhD that I got there in 1982. I wanted to be an academic, and all indications were that I could be a successful one. That's what I was told anyway, and eventually I was given an academic position there doing teaching and research. But not without some difficulty. My department head and my mentor was a big academician, a member of the ruling committee, the Presidium of the USSR Academy of Sciences. Even so, Moscow University did not allow him to hire me after graduate school. He was told that his department already had too many Jews. So he went to his friend who, by the way, was that dean who allowed Jews in the chemistry department in the first place, but he was also director of a big Academy of Sciences institution, the Institute of Biochemistry. And they struck a deal. My department head hired a girl and my department paid her salary, but she actually worked at the Institute of Biochemistry. At the same time, the Institute of Biochemistry hired me and paid my salary there, but I worked at Moscow University. So even a person of my mentor's stature was told, 'no,' he could not hire another Jew. A few years later, my department head at Moscow University managed to hire me properly, but even so, I left four years later.

"Around 1987, a second thing happened in my personal life that was the last straw. There was a big anti-Semitic action in my son's school where the walls were painted with anti-Semitic insults. The director of the school was Jewish, so his door was spray painted with insults. My son was in the fourth grade at the time, and he was not prepared for something like that. The school was great and the teachers were great. We liked the school, and the kids were OK. There had never been anything anti-Semitic. He was not yet introduced to this whole thing of dealing with anti-Semites. After it happened, of course, the school was repainted next day. But it was interesting to watch how a lot of kids had something that they were obviously hiding, and it suddenly came to the surface. They pointed out to my son that he was Jewish, and they educated him about how bad it was. So the week after that was unbelievable. That event exposed a lot of underground anti-Semitic prejudice, and since I was thinking about leaving anyway, I said, 'OK, that's a good enough reason for me to leave Russia.'"

A Director at a leading global asset management firm who came in her teens in the mid-1990s from Moscow related her family's story. It illustrated the mixture of anti-Semitism and suspicion of political unreliability that was a real but unofficial limit on their life prospects in the Soviet Union: "In fact, my mom is actually Russian and my dad is Jewish. The problems started in the beginning when she was getting married. The families were concerned about my parents getting married because they would have to face many difficulties in Russia being a mixed couple. And then my dad, after he graduated from college, had no choice about his future education, whereas my mom had a choice of where she could go."

Semyon Kogan, Founder and CEO of Gen5 Group, brought his family from St. Petersburg in 1989, when he was in his late thirties. He described the discrimination against Jews that four generations of his family had endured, including his young daughter, Nataly: "My parents and my grandparents and I hadn't done anything but study perfectly, work hard, and do as much as we could. Yet we always had been treated as second-class citizens. My parents since I was a kid would tell me, 'If you want to be equal, you have to be five times better than anybody else.' With Nataly it was different. We used to protect her. She really never knew what it meant. I remember she was in a children's dance group that was one of the best in Leningrad. It was taught and run in a pure Soviet style, five times a week, with no smiling, no nothing. If you couldn't do the splits, the teacher would sit on you. So it was torture, but they were perfect. When she was maybe in the fourth grade, the dance group was supposed to tour Bulgaria, which meant going abroad. This was really a big deal because nobody in our family had ever been abroad. The manager called me up and asked me to come over. She said, 'I am sorry, but the authorities wouldn't allow Nataly to go, because ... you know why.' I did know why they would not let her go. So I said, 'Don't tell her.' I bought a big box of candies, went to her doctor, and said, 'Could you write a note that you're not allowing her to travel because of some medical issue?' Nataly obviously was upset and cried, but at least she did not feel that she was treated differently because she was Jewish. My parents never did that with me. Instead, it was the other way around: 'You know that's the way it is, so just live with it.' So, the more I live here in the US, the more I hate the communist authorities for this injustice and humiliation, even though on the surface everything was picture perfect. I graduated from a special school for gifted and talented kids, got a PhD, I was a successful manager, my wife was a piano teacher, we had our own apartment instead of living in a communal one, and our daughter went to a special school. Picture perfect on the surface, but ...."

Kogan's daughter, **Nataly Kogan**, Founder and CEO of Happier, Inc., in a separate interview also recalled how devastated she was at not being allowed to go abroad to perform with her dance troupe, unaware at the time of the true reason. She then mused about another incident during her childhood, noting the irony about then-Soviet President Leonid Brezhnev: "There was a billboard I could see from our apartment window that had a large picture of Brezhnev. When I was a young child, I was completely obsessed with him to the point that when he died, my parents kept the curtains closed for a while because they were actually really afraid of how I was going to react to Brezhnev dying. And when they told me, I asked in disbelief, 'What do you mean?' And this is all very funny. We are Russian Jews and we are constantly persecuted, but here I am obsessed with Brezhnev." Kogan's story illustrates the mixed identity and the impact of contrasting imprinting from formal and informal institutions in developing her personal identity at that time.

The institutional collapse of the mid-1990s also led to more open expression of informal anti-Semitism, as illustrated by the experience of **Leonid Komarovsky**, President of the Boston Russian Media Group, who came in 1995 at age forty-eight. Born in Chisinau, Moldova, he emigrated from Moscow: "Well, for example, I had a country house. We would drive there, and I would see written in big letters on the side, 'Leave, Jew. Go back to Israel.' And people would walk up to me on the street and say, 'You Jew, I am going to beat you now.' In the school that my child went to, one time the director with whom we had a good relationship and had even helped with money, called me and said, 'You know, we caught a person in our school who was writing the addresses of Jewish children in their journal.' And I thought, what's the point of causing a ruckus? And this just happened to coincide with my father going to the hospital. Since he was a big boss, he was in the Kremlin hospital, but they basically killed him there. It was a bad operation, and he didn't survive. But when going to the operation, he had told me, 'Leonid, that's it. Leave.' So when he died, we left."

Fewer interviewees in the Third Wave spoke of anti-Semitism, and it was apparently not as important a reason for leaving as it had been for those who came in the first two waves. Although anti-Semitism was the most common form of discrimination in the former USSR, it was not the only one. **Diana Tkhamadokova,** who was born in Nalchik, in the

Kabardino-Balkar republic in southern Russia, recounted: "We faced discrimination all the time. But it's not like we cared. The unofficial name for our republic is 'Black Hats' because we are from the region where we're kind of naturally darker. But I don't know. I think we also happen to be more educated. People who want to can discriminate against wealth, origin, hair color, skin color, whatever they want." Tkhamadokova, Founder of the online fashion firm, I Style Myself, earned a master's in finance and economics from the University of Manchester in the UK and worked in London at Goldman Sachs, JP Morgan, and Barclays before coming to the United States in 2015 at age thirty-two.

Interviewees from all three waves indicated how their lives in the USSR or former USSR seemed to have been fraught with more difficulties than opportunities for happiness and personal progress. Many lived with discrimination, including anti-Semitism, while others simply faced the hardships of daily life, including shortages of essential goods, along with various reasons to be fearful. Membership in the Communist Party was clearly a prerequisite for career success during Soviet times, almost regardless of level of education or technical or scientific achievement. This was primarily true for parents of our interviewees, with the possible exception of Wave One immigrants. For interviewees who emigrated later to the United States, little was said about their own experience with the Party, but a great many had belonged to communist youth organizations such as the Young Pioneers and the Komsomol. Like their parents, though, membership was virtually expected if one hoped to gain opportunities for leadership, involvement with special projects, and access to better quality goods and services.

Most of these realities for interviewees and their parents virtually evaporated after the breakup of the Soviet Union. As some explained, that period began with high hopes and even jubilation, but those feelings soon gave way to hardships, shortages, lawlessness, civil wars, and wars between former republics, causing widespread feelings of despair. Among Wave Three immigrants particularly, the reestablishment of law and order, especially during Putin's second term, brought about a renewal of hopes for a better future in their homelands. That promise, too, soon evaporated as Putin's third term, after the Medvedev interregnum, gave way to the annexation of Crimea and war in eastern Ukraine. Ukraine, of course, was also hard hit during this period, and other republics lacked the resources and political leadership and will to offer anything like abundant opportunities. By this time, virtually all who might have considered returning to their homelands had given up that idea and seemed to have become more resigned to their new circumstances and opportunities in the US innovation economy, with its highly supportive institutional structure.

3 | Soviet Educational Institutions
Capability for Contribution

Most people interviewed for this book came to the United States with capabilities gained through the Soviet educational system. Many of them contributed to the US technology sector immediately or soon after becoming somewhat acclimated. Their education, coupled with their intelligence and drive, prepared them to join the US technological revolution that took place in the later half of the twentieth century and the beginning of the twenty-first century. Without an overview of the Soviet educational system, it would be difficult to understand the abilities, cultural reference points, and attitudes of immigrants from the former USSR. This chapter will review the educational system and the special institutional and social aspects of that system that helped shape our interviewees through an inevitable imprinting process.

In the early twentieth century, the Soviet government succeeded in raising a largely illiterate population to almost full literacy in a very short time. The educational system was developed to serve the needs of the forced march toward industrialization, which was heavily dependent on developing human capital in engineering, science, and mathematics. Equally important were the requirements of military production and, later, the Soviet space program, both of which also demanded world-class science and engineering.

To meet these demands, the Soviet government began to develop special high schools and elite university programs. Many chose careers in math and science since they were among the least politicized career paths. Jewish students often found their paths full of unforeseen obstacles and sometimes blocked by unspoken anti-Semitism and quotas. So, while the educational system prepared highly trained and skilled professionals, it also provided a stimulus for some to leave when the opportunity presented itself. This stimulus was broadened and deepened by the economic collapse at the time of the dissolution of the Soviet Union, leading to a growing influx of talented and trained professionals to the United States. The strains between the declared ideals of Soviet society and its actual practice created the paradox of high-level Soviet and post-Soviet educational institutions preparing some of their best and brightest to make their contributions to the technological revolution elsewhere.

In this chapter, we describe a system with very strong educational and research institutions developed by the Soviet government with its socialist political and social objectives and mode of operation as a centrally planned economy. When the Soviet Union ceased to exist as of December 26, 1991, the fifteen newly independent countries each had the sovereign authority to define their own objectives and modes of operation, including educational systems. However, the ongoing impact of the heritage of the Soviet educational system remains a dominant influence in most of these countries. The situation will not be as clear in the future as these countries continue to develop their own national systems of education that reflect their diverse societal values. For our interviewees, the Soviet educational system remains a dominant institutional influence; therefore, we focus primarily on that system and its main successor country, the Russian Federation.

Czarist Educational System

Civilian education was very limited under the czars until the beginning of the eighteenth century, when Peter the Great focused on it as part of his goal of modernizing the Russian Empire. He required mathematics expertise in particular for development of the military and armaments, and he founded the School of Mathematical and Naval Sciences in 1701.[1] Moscow University was founded in 1755, but, until around 1825, the only important educational institutions were the cadet schools, military and technical academies to train future officers from the gentry.[2] The Russian Academy of Sciences (RAS) was established in 1724, with the mission to organize and conduct research in the physical sciences and mathematics. Although it had some educational function, it was primarily involved with research.[3]

---

[1] Joseph I. Zajda, *Education in the USSR* (Oxford: Pergamon Press, 1980), 1.
[2] Martin Malia, "What Is the Intelligentsia?" *Daedalus* Volume 89, Number 3 (Summer, 1960): 453.
[3] Yury S. Osipov, "Traditions Generating," in *Academy of Sciences in the History of the Russian State* (Moscow: Nauka 1999). http://www.ras.ru/en/about/history/traditions.aspx.

It was only under Alexander I that the Russian Empire began to pay serious attention to building an educational institutional structure for civilians. After 1803, the Czarist government established five new universities, high schools (gymnasia) were established in almost all provincial capitals, some efforts for elementary education were begun, and diocesan seminaries were reformed and modernized.[4] This system began in the 1830s and 1840s primarily to educate bureaucrats for the increasingly complex administrative tasks of the government.[5] The system thus had limited impact on the vast and populous Russian Empire, which remained dominated by the feudal relationship between a narrow autocracy and the mass peasantry until the end of the nineteenth century. After the emancipation of the peasantry in 1861, there was a gradual development of industry and a more diverse social structure. However, the general population was not educated. In the 1870s, the university population consisted of only 5,000 students.[6] The rate of illiteracy in 1916 for the population as a whole stood at 66 percent.[7] In prerevolutionary Czarist Russia, only 250 private and 30 public kindergartens existed in the entire country.[8] At the time of the 1917 Bolshevik Revolution, the Russian Empire had a small but prestigious university system comprised of 105 universities, half of which were private, and 127,000 university students.[9]

Soviet Educational Objectives

The Soviet educational system is important for our story through two central themes. First, given its origin in the 1917 Revolution, it had the broad goal of creating a new system of morality, culture, and civic behavior in the "Soviet man" or *Homo sovieticus*, which exerts an impact on the identity and frame of reference of our interviewees. The educational system was also designed to provide the human capital for the massive, forced transformation of the Soviet republics from a predominantly peasant, agriculture-focused economy to an industrialized one.

---

[4] Malia, "What Is the Intelligentsia?," 453. [5] Ibid. [6] Ibid., 454.
[7] Zajda, *Education in the USSR*, 11.
[8] "Chapter 1: Principles and Structure of the Soviet Education System," *Soviet Education* Volume 26, Number 9 (1984): 86.
[9] Mervyn Matthews, *Education in the Soviet Union* (London: George Allen & Unwin, 1982), 97.

The Council of People's Commissars created the State Commission on Education on November 9, 1917, under the direction of the Marxist philosopher and literary critic Anatoly Lunacharsky.[10] The Commission quickly brought all educational institutions into one unified system, with a complete separation of religion from the educational process. The result was a secular system of preschool, primary, secondary, and higher education unified under one government ministry.[11] By 1919, government policy on education included free and compulsory education for all between the ages of eight and seventeen, as well as creation of preschools and coeducation.[12] The main tasks of Soviet education in the 1920s were how "to educate 100 million illiterate people and to create a Soviet intelligentsia."[13] After the 1917 Revolution, the RAS was incorporated into the Soviet system as the Academy of Sciences of the USSR.[14]

New Youth Organizations and the Culture of the "Soviet Man"

Along with the structured educational institutions, the Bolsheviks initiated a system of youth organizations to broadly promote their vision of morality, values, and culture. This system provided a unitary experience and set of cultural references for all Soviet citizens. As we will see in interviews referred to throughout this chapter, individuals varied in their acceptance of the all-encompassing nature of this system, but it formed a reference point for everyone who grew up during Soviet times.

The Bolshevik Party did not have any affiliated youth organizations at the time of the 1917 Revolution and was faced with the need to create them as part of their mobilization of supporters during the ensuing Civil War of the 1920s. Afterward, these organizations assumed a critical role in the socialization of young people and in promoting socialist values. These youth organizations also provided extracurricular social activities, along with acting as a political filter for

---

[10] M. A. Prokof'ev et al., "Development of the Soviet Preschool Education System," in *Narodnoe Obrazovanie v SSSR, 1917–1967 [Public Education in the USSR, 1917–1967]* (Moscow: Uprosveshchenies Press, 1967), 3.
[11] Zajda, *Education in the USSR*, 10.   [12] Ibid., 11.   [13] Ibid., 39.
[14] Yury S. Osipov, "Soviet Times," in *Academy of Sciences in the History of the Russian State* (Moscow: Nauka 1999). http://www.ras.ru/en/about/history/traditions.aspx.

the career ladder in the ideologically driven Soviet Union. The first-tier youth organization was the Young Oktobrists, enrolling children from ages seven to nine. The children were divided into small groups of five, with a member of the next age group, Young Pioneers, as their leader. The Young Oktobrist organization provided a variety of social and physical activities.[15]

The second-tier organization, the Young Pioneers, was formed during the Russian Civil War. In Czarist times, there had been an active Boy Scout movement, the majority of which fought against the Bolsheviks. A minority of the leaders of the Czarist Boy Scouts sided with the Bolsheviks and formed the Young Pioneers based on their Boy Scout experience. They borrowed the format and the slogan of the Scouting movement, "Be Prepared" for the new pro-revolutionary organization.[16] The slogan was slightly altered to "Always Prepared," a message that was implanted in the minds of the young people who were members. The Young Pioneers were formally established in 1922, and, by 1925, 11 million children were members.[17] In addition to its political and ideological mission, the Young Pioneers was also an organization for outdoor exercise, camping and adventure trips, arts and crafts, and music and sports. The organization was made up of children aged ten to fifteen. Officially, Young Oktobrists had to apply and then be accepted into the Young Pioneers.[18]

The third tier of the Soviet youth organization system was the All-Union Leninist Young Communist League, or Komsomol, which was established in 1918 as the youth branch of the All-Union Communist Party.[19] It was composed of young people aged fifteen to twenty-eight. Like the Young Pioneers, members were required to apply and be accepted. Ultimately, the Komsomol was the source of members for selection into the Communist Party. For some levels of government, it was considered highly recommended to show political reliability through membership in the Komsomol and then the Communist Party.

[15] "Of Russian Origin: Pioneers," *Russiapedia*, http://russiapedia.rt.com/of-russian-origin/pioneers/ (accessed September 20, 2016).
[16] Ibid.   [17] Zajda, *Education in the USSR*, 22.
[18] "Of Russian Origin: Pioneers," *Russiapedia*, http://russiapedia.rt.com/of-russian-origin/pioneers/.
[19] "Komsomol," *Wikipedia*, https://en.wikipedia.org/wiki/Komsomol/.

Goals of Soviet Education

The goals of the institutional educational system set during the early years after the Bolshevik Revolution survived roughly intact as the guiding principles of the USSR almost until its dissolution. The continuity of Soviet educational philosophy over time can be illustrated by the following Soviet text, translated and published in English in 1984:

> The singleness of purpose of education and upbringing in all areas underlies the basis of the uniform educational system in the USSR. .... This purpose is realized in the teaching of communism, the intellectual and physical development of young people, and their preparation for work and social activity. In other words, the task of developing the well-rounded personality of each member of society is given top priority and is seen as an indispensable prerequisite of preparation for occupational and professional activity and social life. Hence, the educational system helps reinforce the socio-political uniformity of Soviet society, for inasmuch as it creates no distinctions between the representatives of various classes and strata, the balanced educational preparation guarantees the formation of a scientific worldview in each person, conferring basic knowledge and cultural values and inculcating the norms of socialist morality, the principles and rules of Soviet public life.[20]

The fundamental emphasis on extreme egalitarianism in this statement of principles came into conflict with the necessity for producing highly trained specialists and for nurturing those with special abilities. In reality, the post-Stalin period saw an increasing number of special schools and enrichment programs to foster competition and support for gifted students.

Institutions of Soviet Education

The structure of the Soviet educational system was divided into the following components: preschools, general education schools, vocational educational (three types), higher educational institutions,[21] and graduate work in the Academies of Science. These were supplemented by extracurricular organizations like school clubs, as well as special evening and summer schools. Political and civic education took place

---

[20] "Chapter 1: Principles and Structure of the Soviet Education System," 55.
[21] Ibid., 55.

through membership in youth organizations – the Young Oktobrists, the Young Pioneers, and the Komsomol described earlier.

By 1934, the structure of the Soviet educational system was settled by government decree as a ten-year secondary general education. After four years of primary school, students had the option of pursuing seven or ten years of secondary school.[22] The number of years in each component of the system was a matter of debate and change throughout the 1960s. Since citizens had the legal right to study in their native language, the number of years was sometimes expanded in different Soviet republics to account for language differences. Some schools included all levels in one building, and some were in separate buildings depending upon local conditions. All schools had a common curriculum and textbooks. This common curriculum and set of textbooks formed a cultural reference point for all Soviet students, as noted in our interviews.

In terms of formation of a value system, it is important to include the influence of family traditions, history, and values discussed in Chapter 2. Although not considered within the formal educational system, they constituted a critical source of the moral values and worldview of our interviewees. In a number of cases, this influence proved stronger than the official ideologically oriented education.

Development of Special Schools

If uniformity and egalitarianism were the full story, then the impact of immigrants from the former USSR on the US high-tech sector would have been quite different. The vast majority of our interviewees were the product of the special schools that developed in the 1960s. In reality, the uniformity of secondary general education schools was eroded by a combination of factors that led to increasing numbers of special schools in which one subject, like foreign language, math, physics, or biology, received special attention and advanced teaching. Ultimately, the Soviet system included special schools for the arts; sports and circus schools; foreign language schools; biology, mathematics, and physics boarding schools; and others. In mathematics and

[22] "Secondary General Education School," in *The Great Soviet Encyclopedia* (Moscow: Soviet Encyclopedia Publishing House, 1979), excerpted in *TheFreeDictionary*, http://encyclopedia2.thefreedictionary.com/Secondary+General-Education+School.

physics in particular, these new special schools, as well as reforms in education, had a huge impact on the lives and training of many of the immigrants interviewed for this book, particularly for those from Moscow and St. Petersburg, along with a few regional centers of education like Novosibirsk.

In a society that philosophically did not construct differences and boundaries between school life and private life, the system of education for talented young people included an array of extracurricular activities that should be considered as part of the educational process. These activities included local, regional, and national competitions or Olympiads, after-school clubs and activities, summer camps, and special-interest magazines and journals targeted at young people. It was a very competitive environment, but one that supported and valued excellence in science and math from an early age. The impact of that approach is seen clearly in the quotes of our interviewees.

Alex Polinsky, CEO of OncoTartis and Everon Biosciences, came from Moscow in 1988 at age thirty-two. He discussed how he convinced his parents to look into a special school for him to increase his chances of getting into college and avoiding the army: "I got them thinking about it and they started talking to friends. They realized that it was time to put me in some better school that would prepare me for something. After a short search, they put me into an experimental school, where, for the last two years, they divided children into classes with different specialties. We had seven different classes: two with math emphasis, two with physics, one with biology, one with humanities, and the last one was chemistry. I don't know why, just probably because my father was a chemist, I decided to go into chemistry. So I went to the chemistry class and that was very important for two reasons. One is that it brought me to a completely different circle of boys and girls. I had never seen anything like them before. They were smart, and they all knew more about chemistry than I did, which didn't sit well with me. So pretty soon I knew exactly as much as they did."

Alexey Bulychev, Principal Scientist at Moderna Therapeutics who came from Moscow in 1992 at age twenty-four, had the unusual experience of starting out at one type of special school and later switching to another: "I went to a school specializing in English in the second grade. Then, after eighth grade, I transferred to a school specializing in chemistry. So I spent the last two years of school taking advanced chemistry classes. The school was one of a kind in the country: except for chemistry, all other classes were taught according to the standard program. But chemistry was taught by recognized university faculty, adding a notable amount of hands-on lab experience. Although my career aspirations were never up for serious discussion, such a pedigree assumed that I would go to college for chemistry. Hence, it was really strange and confusing for me to realize that I might have a greater passion for electronics than chemistry. As it was right around the time when I had to apply for college, and the Chemistry Department of Moscow State University was my first choice, I almost panicked. But in the end, reason took over and the rest is history."

If we can borrow an expression from today's lexicon, it would be correct to state that there was an entire ecosystem of supporting institutions built around mathematics. This ecosystem included cultivating interest in mathematics by talented young people at an early age. At the beginning of the 1960s, four leading universities established boarding schools for talented high school students. University professors, as well as specially selected teachers, gave lectures in these schools. The best students from all over the country were gathered together in these schools. Special effort was made to bring in students from small towns and villages. Later, this network of boarding schools was expanded and included specialized math schools in large cities. In 1970, the monthly magazine, *Kvant*,[23] was founded to explain complex concepts in mathematics and physics to high school students. *Kvant* ultimately had 300,000 subscribers.[24] Many interviewees included comments about these schools.

Leonid Raiz, Cofounder of Rize, came in 1980 at the age of twenty-nine and was one of the very early members of the industry-changing engineering software company, PTC: "I was born in Moscow and moved to the Far East. Afterwards, we lived in the former East Prussia, in the Kaliningrad region. An important event in my life took place after the eighth grade. One of my friends told me that there were people interviewing kids for boarding school back in Leningrad, which was affiliated with Leningrad University and specialized in science and math. It was in 1966, and I ended up going to that special school. It was

[23] *Kvant* means "quantitative."
[24] Sergei Lando, "Mathematical Education in Universities in the Soviet Union and Modern Russia," *Higher Education in Russia and Beyond* Volume 4, Number 6 (winter 2015): 6.

quite a transformation for me because, in the first years, I was almost expelled for bad behavior and bad grades at the same time. By the second year, which was the graduation year, I was already doing quite well. By that time, I was indoctrinated with the idea that I had to enroll in Leningrad University, which was, if not the number one, the number two university in the Soviet Union. I had to get there and I had to become a mathematician."

Another graduate of a specialized school was **Sergei Burkov**, Founder and CEO of Alterra, who came from Moscow in 1990 at age twenty-one. He also put the Soviet specialized school in a personal context as well as in a comparative context with a US analog: "The most formative experience for me was this mathematical magnet school in Moscow. It's called School Number Two, and at that time it was the best. Now there is a better one called School 57. And that's high school, so it's from fourteen years old and up, and there is an entrance exam. In some sense, it's similar to the Bronx School of Science and also to Stuyvesant that are in New York – the same idea. It's a high school with entrance exams and a focus on mathematics and physics that was fantastic. That was probably the best three years of my life, and I learned more than in university."

Many Second Wave interviewees had very similar experiences with the formative influence of specialized secondary schools. **Alexey Wolfson**, CEO of Advirna, who came from Moscow in 1997 at age fourteen, recounted the profound influence of evening school on his life: "I went to a special English school and basically started learning English in second grade. It was usually started in seventh grade, but it was intensive, which was definitely felt. It just so happened that at the age of twelve, I was introduced to an evening school for kids in biochemistry, and I met there an absolutely wonderful professor, Kirill Gladilin. Basically, I fell in love with him right away. And then I started to work there and do biology, biochemistry, and chemistry, and I loved it. I didn't know I was supposed to become a scientist until I met him, and in a half hour he convinced me that science was the best thing for me."

Max Skibinsky, serial entrepreneur and investor who came from Moscow in 1996, followed the path of specialized school coupled with extracurricular enrichment: "I was at math and physics School 625, which is somewhat known in Moscow, not as well-known as School Number 52. Through that school, I participated in the

Russian Olympic math and physics competitions. I participated in all those Olympics in high school, and then went to do physics at Moscow State University. I moved to the United States after graduating."

Dimitri Popov is a serial entrepreneur who came in 2014 at age forty from the electronics hub of Zelenograd near Moscow. He described his experience with the rare computer technology that was available in his high school: "It was the first school ever in the Soviet Union that installed PCs. There were twenty-four of them, and you could come over and just work on them. They were green, darker screen Soviet PCs that weighed probably as much as a washing machine."

The educational system also forced difficult choices of specialization at an early age. As with the general school system, the main impacts from extracurricular activities did not differ significantly from wave to wave due to the continuity of the educational system, even after the breakup of the Soviet Union. **Dmitri Krioukov**, Associate Professor of Physics, Mathematics, and Electrical and Computer Engineering at Northeastern University, who came in 1994 at age twenty-four from St. Petersburg, spoke about the difficulty of making career choices at a young age and the stakes involved in choosing or not choosing to attend a specialized school: "I play classical guitar, but after the eighth grade, you had to decide: you either go to this professional school or you continue two more years in high school or you continue to university. So after finishing the eighth grade, I had to decide if I wanted to continue to study guitar playing, then go to music school for two years and then maybe go to the conservatory. The other choice was to go to this school in physics and mathematics, which I also liked. So it was quite a hard decision for me, and at that moment nobody tried to influence my decision but left it for me to decide. And I decided to go to physics and mathematics school, and I'm happy I did because I still think it gives you a lot more freedom. This science career gives you a lot more freedom. More than art, I still think."

Third Wave interviewee **Yury Lifshits**, Managing Director of Entangled Solutions, who came from St. Petersburg in 2007 at age twenty-three, reinforced the continuity and ongoing influence of the Soviet system in the school and extracurricular system in Russia in the 1990s. He described the rigorous training he received in his extracurricular math circle that propelled him to win two gold medals in the International Olympiads in mathematics: "I had a teacher who was working with me in mathematics from the sixth to eleventh grades. He started with about seventy kids in sixth grade, ending up with about eight to fifteen in eleventh grade. Of these, four became members of the national mathematics teams out of a total of six for the entire Soviet Union. So, in the whole country, he took a random seventy kids and got four of them out of six onto the national team. Definitely the sample he was working with was extremely small. That the rest of the country put just two kids on the team and he put four kids definitely demonstrates that it's more about the teacher than about the kids."

Another International Olympiad participant, serial entrepreneur **Max Polyakov**, won second place in biology in 1995. Originally from Zaporizhia, Ukraine, he came to Silicon Valley in 2012 at age thirty-five. He explained that in the regional Olympiads he had won prizes in the biology, physics, and mathematics competitions, crediting his parents who were both rocket scientists. He continued: "But then for the All Soviet Union Olympiads you had to choose only one subject, so I chose biology. And after winning the biology International Olympiad, I decided to go to medical school because it didn't require any exam because of my gold diploma and my medal, and I could go to any medical school in the Soviet Union. But we didn't have money for a train ticket to Moscow or to support me there. I cried because I obviously wanted to go to Moscow for medical school and genetics. So that's why I stayed in Zaporizhia. It was a very poor situation."

Thanks to the centralized, unitary government system of the USSR, people from all fifteen Soviet republics shared common aspects of special schools. They also experienced the differences that come from being either far from the capital or in a linguistically different and more complex environment. After the Soviet Union broke up, the educational institutions in the newly independent countries went through very hard times. **Anton Rusanov**, a software engineer at Google who came to the United States in 2012 at age twenty-eight, had moved from his home in Kazakhstan to Russia for high school and also experienced the highly competitive nature of the Russian school system: "I studied in Karaganda, Kazakhstan, until middle school, and then I left for Russia to enter high school. It was a boarding school, where I lived away from my parents for a year. The school was affiliated with the university and specialized in math and physics. I was the best student in my class in Karaganda, one of the three best students among the eighty or so students in my age group. But when I moved to Novosibirsk, I found out that I was just average because there were a lot of more talented students. It was hard."

Alex Petetsky, Director of Software Development at PatientKeeper, gave his interpretation of the educational opportunities in different Soviet republics, including the highly specialized training he received at an institute that was a feeder of engineering talent for a major enterprise. Originally from Berdychiv, Ukraine, he emigrated immediately after college in 1991 at age twenty-four: "I decided to leave Ukraine and go to Minsk in Belarus because it was hard being Jewish in Ukraine, and I had family in Minsk. I was a little afraid, a little intimidated to try Moscow or St. Petersburg. Coming from a small town, I just didn't think that I was good enough. So I went to the Belarusian Institute of Technology and studied furniture technology, everything wood-related and the manufacturing machinery involved with that. A town near Minsk had a huge manufacturing facility that made all kinds of stuff from wood – from matches to guitars. The program was quite similar to what you learn here in mechanical engineering. For the first one and a half years, it was algebra, physics, and chemistry." When asked if he got a good education, he replied: "I think so. Probably not as good as some of the elite schools in the country, but I think I got a decent education."

David Gamarnik, Assistant Professor at the Sloan School of Management at MIT, came in 1990 at age twenty-one from Tbilisi, Georgia. His experience illustrates how the Soviet system dominated the entire country and brought with it the emphasis on math and science: "I went to a Russian school and actually grew up not knowing Georgian at all. So up until seventh or eighth grade, I was reasonably good in math. I thought I wanted to become a scientist, but I really wasn't sure. My mother told me, 'There's a math and physics school that is very selective, and so you have to take tests and so on to get in. You should go there, you should try there.' I thought, OK, that sounds good. So I was admitted to this school, which was very selective using those tests to pick a bunch of talented, promising young kids. Only then did I develop a love of mathematics when I realized that mathematics is really a fascinating thing. So on top of regular classes in mathematics, I had what is called 'math circle.' It was a math club, where we were doing all kinds of recreational, Olympiad mathematics problems. It just blew me away. Being surrounded by kids interested in mathematics makes you more interested, and we sort of rubbed off on each other and became very engaged in math subjects. In this school, the focus was math and physics. Some kids fell in love with physics and continued, becoming physicists. I decided that math was my thing."

Shalva Kashmadze, also from Tbilisi, came in 2011 at age twenty-five, after Georgia had gone through the rocky early years of independence. Senior Project Manager at Pocket Gems, he described the difficulty of getting current textbooks and information at Tbilisi State University: "My major was finance and banking, and the problem was that there was an absence of good textbooks in Georgian. The banking books, finance books, they made no sense. In many cases, there were people who were teaching us something that they didn't know themselves. That was true for many, many disciplines. People were always trying to kind of get their education in Western countries because of that, because they could not get a good education in Georgia. There were no good professors, no good books. And they were trying to learn something on their own, trying to get those books, bought from Amazon or from somewhere else, and then to read up on their own. In my undergrad days, I was really curious about derivatives. I was curious about more complex financial instruments, and they didn't teach me that. There is a famous investment book by Merton Miller, and I bought the Russian edition. So I was reading and trying to learn those things."

Armenia also experienced great hardship after the breakup of the Soviet Union. **David Gukasian**, Senior Systems Administrator at Battery Ventures, who came from Yerevan in 1995 at age seventeen, described the impact of the economic problems on his education: "There were some interruptions, but I was going to a private school back then. Right after independence, they opened some private colleges, so I transferred from public school to a private college. In the private schools, the biggest challenge was in the winter. They did not really have the sustainability to heat the classes. We were asked to just come to class once a week, receive our homework, and then go back home for the whole week. We would just do the homework at home and then show up once a week. They would heat up the stove and then sit down to discuss the homework material. That was pretty much it."

Another interviewee from Yerevan, **Nerses Ohanyan**, Director of Growth Analytics at Viki, who came in 1998 at age fourteen, recounted the language difficulties in Armenia after the breakup of the USSR:

"My school was actually traditionally a Russian language school. It was called Chekhov Number 55. When my aunt went to that school, all the curriculum was in Russian. The year I started is when everything changed. So when I was in first grade the school became an Armenian-language school. Russian was still very strong, because obviously the teaching staff there was very heavily Russian. The Russian teachers were very well-versed in Russian, but all the curriculum changed to Armenian, which made it kind of hard for some teachers because they were used to teaching it in Russian, and they had to change it all to Armenian."

Anna Lysyanskaya, Professor of Computer Science at Brown University, who came from Kiev, Ukraine, in 1993 at age seventeen to attend Smith College, noted the impact of specialized education: "When I was in high school, my favorite subject was English, but that's just because I went to a high school that specialized in English. So in Kiev, much as throughout the Soviet Union, there were certain high schools that were considered elite. It wasn't easy to get in. You needed to have connections. These schools would specialize in languages, and that is where the elite parents, who were allowed to travel abroad when most people weren't, would send their kids. They wanted their kids to learn foreign languages so that they could have a better life than everybody else. And I went to that school not because my parents had this kind of background, but because, a long time ago, before that school had acquired this kind of status, my grandmother used to work there. I got all A's in my junior and senior years and should have gotten a gold medal, but they changed the rules retroactively in order to help somebody else." Lysyanskaya's story shows not only the positive side of the informal institution of networks in the educational system, but also the negative side.

Stratification and Hierarchy of the University System

As noted earlier, many of the people interviewed for this book graduated from special secondary schools, and many are also are graduates of a small number of leading universities in the former USSR. Soviet society had an official ideology of equality, but an actual practice of stratified status. Within the university and scientific worlds, the pecking order of Soviet universities and leading institutes of the Academy of Sciences was very strong and influential. This status hierarchy was created by a combination of history and government policy during the Soviet period. It remained largely in place even after the breakup of the USSR and has provided valuable background for the qualifications and attitudes that these immigrants brought with them to the United States. Their university affiliation remains a reference point toward which they have deep loyalty even after the many years that have passed, not unlike alumni of US universities.

The very limited number of universities created in Czarist times was primarily associated with the needs of the state, and these institutions were mainly located in St. Petersburg and Moscow. These included St. Petersburg State University, founded in 1724; Moscow State University, in 1755; followed by the first technical university, Moscow Craft School, established in 1830, which was renamed in Soviet times as the Bauman Moscow State Technical University. There was a continuing and slow expansion of the establishment of additional universities in parallel with the expansion of the Russian state boundaries, with Kazan and Kharkov founded in 1804, Kiev in 1834, Odessa in 1864, Tomsk State in 1878, and Tomsk Polytechnic in 1896.

At the time of the 1917 Russian Revolution, the new Soviet state made expansion of the educational system a priority, including rapid expansion of universities in the Russian regions and within the new nationality-based republics that composed the political organization of the new Soviet Union. Between 1916 and 1923, new universities were founded in Nizhny Novgorod, Perm, Dnepropetrovsk, Smolensk, Voronezh, Yerevan, Samara, Yekaterinburg, Minsk, and Tbilisi, not to mention a number of new technically oriented or training universities in Moscow. The new tasks facing the Soviet State as a result of World War II, known in the USSR as the Great Patriotic War, led to the founding of the Moscow State Institute of International Relations (MGIMO) in 1944, and the Moscow Institute of Physics and Technology (PhysTech) in 1946. Both became prestigious institutions in their respective fields.

Under Nikita Khrushchev in the 1950s, and later Leonid Brezhnev in the 1960s through the mid-1980s, there was further expansion of universities into the regions and republics. Of particular importance was Khrushchev's project to mimic Silicon Valley, developed after his historic trip to the United States in 1959. In 1958, he had already approved a plan to create a new eastern division of the RAS, to be located outside Novosibirsk in the new district of Akademgorodok.

Ultimately, nineteen scientific research institutes were located in that area. As part of the plan to boost the scientific human capital of the Soviet Union, Novosibirsk State University was founded in Siberia in 1959, in conjunction with new research laboratories. This university quickly became a prestigious center for scientific research, particularly in mathematics. Many interviewees confirmed the importance of the leading universities in preparing them to become valuable contributors to the US tech sector.

Particular respect and attachment were shown toward Moscow State University, St. Petersburg State University, the Moscow Institute of Physics and Technology, and Novosibirsk State University. **Vadim Gladyshev,** Professor of Medicine at Harvard Medical School and Director of the Center for Redox Medicine, came in 1992 at age twenty-six: "I went to Moscow State University in the chemistry department. Being from Orenburg, it was not an easy transition because Moscow State is the best university in Russia, and only very few people can go there. The entrance exams were very competitive, but I was accepted. I was lucky, I guess. Being around many smart and creative people was a big change for me. In my high school I was the best, but at Moscow State I initially found myself more or less in the middle and had to work hard to change that. It was really an interesting feeling, I still remember."

Among our interviewees was a significant cluster from the Moscow Physical and Technical Institute, known as PhysTech, who tended to stay in touch with each other. All expressed a great deal of allegiance to their university and the education they received there. **Mike Sandler,** Founder and CEO of Epiphan Systems, was one such alumnus. He came to Canada from Moscow in 1991 at age twenty-six and later established an office in Palo Alto, California: "I'm educated as a physicist. I graduated from the Moscow Physical and Technical Institute, PhysTech. It's considered the best in Russia and the best in the world, and obviously for me it is." **Sergei Burkov,** Founder and CEO of Alterra, quoted earlier regarding his specialized high school education, is another PhysTech graduate: "Yes, I graduated from PhysTech, where I worked at the Landau Institute for Theoretical Physics and then left for the US. It was a very good education, a great curriculum, great professors, and the best thing was that starting from year three, you were working with research physicists, not professional educators. So in Russia, as you know, the situation is different from the

US, at least at that time. There were universities that were education-only, and all the research work was done at Academy of Sciences institutes, which are the equivalent of national labs here. So, PhysTech was different in the way it had collaboration with those institute research labs. After two years of basic training in physics, you were transferred to one of those labs and started working and getting education from people who were practicing, cutting-edge physicists. This is kind of normal in the US, because here the professors are researchers at the same time. In the Soviet Union, and in many other places, the professors were educators only. They didn't do practical work, but at PhysTech it was very different."

Another PhysTech graduate is **Shamil Sunyaev**, Professor of Bioinformatics and Medicine at Harvard Medical School, who came in 2002 at age thirty-one. He attended from 1988 to 1994, and then spent three additional years there to obtain his PhD in biophysics. Like others, he spoke about obligations for military service in Russia and how being in university provided exemption from it: "The situation with the draft was that two years before I entered PhysTech, no college, including PhysTech, provided temporary protection from the draft. When I started, PhysTech did provide a temporary shelter, and later it was extended to graduate school."

David Boinagrov, Research Scientist at Sciton, who came in 2008 from Moscow at age twenty-two as part of the Third Wave, evaluated his experience at PhysTech: "My opinion of that experience has been changing for some time after arriving here. The undergraduate education I had was very good. My college was good academically, but maybe not good in terms of the social life. In the college curriculum, if you counted up all the hours that you had to do homework and the hours for lectures, it was around thirty hours a day. You didn't have much time to do anything else besides studying, if you could keep up with that. It was a pretty good education. However, I wouldn't say that graduate school there was the best because, in order to do research, at least experimental, a lot of money needs to be invested in up-to-date laboratory equipment. Big science needs big money. When you are an undergraduate, you might still need some resources for lab equipment, but it's not as crucial." Additionally, Boinagrov's hard work in high school paid off: "I had won first place in the national physics Olympiad of Georgia in the ninth, tenth, and eleventh grades, and then I got onto the Georgian national team for the international physics Olympiads in Taiwan, where I received a silver medal."

A number of interviewees showed the same kind of loyalty to universities in St. Petersburg and Novosibirsk. **Michail Shipitsin**, Senior Associate Director of R&D at Metamark Genetics, came from Novosibirsk in 1997 at age twenty-two. He explained why his education at Novosibirsk State University was so valuable: "In my university there was the system where you studied for three years and then two years you worked in a lab and took some special course. So Novosibirsk University was really geared toward producing scientists as a primary goal. It was a great thing, actually."

Prior to the introduction in Russia of the Unified National Exam in 2009, students had to apply in person at the university where they wanted to study and could not apply to multiple institutions. Furthermore, each institution had its own admission process and criteria. The result was that admission to the most prestigious universities in Moscow, St. Petersburg, and Novosibirsk was very difficult for students in the different regions of the vast Soviet Union. So, in practice, most students looked to their regional university or the university that had been established in their republic as the most likely location for their education. Nonetheless, these regional and republican universities also followed the same patterns of competitiveness as described for the more recognized universities.

Danil Kislinskiy, Founder and CEO of IB Consulter who came in 2002 at age eighteen from Yekaterinburg, Russia, illustrates the movement from smaller cities to regional university centers: "I transferred from the Pedagogical University to a university in Yekaterinburg. It is a different city absolutely. Yekaterinburg has a population of one and a half million people, and it is way more advanced than Nizhny Tagil, where I am from. It is just a different environment. So I started studying, learning different things, and being independent – becoming an adult."

The experience of **Margarita Hunter-Panzica**, Global Product Marketing Leader of BioProcess Cell Culture at GE Healthcare, who came in 1998 at age twenty-two, illustrates the competitive nature of the regional universities, having come from the far western Russian enclave of Kaliningrad: "I went to Kaliningrad Technical University. I'm a manufacturing engineer by education, however that was not my initial goal. I actually wanted to study Russian literature or foreign languages, and English was a passion of mine. I couldn't get into that program because I missed one score. In Russia, English was the most competitive major at that time, as my generation was emulating the US. I was a top student, graduating with a gold medal, the highest award in Russian schools, so I didn't need to pass four or five exams. If I didn't get an A on the first exam, I would have to take all of them. And I failed to get into the English program because I got one B after two A's, one score short of the threshold. I quickly reapplied to the mechanical technology faculty and got in right away."

Michael Fayngersh, Vice President of Quality and Test at power transformer manufacturer Delta Star, who came in 1990 at age thirty-six, received a highly specialized education that served a major industry in his home city of Zaporizhia, Ukraine: "I graduated from the local technical university that was preparing people for particular enterprises. In my case, they were preparing engineers for the local transformer factory that was the largest one in the Soviet Union, and at the time it was also the biggest transformer company in the world."

The Special Role of Mathematics in Soviet Education

Mathematics plays a special role in our narrative. After the dissolution of the Soviet Union, more than 1,000 mathematicians emigrated, many of whom came to the United States.[25] They had a dramatic impact on academia, displacing many American academics.[26] Outside of academia, many others used their strong Soviet mathematical training in their new careers in US high-tech companies, as was evident in our interviews. Mathematics in the Soviet Union developed in almost complete isolation from the rest of the international mathematics community. In these conditions, they developed expertise in certain highly advanced areas, such as integral and differential equations. Later, they became heavily involved in mathematical modeling for trading software as the financial industry's technological revolution began.[27]

[25] George J. Borjas and Kirk B. Doran, "The Collapse of the Soviet Union and the Productivity of American Mathematicians," *The Quarterly Journal of Economics* (2012): 1146.
[26] Ibid., 1143.
[27] Leonid Bershidsky, "Russia's Math Geniuses Work Mainly in the West," *Bloomberg* (January 21, 2016) (https://www.bloomberg.com/view/articles/ 2016-01-21/how-russia-has-lost-its-mathematical-minds-to-the-west).

Given the importance of mathematics and mathematics education in our story, it is worth a short examination of the specifics of this particular field. The path to the United States for talented mathematicians begins with a very advanced culture of Soviet mathematics education that was later hamstrung by politics and anti-Semitism, providing part of the stimulus for the waves of immigration described in this book. A recent Russian analysis of the reasons for the success of Soviet mathematics pointed to several factors. The first was the significant support from the government and the high prestige of science as a profession. Both factors are related to the rapid industrialization efforts of the Soviet government. Second, doing research in mathematics or physics was one of the very few intellectual activities that had no mandatory ideological content. Many would-be historians, philosophers, or economists, and even artists, musicians, and computer scientists, became mathematicians or physicists. Another factor was that the Iron Curtain prevented international mobility. An additional feature of Soviet mathematics was a relatively high percentage of Jews, who traditionally gravitated toward intellectual professions.[28]

The years between Stalin's death in 1953 and approximately 1968 are considered the golden years of Soviet mathematics.[29] During that period, a rich environment combined university mathematical instruction and research with a network of supporting institutions and publications. At the start of that period, the most talented mathematicians in the Soviet Union were concentrated at Moscow State University in the Faculty of Mathematics and Mechanics, known as "MekhMat."[30] Other key centers of mathematics were in St. Petersburg and later at Novosibirsk State University. This expertise is illustrated by the personal experience of **Alexis Sukharev**, Founder and CEO of Auriga, who came in 1990 at age fifty-four: "I went to high school in Grozny, Chechnya, in southern Russia and graduated in 1963, and then I became a student at MekhMat. It was the number one mathematical school in the world. There are good mathematicians in other schools, in America and other countries, but the scope of MekhMat was much

---

[28] Vladlen Timorin, "Reasons for the Success of the Soviet Mathematical School," *Higher Education in Russia and Beyond* Volume 4, Number 6 (winter 2015): 9.
[29] Ibid.
[30] Lando, "Mathematical Education in Universities in the Soviet Union and Modern Russia," 6.

larger than any of the prominent mathematical departments in the US, believe me. Prominent professors, especially in mathematics and mechanics, numbered in the dozens and dozens and dozens. And there were schools and different directions of mathematics, very active, and I was happy to be part of MekhMat of the 1960s. I graduated in 1968 from the Department of Probability Theory, and then I entered graduate studies right away at Moscow State University. It was a different department, and I specialized in the theory of games from the very beginning and got my PhD in 1971."

Anti-Semitism and Elite Mathematics Education

The Golden Age began to be undermined by political events in 1968. The defining event for the end of this period was the publication that year of the "Letter of 99 Mathematicians."[31] The letter, signed by ninety-nine Moscow mathematicians, was addressed to Soviet authorities asking for the release of Alexander Esenin-Volpin, a prominent mathematician, poet, and dissident who was confined in an asylum as punishment for his anti-Soviet activities. Against the wishes of the signers, this letter was published in the West and brought political reprisals. When public manifestos like that letter appeared, the period of benign neglect of political oversight of mathematics came to an end. The administration of MechMat was changed and the Communist Party took a key role in deciding admissions, with an unstated rule against admitting Jews into the leading math schools and in faculty hiring practices.[32] Talented Jewish students interested in mathematics were channeled to other universities, like the Institute for the Petrochemical and Gas Industry, also known as the Gubkin Institute.[33]

Leah Isakov, Senior Director of Biostatistics at Pfizer, came in 1992 at age twenty-six, from Moscow. She described her experience with the channeling of Jews interested in mathematics away from Moscow State University to the Gubkin Institute: "I was trying to go to Moscow State University, but it was kind of known up front that it was no place for

---

[31] Timorin, "Reasons for the Success of the Soviet Mathematical School," 9.

[32] Lando, "Mathematical Education in Universities in the Soviet Union and Modern Russia," 6.

[33] Mark Saul, "Kerosinka: An Episode in the History of Soviet Mathematics," *Notices of the AMS* (November 1999): 1217.

Jews in the 1980s. So I failed, of course, and in the same year I was accepted to the Gubkin Institute. My department was applied math, but the way higher education was organized then, it was by industry. So this Gubkin Institute was supported by the petrochemical and oil industry, not in computer science and information systems."

Katya Stesin, serial entrepreneur and Founder of the IoT (Internet of Things) fashion startup, Fit-Any, came in 1991 at age twenty-three from Moscow: "I tried to go to Moscow State University, they but didn't take me in because I had a Jewish last name. They asked me in the oral exam whether my name was Jewish-related, and when I said, 'yes,' they failed me on the exam. I just wanted to go to university. For somebody who was seventeen years old, it was a tough blow. So instead I went to the university where my father was teaching, and I got straight A's. It was very easy to do that, and I was pretty bored at that time, so I was trying to do other things. Later, I went to Stanford so it turned out OK."

Simon Selitsky, Cofounder of Coingyft, who came from Moscow in 1991 at age twenty-one, described his early years and the impact of anti-Semitism on his university studies: "I started at a regular elementary school in the suburbs of Moscow where I grew up. I decided to go to a specialized math school in eighth grade, and it was a ten-grade system at the time. I had to pass some exams, and I was accepted into that specialized school. And then I got really involved in physics and math Olympiads, all sorts of Olympiads. I wanted to go into physics, and I had a physics tutor, and I won a bunch of Olympiads. And I wanted to go to MPT, which was a school outside of Moscow. And my tutor told me, 'No way; you don't want to go there.' He said he was the last of the Mohicans. He was Jewish himself and graduated from MPT in 1973, and he said that after him no Jews were accepted. That was not entirely true, but at the time I believed him because he didn't want to take the responsibility of steering me away from actually applying, which was a huge mistake that I now regret. So I tried to get into Moscow State University instead. You know how entrants were treated at Moscow State University and what the procedure was that they used in 1978. So I was not accepted, even though I was the winner of the All-Soviet physics Olympiads and a bunch of math Olympiads. I solved all the problems, but I was denied. In my Russian-language composition, I made one mistake. It was actually not a mistake, and I proved it during the appeal. Because there was some ambiguity, you could spell it two different ways. So I got a D, a 2, with the comment that the topic was not sufficiently developed. What a way to put it. And once you get a D, you are automatically denied entrance, even though I'd gotten A's in the other subjects. So I ended up at pretty much the only place that took Jews. It was the Moscow Road Automatics Institute. It was an engineering school where you learn how to build things like big machines and cranes and the switches used to control stuff. But automatics was actually before electronics took a strong hold in the industry. So that's where I started, automatic control systems."

Wave One 1972–1986: Anti-Semitism and Emigration

The unspoken, but not unknown, anti-Semitism of this period led to increasing numbers of Jewish mathematicians leaving the country in the First Wave of the 1970s, when the Soviet Union began to allow emigration through an Israeli visa. Many leading physicists and mathematicians were of Jewish origin, including five of the eleven Soviet Nobel Prize winners in physics.[34] These policies were the catalyst for the patterns of emigration from the Soviet Union during the First Wave. The emigration of leading mathematicians reached even greater heights around 1988, as the Second Wave began to grow. It is estimated that more than 300 Soviet mathematicians found permanent positions in American universities, and, in total, more than 1,000 moved to the United States, in some cases replacing native-born American mathematics professors and disrupting the profession.[35]

Wave Two 1987–1999: Breakup of the Soviet Union and the Brain Drain

With the dissolution of the Soviet Union, the educational system in the newly independent countries lost its privileged role and caused a much broader push for emigration. University and research establishments were in deep crisis. To illustrate the depth of the crisis in Russia alone, between 1992 and 1998, spending on higher education declined as a proportion of gross domestic product (GDP) from 1.2 percent to

---

[34] Lando, "Mathematical Education in Universities in the Soviet Union and Modern Russia," 7.
[35] Ibid., 8.

0.4 percent.[36] The collapse in government support of both universities and the Academy of Sciences research institutes provided additional stimulus to emigrate and contributed to the exodus of scientists and researchers from the former Soviet Union during this period.

American researcher Loren Graham and Russian researcher Irina Dezhina have presented a comprehensive analysis of these dramatic changes in their book, *Science in the New Russia.*[37] It is a widely recognized fact that there was a significant brain drain from Russia and Ukraine during this period. The concept of "brain drain" consists of external emigration, an internal shift by scientists to business or other nonscientific occupations, and the decline in demand for highly trained researchers in the military–industrial complex.[38] There was a significant movement out of science and academia into business during this period. Some who made the switch became the pioneering founders of Russian, Ukrainian, and Belarusian software outsourcing companies and later became part of the US tech sector.

Data reported by Graham and Dezhina indicate the depth and strength of the push for emigration or the alternative of leaving science entirely during those years. In the first years of the new Russian Federation, financing for science collapsed. Graham and Dezhina note: "The real level of financing of science in Russia in 1994 was 20 percent of that of 1991."[39] Additionally, "between 1990 and 1993, the share of science in Russian national income fell from 5.6 percent to .52 percent."[40] For many researchers, this meant no salary at all and no material support for continuing scientific work. Scientists who had been a relatively favored group in the Soviet Union, with privileged access to food, cars, and summer homes, could not provide their families with even the basics of normal life. As a result, many researchers sought to emigrate so they could both support their families and continue their research.

---

[36] Daria Luchinskaya, "RUSSIA: Modernizing the Higher Education System," *University World News* Number 192 (October 9, 2011) (http://www.universi tyworldnews.com/article.php?story=20111007130249809).

[37] Loren Graham and Irina Dezhina, *Science in the New Russia* (Bloomington: Indiana University Press, 2008).

[38] Jeffry L. Roberg, "Is the NIS Brain Drain Exaggerated?," *Demokratizatsiya* Volume 2, Number 4 (fall 1994): 595. https://www2.gwu.edu/~ieresgwu/assets/ docs/demokratizatsiya%20archive/02-04_roberg.pdf

[39] Graham and Dezhina, *Science in the New Russia*, 19.

[40] Roberg, "Is the NIS Brain Drain Exaggerated?," 599.

One research scientist at a major multinational corporation explained how, in the 1990s, he had come from another republic to study at Moscow State University and had a dream of pursuing pure science for his career. But, during the 1990s, funding for scientific research plummeted and people had to resort to finding ways to simply survive: "A lot of people of my generation, if they were to pursue science, had to find a way to pursue it elsewhere. I thought long and hard. I dreamt of contributing to some scientific work. I dreamt of continuing in science, and it would be regrettable for me at the time to have spent quite a few years moving toward my dream, and then just go and get into the business of buying and selling. So I had two choices, and I ended up coming here to the States for my PhD."

Sergei Ivanov, CEO of Optromix, expressed his motivation for leaving in 1992 at age twenty-six. Born in Murmansk, in the Russian far north, he, too, moved to Moscow for university: "I went to the Moscow Steel and Alloys Institute, and it turned out to be a pretty decent school. I was in physical chemistry. So I entered that school, and I moved to Moscow, and spent almost six glorious years there. They had a model where the students from the third year could spend increasing time at one of the basic research institutes of the Russian Academy of Sciences. I decided to take up on it because it was kind of going in the direction that I wanted to go. You know, physics, science. We were spending time in Chernogolovka, which is an academic center outside Moscow. But by 1992, I had started already falling on hard times. The reason why I came here was because I wanted to do science. I didn't want to go into business. In 1989 or 1990, a few of my friends were going into business, and I could have been working for them easily. But I went there and sort of hung around, and I didn't like it. I was still an idealist. I didn't like the atmosphere. I didn't like the dirty money. So I was looking for ways out, and I think it was around 1990 or 1991. Chernogolovka was still a pretty advanced research center. Lots of people from there left, and sometime later they would send emails. There was no Internet, but there was email. I got a message somehow from my good old friend from college who had emigrated to Israel. I started to send messages, and I started getting some replies with admission forms to graduate school. I couldn't really pay the processing fees, but they were willing to take an application. Essentially, I went to the first one that accepted me, and it happened to be Clark University in Worcester, Massachusetts. The reason that they took me was there was a new professor there who was setting up a pulsed magnetic field lab, and I was doing that type of research."

Other interviewees were essentially marooned in foreign countries as graduate or postgraduate students, with no ability to return home due to lack of scientific employment opportunities. Still others came to the United States through academic exchange programs that led to gradual assimilation into the US technology sector. While some were Jewish, the political and social impetus of the earlier period of emigration was replaced by a predominantly economic motivation. While many of these people never intended to become part of the high-tech sector, they made this transition through their experiences in the United States. In the process, they brought world-class scientific talent to the US innovation economy. One such interviewee was **Anastasia Khvorova**, Professor at the RNA Therapeutic Institute at the University of Massachusetts Medical School, who came in 1995 from Moscow, at age twenty-six. She was blocked from returning to Russia to continue her research due to economic hardship there and had to decide between working in Europe or the United States: "Essentially, when I was in France a second time, I decided that I would like to get a permanent position in the French Academy of Sciences. When I talked to people over there, they told me that if this is what I wanted to do, then I would have to do an American postdoc. That would ensure that my English would become more fluent. And second, I would learn how science is done in a leading innovation country. This was a tough requirement, to learn that I would have to have an American postdoc to qualify for certain jobs. After staying in France for another three months, I came back home to Russia and decided that I had to find an American postdoc, taking my daughter and flying 5,000 miles to the US."

In a somewhat different vein, another interviewee from Russia was encouraged by her parents to study at Stanford in the late 1990s, although she was reluctant to do so because it was not in her direct field of interest. She was accepted, and with two weeks' notice from the university, started her journey that eventually led to her becoming a US citizen and playing key roles in the Silicon Valley technology landscape.

Putin Era and Reassertion of the Government's Role in Russian Universities

We provide the example of Russia to give insight into some of the educational reforms that took place in the various independent countries of the former USSR. In the early part of the twenty-first century, the Russian government began to reassert its role in defining the role and purpose of universities. It instituted significant reforms of the Academy of Sciences, and the changes had a significant impact on the educational system in addressing the brain drain of the 1990s. Having lost their defining purpose as providers of human capital for a centrally planned economy, universities in Russia and in the other newly independent countries of the former Soviet Union struggled to establish a new purpose and some kind of business model to sustain them. The RAS was also struggling to maintain its ability to continue with fundamental research and preserve its privileges from the Soviet period. Also during these years, there was rapid growth of private higher educational institutions of varying quality that enrolled increasing numbers of students as many people were seeking their place in the new, changed world.

Beginning around 2005, the Russian government began a process of redefining both the universities and the Academy of Sciences to change the role of each in promoting practical technical research to modernize the Russian economy, an area previously reserved solely for the institutes of the Academy of Sciences. The government set the foundation for major institutional change in reforming the university structure by creating a system of national research universities and federal universities. Led by the Ministry of Education and Science, a program created national research universities aimed at integrating science and education with the high-technology sector of the Russian economy. This was a dramatic shift in the historical mission of Russian universities away from a purely pedagogical role. The first two of these new national research universities were designated by presidential decree in 2008. Then, in 2009, twelve more universities were selected through a competitive process and another fifteen by a second competitive process completed in April 2010. Currently, there are twenty-nine such national research universities. The main goal of establishing national research universities was to link universities with the real economy and to turn them into centers of applied scientific knowledge.

Being designated as a national research university brought substantial new funding from the Russian government. Each university was slated to receive approximately $110 million of additional funding over five years, doubling the overall budget for many of them. The majority of funds were to be spent on creating state-of-the-art labs, with the remainder spent on training and new programs in science, technology, and engineering. In return, universities were expected to show results in the form of new curricula, publications in international journals, and creation of new companies or licensing based on university intellectual property. Their status as national research universities and the associated funding could be revoked it they did not show results.

A critical component of the reform of Russian universities was the enactment of Law 217-FZ, which was intended to be the Russian analog to the US Bayh-Dole Act. The Russian legislation is intended to empower universities to commercialize basic science and duplicate the success that Bayh-Dole engendered in US universities. The reforms of Russian universities were intended to assist with the transformation of the Russian economy from its dependence on natural resources and move it toward a knowledge-based economy. The objective was to remake some portion of the leading universities into analogs of the innovation-generating institutions of the United States and Western Europe. A by-product of these efforts was additional funding and the creation of challenging opportunities for scientific researchers that helped stem the outflow of the country's technological elite.

Wave Three 2000–2015: Educational Opportunities and Intellectual Recruitment

As government funding of universities and research increased dramatically, along with improving economic opportunities in Russia related to rising world oil prices, the motivation for emigration to the United States shifted even more decisively toward taking advantage of educational opportunities there, as well as in response to recruitment of talented people to the US tech sector. In Wave Three, the number of educational exchanges and opportunities for study in the United States continued to expand for people from the countries of the former Soviet Union. And while education in the States continued to be an escape hatch for some from the difficult economic conditions in their countries, for others it was more a search for the highest quality education that they could find. Whatever the reason, many interviewees came initially for education and then stayed. For example, the University of Massachusetts, Amherst, had an active program of exchange with the Perm region of Russia near St. Petersburg. The university sponsored a number of students who constitute a small network of people from that region who later participated in the US tech sector. **Natalia Goncharova**, Founder and CEO of Finance Alpha LLC, was originally from Tashkent, Uzbekistan, and came in 2001 at age twenty-five, through a University of Massachusetts program: "I applied to three schools and all of them accepted me. UMass Amherst gave the best terms, so I went there because all my expenses were completely covered. I came fifteen days before September 11, 2001."

Jane (Evgeniia) Seagal, Senior Scientist III at AbbVie, who was originally from Novosibirsk, Siberia, came to Harvard Medical School as a postdoctoral fellow in 2004 at age thirty-two, after receiving her PhD in immunology at Technion in Israel: "Dr. Klaus Rajewsky is one of the stars, one of the top scientists in the field. If you decide that you want to continue to stay in this field, you get your education and you continue to try to get to the best places. So, right after my PhD, I sent him an email saying who I was and what I was interested in. He invited me for an interview. There were a few other places to which I applied here in US, and they all invited me for an interview. I came, and it was interesting. He took a real personal interest, and I really felt like he wanted me to come. So that's how I decided to join his lab."

Alexei Dunayev, Cofounder and CEO of TranscribeMe, was originally from Kiev, Ukraine, and came to the United States via New Zealand in 2007 at age twenty-six: "I came here to do my MBA at Stanford. I wasn't interested in coming to the US prior to that time. I first received a Fulbright scholarship, and then I decided to apply to MIT and Stanford. I chose Stanford after getting into both, because it was the tech hub of innovation and startups."

Yury Lifshits, Managing Director of Entangled Solutions, explained his path to the United States, coming in 2007 at age twenty-three, after receiving his PhD in computer science from the Steklov Institute of Mathematics in St. Petersburg: "Half a year before my PhD was scheduled to be defended, I started looking for postdoc positions. I was looking for open application positions because most postdocs are kind of invitation only. The professors identify a person they want to work with and that settles it. At that time, I was close to moving from pure mathematics to applied mathematics and computer science. I was interested to go to the United States. Actually, that year there were only two of those positions, and one of them for some reason was not ready to accept applications. The other one was Caltech that had a big grant from Gordon Moore, the Founder of Intel, and they were accepting postdocs. They were open applications and anyone could apply. You just had to send them a resume and cover letter, something like that. And I applied. They had a list of faculty members who go through these, say, 200 applications and pick seven or so people. One of the professors, Professor Jehoshua Bruck, saw the name of my advisor who was a very famous scientist, on my resume. He put his weight and his reputation behind selecting me. That was amazing. And then he wrote me an email saying, 'Please come.' And that was even before my PhD graduation, like three months before. So I got that offer and came to Caltech."

Others were exposed to life in the United States through exchange programs that led them to want to return later, even in the case of fashion blogger and consultant **Natasha Lavrishina,** whose exchange experience in 1998 at age twenty was a major disappointment. Coming from Baltic State Technological University in St. Petersburg to a small college in a small town in Missouri was a difficult adjustment: "I won a grant from the US State Department which allowed me to come here for a year as an exchange student. I was in Liberty, Missouri, a small town near Kansas City, in a private Baptist school. It was obviously a very different world, and I was astonished at the poor level of understanding of the world and Russia there at William Jewell College. It was 1998, but they still lived as if it was 1979. They knew nothing, they didn't know whether it was still the Soviet Union. They would ask me whether we had tomatoes there or cars or stone buildings. The mentality shocked me as a narrow-mindedness in so many respects. I had set my expectations very high in going to the States. Spending a year in America, nothing was met. I thought it would be so much more than what we had in St. Petersburg or in my town in terms of understanding, acceptance, willingness to see, to hear, to accept." Despite that disappointing experience, Lavrishina returned to the United States fairly soon afterward, settling in California, where she found a better fit, and eventually became an American citizen.

Igor Gonebnyy, Cofounder of Coursmos, who came for his business in 2014 at age thirty-six, fondly recalled his exchange program about a decade earlier: "My first experience with the United States was about ten years ago. I was born in Siberia, in Krasnoyarsk. We have a sister city relationship with Oak Ridge National Laboratory in Tennessee. We had a summer student exchange program, and I moved to the United States for the summer. I lived in Nashville, actually in Maryville. I worked at a Japanese factory for the whole summer. It was a pretty interesting time. I was really excited about the country and what I saw. I definitely wanted to move to the United States."

Danil Kislinskiy, Founder and CEO of IB Consulter, came for the first time in 2002 at age eighteen from Nizhny Tagil in the Urals region of Russia. His initial exposure to the United States was also the result of a summer program called Work and Travel: "It was an exchange program in work and travel. I was wanting so much to get to the US. The US at that time, for me, was a dream come true to go and visit. I think this was because of the American films I saw." He returned to the States in 2004 for four months in the Work and Travel program and then decided to immigrate in 2013. "And it was a decision to come here, so I had to quit my job. I had a really good job, good career, good salary, and I was respected. I came here as a student, an international student with no connections and away from the good life I had back home. But it was Silicon Valley, a nice place and a new challenge. It was a really hard decision. So my wife and I talked it over together and decided to go. It was hard. I wouldn't say it was easy."

One interviewee presented a somewhat different story from most others about her motivation to study in the United States, expressing dissatisfaction with her education in Russia. **Sophia Kovaleva** came from Moscow in 2013 at age twenty-two as a PhD student in electrical and computer engineering at Carnegie Mellon University West in Silicon Valley: "I guess the formative moment was when I first took online classes in Russia from Stanford on the machine learning artificial intelligence and databases. I was really impressed by that because in our university some courses and projects were maintained by PhD students and young professors who were doing that just for their own fun or their own interest, while the rest of the classes were taught by a very old professor. So I was really impressed by the fact that there were people with credentials and with a lot of experience who were teaching things that were really relevant and up to date. So that was really interesting for me and I decided that, OK, maybe I want to take a look into some of the top US schools. And again, I liked quite a bit the way the academic system is organized and the way the academic culture is compared to what I saw in Russia." Kovaleva also had her own way of dealing with having to learn Marxist philosophy even in the 2000s, long after the fall of the Soviet Union: "Undergrads in Russia are required to take philosophy classes, and the thing is that these classes are taught usually by all the Marxist professors who the administration didn't have the heart to fire after the Soviet Union collapsed, so they continued teaching the same thing with the same style of teaching. And what they taught made me really, really mad. So just to spite them, I read a lot of other books on philosophy, including Karl Popper's criticisms of Marx's works, just so that I could make comebacks with them. And about the same time, I actually read *Atlas Shrugged*[41] and that also helped, not that I really bought the ideas from there, but just throwing these ideas around the old style system was a really fun experience."

One interviewee chose the United Kingdom as the country in which to do his PhD, and then saw the United States as the ideal place to launch his career as an academic. **Ilya Strebulaev**, who was originally from Moscow, joined Stanford Graduate School of Business as a finance professor in 2004 at age twenty-nine after obtaining his PhD at London Business School: "The reason is very simple: I wanted to be in academic research. LBS was the best school in Europe, but you can't stay in the same school that you graduated from. That is not professional. So once LBS was out of the picture, then you really had to go to the United States. There are some other very decent European schools, but you can't really compare them with the top like Stanford. I did apply to a number of European schools as well, but my eyes have always been on the United States. Russia has not been an option at all because, even at that time, there were few business schools, and therefore I felt my qualifications wouldn't be needed."

---

[41] Ayn Rand, *Atlas Shrugged* (New York: Penguin, 1957). Rand (1905–1982), who emigrated to the United States from St. Petersburg, developed the philosophy of objectivism that advocated reason, individualism, and capitalism.

Russian Outreach to the Diaspora

We again refer to the case of Russia as an example of attempts by the various independent countries to encourage their highly educated former citizens to return home. By the mid-2000s, the Russian government began efforts to convince leading members of the scientific diaspora to return home to assist in rebuilding Russian science. There was a lot of internal discussion about how to turn the "brain drain" into a "brain circulation."[42] In 2010, the Russian government established tenders for mega-grants of up to $1 million dollars for leading scientists to set up their research in Russian universities, institutes, and research facilities. These grants went primarily, but not exclusively, to scientists who had emigrated. In the period from 2014 to 2016, some twenty-three research projects were carried out in Russian universities that were supported by these mega-grants.[43] Another effort to attract scientific researchers to return was the founding of a new research university, SkolTech – Skolkovo Institute of Science and Technology – in 2011. The university was founded through a three-year collaboration with MIT and was intended to provide world-class research and living conditions for international and Russian faculty. The programs, along with other exchange programs, have had some limited success in bringing scientists back to Russia for temporary or short-term assignments but have yet to produce a substantial number of returnees. One of main themes of the St. Petersburg Economic Forums in June 2015 was the brain drain and the impact of Ukraine-related sanctions imposed on Russia by Western governments.

Subsequently, some Russian experts expressed the opinion that they expected some 10,000 to 15,000 researchers to return home, based on initiatives begun under Russia's National Technology Initiative (NTI).[44] This opinion was widely disputed in US émigré social media

[42] Andrei Korobkov, Dmitry Polikanov, Michael Spaeth, and Sergei Sumlenny,"From Brain Drain to Brain Gain," *Russia Direct Report* Number 5 (May, 2014). http://www.russia-direct.org/system/files/journal/RussiaDirect_QuarterlyReport_BrainDrainBrainGain_May2014.pdf.

[43] Andrey Melnikov, "Russian Government Supporting International Cooperation in the Big Data Field," *Global Statement Blog* (2015). https://globalstatement2015.wordpress.com/2015/09/02/russian-government-supporting-international-cooperation-in-the-big-data-field/#more-219.

[44] Pavel Koshkin, "Russia Needs a Bold New Plan to Reverse the Brain Drain," *Russia Direct* (May 26, 2016). http://www.russia-direct.org/qa/russia-needs-bold-new-plan-reverse-brain-drain.

commentaries. NTI is a broad government-initiated program to develop priorities in the research community, ensure world-class research through globally competitive university centers, and enhance the prestige of Russian universities by 2035. Part of the program is focused on creating university innovation hubs in targeted technology sectors.[45] The goals for these NTI-related universities include having an average of 50 percent of professors being either foreigners or Russians who have returned to the country after working abroad. They further seek to have up to 30 percent of students from outside Russia.[46] This activity would represent a significant integration of leading Russian universities into international scientific and technical ranks. NTI is providing funds for a select group of Russian universities to take measures to improve their ranking in worldwide university rankings systems.

None of these initiatives has yet shown significant success in bringing back members of the diaspora, but all of them provide room for talented entrepreneurial researchers to seek achieving their goals in Russia and have somewhat reduced the allure and incentives to emigrate.

Ongoing Reform of the Russian Academy of Sciences

In the current era of reform of the university system, the RAS has been challenged in its prior role as the exclusive focus of fundamental and practical technical research in Russia. It has resisted these changes, but still has itself undergone significant reform. After almost a decade of conflict with the Ministry of Education and Science, significant changes have been imposed on the RAS.

For a number of years, the Russian Ministry of Education and Science has been attempting to shift the balance of research toward the national research universities. Only a very few Russian universities receive state funding for basic research, but, in 2013, there was a major change in Russian funding mechanisms for competitive awards to make universities and institutes compete by submitting cooperative

---

[45] Alexandra Engovatova and Evgeny Kuznetsov, "A Plan for the Growth of the Knowledge Economy in Russia," *Russia Direct* Volume 4, Number 8 (2016): 9. http://www.russia-direct.org/system/files/journal/RussiaDirect_Report_FromU niversity1to4_August_2016_0.pdf.
[46] Ibid., 9.

proposals.[47] This change in policy has understandably been opposed by the RAS. Its institutes still receive the majority of research funding, but the trend is toward a declining share, with their share in overall Russian R&D funding declining from 59.3 to 47.7 percent from 2005 to 2013, with the university share increasing from 11.4 to 18.6 percent in the same period.[48] One of the main reforms imposed in the latest round for the RAS involves increasing the role of competitive funding as well as assessment that is based on actual performance.[49]

In addition to protests by RAS scientists, some of those who did not expect to receive funding on a competitive basis began looking for opportunities abroad. According to Professor Irina Dezhina, a leading Russian social scientist, as the certainty of this direction of reform became obvious, the brain drain began to increase from the RAS. She observed that "over the first eight months of 2014, the outflow of human resources from this country exceeded the corresponding index for every full year over the last one-and-a half decades. The bulk of these emigrants are scientific researchers and entrepreneurs."[50] We have no statistical information on how many of these people are coming to the United States, so it is not possible to assess their impact on the tech sector at this point. However, their emigration does have a potential future impact on Russia, as well as on countries like the United States to which they emigrate. To a lesser extent, the same could be said of the other countries created after the dissolution of the USSR.

Potential for Future Brain Drain

The countries that have emerged from the Soviet Union are divided in their ability to continue to independently support higher education, scientific research, and research and development activities depending

---

[47] Maria Yudkevich, "The Russian University: Recovery and Rehabilitation," *Studies in Higher Education* Volume 29, Number 8 (2014): 1466.

[48] Mikhail Gershman and Galina Kitova, "Assessing Government Support for Research and Innovation in Russian Universities," *Journal of Knowledge Economics* Volume 8, Number 3 (September 2011): 1067–1084.

[49] Vladimir Pokrovsky, "Russian Researchers Protest Government Reforms," *Science Magazine* (June 8, 2015) (http://www.sciencemag.org/news/2015/06/russian-researchers-protest-government-reforms).

[50] Irina Dezhina, "Restructuring of the Scientific Research Institutes Formerly Subordinated to the Russian Academy of Sciences: An Assessment of Change," *Russian Economic Developments* Volume 12 (2014).

upon whether they are energy producers or not. Russia and Kazakhstan, as major oil and gas exporting countries, have had the resources to reform higher education and renew support of scientific research and commercialization. Ukraine, Belarus, Armenia, and Georgia, by contrast, being importers of oil and gas, have not found alternative sources of revenue in world markets and are thus much less able to support educational reform and scientific research.

In Russia, the reforms of the Putin era have had two potentially different impacts on the topic of emigration of technical professionals under consideration in this book. On the one hand, the transformation of national research universities and supplemental funding provided for innovation and technology initiatives has undoubtedly removed some of the economic motivation for the 1990s brain drain. There also has been some progress made in the internationalization of Russian universities and improved relations with the Russian-speaking scientific diaspora. However, the imposition of Ukraine-related sanctions has limited further internationalization of Russian universities and has encouraged a new surge of emigration by young entrepreneurs. At the same time, there is some anecdotal evidence that many young researchers who might have become entrepreneurs in an earlier period have moved into more secure research positions within the Russian military research and development sector, which has become increasingly high tech.

A counterpressure pushing toward emigration comes from the reforms of the RAS, as described earlier in this chapter. Thus, it is rather difficult to evaluate the potential impact of this new rush to the exit without further information. First, we do not know where these researchers and scientists are going. Second, we do not know their age or scientific relevance to the US tech sector. What we do know is the general picture of increasing numbers of people emigrating from Russia over the past several years, with 186,382 leaving in 2013, 310,496 leaving in 2014, and 353,233 in 2015.[51]

We now turn to the situation in some other former Soviet countries. Kazakhstan has a complex pattern of brain drain, with a large number of scientists and engineers of Russian and German ethnic backgrounds having left the country in the 1990s. After 1998, the economy began to stabilize, and, as revenues from oil and gas have increased, the

---

[51] "University 1.0 to 4.0," *Russia Direct* (2016): 5.

government of Kazakhstan has focused on developing a new intellectual elite. The existing school system continues to reflect some of the strengths of the Soviet system, particularly in mathematics, where, for instance, the country scored as high as fifth in international mathematics competitions in 2010.[52] At the university level, Kazakhstan has established an extensive program to send students to foreign universities through the Bolashak program, but requires them to return home to engage in government work for five years after completing their studies.[53] The picture in Kazakhstan is more reflective of a model of brain circulation than of a continuing brain drain.

Ukraine has stagnated economically and politically as it has struggled to define its political identity and basis of economic modernization in the post-Soviet period. The number of researchers in post-Soviet Ukraine declined dramatically from 1991 to 2013. This was reflective of a dramatic decline in spending for research and development, from constituting 2.44 percent of GDP in 1991 to only 0.76 percent in 2013.[54] Increasing numbers of scientists and technical people are leaving the field of science due to lack of funding, creating ongoing pressure to emigrate for those who can take their training to new positions abroad. According to the Ukrainian State Statistics Committee, there were significant losses of Doctors of Science and PhDs in the fields of biology, physics, mathematics, and medical sciences.[55] These losses continue to be substantial, with the number of researchers in the Ukrainian Academy of Sciences having decreased by 13 percent from 2014 to 2015, and only 20 percent of the PhD researchers in the Academy are younger than thirty-five years.[56] The bright spot for Ukraine is in IT outsourcing, with 4,000 companies – more than any other country in Eastern or Central Europe – whose combined revenues are more than $2 billion per year.[57]

---

[52] Naila Mukhtarova, "Brain Drain of Kazakhstan in 1999–2008" (Master's Thesis, Charles University, Prague, 2010), 72.

[53] Ibid., 61.

[54] Halnyna Mokrushyna, "The War on Ukrainian Scientists," *CounterPunch*, www.counterpunch.org (April 29, 2016).

[55] N. Parkhomenko, "Problems of Intellectual Migration of Ukraine," *Ukrainian Studies* Volume 17 (2014): 38. http://ukrbulletin.univ.kiev.ua/Visnyk-17-en/Parkhomenko.pdf.

[56] Alya Shandra, "Reversing Ukraine's Brain Drain: Mission Impossible?" *EuroMaidan Press* (July 18, 2016). http://euromaidanpress.com/euromaidan/.

[57] Ilya Greenburg, "A Ukrainian Brain Drain," *The New Yorker* (June 4, 2014). http://www.newyorker.com/business/currency/a-ukrainian-brain-drain.

Belarus, which is in an analogous situation to Ukraine, has also stagnated economically. The declared aim of reforming university education to conform to European standards through the Bologna Process has proceeded slowly, with a continuing mismatch between education and the real needs of employers.[58] Belarus has significant demand for its software engineering companies, as well as for its firms in the software entertainment sectors.[59] Increasing numbers of university students are also studying abroad.[60]

Regarding Armenia, there have not been many studies of university education and brain drain in that country. However, a 2009 study by the International Labor Organization noted the ongoing problem of mismatch between professional education and actual job demands that creates a surplus population of university-educated people.[61] The unemployment rate for people with higher education is increasing, with the exception of those in the IT industry, and many professionals have had to leave the country to find work.[62]

Similarly, Georgia struggled to reform its education system from 1991 to 2004, to make it conform to real-world requirements and to root out corruption in university admissions.[63] Some observers claim that recent reforms have made a substantial difference in supporting professional education and matching university education with job requirements.[64]

All of the countries of the former Soviet Union have struggled to define what to save and what to reform from the Soviet educational system. The legacy of that system continues to shape their expectations

---

[58] Alena Spasyuk, "Education in Belarus: Between the Bologna Process and the USSR," *ODR, Russia and Beyond* (April 25, 2017). https://www.opendemoc racy.net/od-russia/alena-spasyuk/education-in-belarus-between-bologna-pro cess-and-ussr.

[59] Ibid.   [60] Ibid.

[61] Lilit Avetikyan, "Brain Drain in Armenia: The Impact of Education on Migration Intentions" (Master's Thesis, American University of Armenia, 2013), 14.

[62] Ibid., 28.

[63] Maia Chankseliani, H.-D. Meyer, E. P. St. John, M. Chankseliani, and L. Uribe, eds., "Higher Education Access in Post-Soviet Georgia: Overcoming a Legacy of Corruption," in *Fairness in Access to Higher Education in a Global Perspective: Reconciling Excellence, Efficiency, and Justice* (Rotterdam: Sense Publishers, 2013), 171–188.

[64] Tamar Todria, "Education Reform: Post-Soviet Georgia," *Peace Child International* (January 26, 2017). https://peacechild.org/education-reform-success-post-soviet-georgia/.

for high standards in mathematics, science, and engineering. The extent to which each of them will have pressures for emigration vary depending upon the economic and political stability of each individual country. It will take time for these eventualities to materialize, and thus as noted earlier, it is not feasible to predict the extent of emigration that transpires nor the potential effects on the US innovation economy.

4 | Migration from the Former Soviet Union to the United States
Three Waves 1972–2015

The previous chapters introduced the three waves of immigration, focusing on the institutional impetus or push to leave the Soviet Union as well as the education and other preparation in their homelands that enabled immigrants to contribute to the US tech sector. This chapter will provide a detailed description of the context and the immigration experience of each wave. It presents an overview of the forces shaping each wave, including the legal and administrative framework of both the United States and the Soviet Union and then the post-Soviet successor countries. It also describes the geopolitics that led to the on-and-off flow of immigration and the experiences, desires, and actions of the immigrants themselves.

One of the key results of the collapse of the Soviet Union was the removal of barriers between the two separate systems of science and technology that existed in the West and in the Soviet Union during the Cold War.[1] Each side competed with the other, particularly in military and space technology, and each had some general idea of the scientific research that was being conducted in the other system. However, actual communication and exchange of knowledge were extremely limited.

With the collapse of the Soviet Union, the barriers between the two worlds of science were broken down and a period of integration of science and high technology began based on the US model of science, research, and innovation. Prior to this breakdown, the Jewish exodus from the Soviet Union in Wave One added marginally to the growing internationalization of the US tech sector that started around 1965 and began to show the potential for contributions of Soviet-trained people to the US tech sector. The brain drain during Wave Two dramatically increased the impact of scientists, researchers, and new entrepreneurs

---

[1] Ina Ganguli, "Immigration and Ideas: What Did Russian Scientists 'Bring' to the United States?" *Journal of Labor Economics* Volume 33 (July 2015): 262–263.

from the former USSR. In Wave Three, the causes and nature of ex-Soviet immigrant flows to the US tech sector changed again with the maturing globalization of high technology and diversification of the economic and political systems of the countries of the former Soviet Union. However, a new era began in 2014, when economic and political sanctions imposed by Western powers on Russia resulted in a growing re-isolation of Russian science and technology as part of the new Cold War. Meanwhile, most of the other successor states remain in the orbit of the international technology sector, primarily as suppliers of intellectual talent.

Globalization, Immigration, and Innovation

The stories of the people described in this book are part of a broader picture of a worldwide flow of intellectual talent to the United States in the years after enactment of the Immigration Act of 1965. In her seminal study of the innovation ecosystem of Silicon Valley, Professor AnnaLee Saxenian extensively documented the phenomenon of Taiwanese, Chinese, Indian, and Israeli technology industries that were seeded by members of their national diasporas schooled in Silicon Valley.[2] What went unsaid in her book was that this phenomenon would not have been possible without changes in the policy of exclusion of Asian immigration until 1952 and only very limited Asian immigration until the dramatic rewrite of the law in 1965. The information and analysis provided in Saxenian's book and other works illustrate how the United States has benefited from recruitment of the "best and brightest" drawn from around the world and demonstrate how the immigration of talent and the growth of the high-tech sector in the United States have gone hand in hand.

Three Waves of Migration

As described in the Introduction, our research and interviews found three distinctly separate waves of migration from the Soviet Union and its successor countries from the 1970s to 2014. These three waves have had different influences stemming from institutions, imprinting, and

---

[2] AnnaLee Saxenian, *The New Argonauts: Regional Advantage in a Global Economy* (Cambridge, MA: Harvard University Press, 2006).

identity that will be explored as key themes of this book. These three periods are as follows:

Wave One took place roughly from 1972 to 1986. Before 1972, the Soviet Union had very restrictive and tightly controlled travel and emigration policies. Soviet regulations allowed for exit from the country for very limited reasons. As part of the Cold War contention between the United States and the USSR, emigration rights for Soviet Jewry became a bargaining point between the Cold War rivals. Emigration also began to work as a kind of safety valve to push dissidents out of the country when policy required, at times leading the Soviet government to loosen restrictions for a short period, then revert to limiting the number of Jews who were allowed to emigrate. This First Wave of emigrants went primarily to Israel and some to Germany, while a small number chose to come directly to the United States, and some later emigrated to the United States from Israel.

Wave Two covers the period from 1987 and 1999. Under Mikhail Gorbachev, there was a substantial loosening of exit controls connected with his broader political agenda of opening to the West, as well as re-establishment of diplomatic relations with Israel. Emigration of Soviet Jews became an agenda item for the series of summits between US President Reagan and Soviet General Secretary Gorbachev, and the number of Jews allowed to leave in those years increased dramatically. With the collapse of the USSR, new legislation in the successor states substantially broadened the right to travel for the general population. With the economic chaos of the 1990s, a surge in emigration on a much broader scale went beyond Soviet Jews. This is the brain drain period, with a dramatic exit of scientists and researchers from the former Soviet Union, including a significant number who found new roles in the US technology sector.

Wave Three covers the period from 2000 to 2015. The drivers and catalysts for emigration began to diverge among the countries that emerged from the former Soviet Union. The main dividing line was between the energy producers, Russia and Kazakhstan, whose economies began to stabilize and expand; industrial producers, Belarus and Ukraine, whose economies stagnated; and, primarily agricultural producers, Armenia and Georgia, whose economies also stagnated. The Baltic republics of Latvia, Lithuania, and Estonia followed a very different path through gaining membership in the European Union. In these years, a process of integration took place of the scientific and technology sectors of the former Soviet Union into the global high-tech sector. Emigration continued, but, for those in the tech sector, the reasons for coming to the United States were focused more on specific educational, scientific, or business opportunities. As mentioned, for Russia, this period came to a close with Russian intervention in Crimea and eastern Ukraine and enactment of sanctions by the United States, the EU, and their allies. Russia had again entered into a policy of economic autarchy, mimicking the Soviet period, with uncertain impact on the technological intelligentsia.

Tracking Each Wave of Migration

The statistical record for each wave is difficult to reconstruct with precision but can be presented with indicative numbers. One problem is that the US government statistics up until 1990 include all of the republics of the Soviet Union, even though the country had begun to break apart. The statistics for most purposes from 1990 to 1999 include all of the former republics of the Soviet Union that became separate countries, except that the Baltic countries of Estonia, Latvia, and Lithuania are not included. Some statistics for certain immigration information began to be collected by country during the 1990s, and, beginning in 2000, separate statistics were kept by individual country. The other difficulty in presenting precise numbers has to do with reconciling the different databases kept by the Department of Homeland Security and the State Department, with the nuances of differences between them. For instance, there is a difference between the number of initial admissions into the United States and the number of those holding lawful permanent resident status.

Within the overall waves of migration, our focus is only on those involved in the US tech sector. This is a much smaller group than the overall volume of immigrants in each wave, but are people subject to the same forces creating, shaping, and limiting this flow from the Soviet Union and its successor states. As a result, we will describe the legal and political forces that shape the three separate waves of migration and will present indicative numbers for the relative size of each. Since refugee status was the principal route to legal residency in the United States in Wave One and was also one of the key routes in Wave Two, there will be a particular focus in this chapter on the refugee immigration flow.

As an indicator of the relative size of each wave, we have drawn from information contained in the 2013 Yearbook of Immigration Statistics published by the US Department of Homeland Security. That document includes the number of persons obtaining permanent resident status in the United States by country of last residence. These statistics indicate that, between 1970 and 1989, a total of 91,615 people from the Soviet Union obtained permanent residence status. During the years roughly corresponding to our Wave Two, some 433,427 persons from the former Soviet Union, excluding the Baltic countries, obtained this status. This means that over the ten-year period of the Second Wave, more than four times as many people received that status than during the previous twenty years. While it is difficult to present an exact distribution by country of origin during this period, we provide indicative information. Wave Three is more difficult to track because of the different countries involved, but we will provide indicative numbers for this wave as well. As we examine the forces shaping each wave of migration, we begin with an overview of the complementary interplay between US immigration law and exit policies of the Soviet Union and successor countries that act as the levers of control for the emigration/immigration process.

US Immigration Law: Regulator of Migration of Human Capital to the United States

The immigration policy of the United States has been tied to complex political and economic factors over different historical periods. As a country whose identity was formed by immigrants, the United States has constantly revised its definition of its cultural and ethnic basis. One popular view of the United States is that it is a "melting pot of nations," with a national identity formed out of diverse immigrant communities, primarily Europeans. However, the formation of US national identity is also built on the conquest and removal of Native Americans, forced migration of African slaves, the annexation of Spanish-speakers through wars with Mexico, and the indentured servitude of Chinese laborers in the West. The struggle to define identity has overlapped with the labor market demands of different periods of US history and with the then-current definition of who is an "American." The changing notion of the desired composition of the country is reflected in immigration law as the control mechanism for entry into the country and is a reflection of political preferences.

Throughout most of its early history, the United States had no restrictions on immigration. One could argue that the first US legislation regarding immigration was the Act Prohibiting the Importation of Slaves of 1807, although clearly this was not regulation of free immigration. The beginnings of real immigration policy began with the Chinese Exclusion Act of 1882, which legislated the total exclusion of all Asian immigration into the United States. Policy then vacillated between virtually open immigration for Europeans, who fueled the industrial work force of the late nineteenth and early twentieth centuries, followed by outright discrimination against eastern and southern Europeans with the Quota Act of 1921 and the Immigration Act of 1924.

The Acts of 1921 and 1924 limited immigration in general and initiated limited access by national quota formulas favoring northern Europeans. As a result, a ceiling of 150,000 immigrants per year was allocated according to a quota of 2 percent of the number of foreign-born individuals from each nationality in the census of 1890.[3] These quotas did not apply to Asians, who were totally excluded, nor to persons emigrating from the Western Hemisphere, for whom there were no quotas. The census of 1890 was chosen as the base year for quotas because it predated massive immigration from eastern and southern Europe and thus favored northern Europeans. Under the new national origins quota system, only 2,248 people from Russia, 5,982 from Poland, and 3,845 from Italy would be allowed legal entry annually.[4] This legislative framework was created to manage the ethnic and religious composition of the country, specifically to limit the number of Italians, eastern Europeans, and Jews who had been the dominant immigrant groups in the early decades of the twentieth century. These Acts also reflected a dominant cultural and ethnic definition of the American national identity that remained in place as US immigration policy until 1965, when the national quota system was abandoned and the face of America began to change.

---

[3]   Libby Garland, *After They Closed the Gates* (Chicago: University of Chicago Press, 2014), 14.
[4]   Ibid., 15.

US Immigration Act of 1965 (Hart-Celler Act)

The US Immigration Act of 1965 dramatically changed the more than forty years of previous immigration legislation. It was enacted in an era of significant US legislative change in response to rising opposition to segregation domestically and to increasing competition with the Soviet Union internationally in the so-called Third World. In the court of world opinion, it was politically embarrassing for the United States to continue relying on racial categories from the 1920s in its immigration policy.

The Immigration Act of 1965, the Hart-Celler Act, eliminated the 1920s quota system and replaced it with a new system of annual numerical limitations of 170,000 for the Eastern Hemisphere and 120,000 for the Western Hemisphere. Within these overall limits, there was a new system of preferences. Priority was now given to "family reunification," followed by an order of preference for admitting immigrants as follows:

1. Unmarried children of US citizens under 21 years
2. Spouses and unmarried children of permanent residents
3. Professionals, scientists, and artists "of exceptional ability"
4. Married children of US citizens over 21 years of age and their spouses and children
5. Siblings of US citizens and their spouses and children
6. Workers in occupations with labor shortages
7. Political refugees[5]

Category 3 and Category 6 eligibility is subject to oversight and verification by the Department of Labor. Political refugees were admitted under the Refugee Conditional Entrants Act of 1965, and then later under the Refugee Act of 1980.[6] Within this new system, each country had a cap of 20,000 per year. Within the cap, the highest priority went to people with family in the United States (65 percent) and those with needed labor skills (20 percent). Admission of immediate family members of US citizens was not included within the quota numbers.

---

[5] "The 1965 Immigration Act: Asian-Nation: Asian American History, Demographics, and Issues," Asian Nation website, http://www.asian-nation.org/1965-immigration-act.shtml.
[6] Ibid.

Reform of immigration was a key element in the high-technology sector's rise to prominence through enabling the recruitment of members of the leading scientific intelligentsia from much of the world to the United States. The door for Asians was cracked open slightly by the McCarron-Walter Act of 1952 and then opened much more widely by the Immigration Act of 1965. It is immigrants from these countries who are highlighted in Saxenian's *The New Argonauts* as having assimilated the innovation culture of Silicon Valley and then distributed it back in some form to their respective home countries. The influx of talented people from Asia who have enriched Silicon Valley and other US technology centers would not have taken place without these reforms in immigration policy. In general, the move away from national quotas facilitated the recruitment of the scientific and technological elite to the United States and helped fuel the leading role of the US in global high-technology. The talented people highlighted in this book became part of this general movement as the Soviet Union began to allow limited emigration in the First Wave and then witnessed the full-scale brain drain of the Second Wave.

Political Refugees in US Immigration Law

Supplementing the regular immigration processes, the United States has been flexible in allowing for political refugees to enter the country in response to certain crises in the post–World War II period. This flexibility comes through legislation allowing for specific groups to bypass the quotas by entering as political refugees. Since almost all of the Jews who came to the United States in the First Wave and a smaller number in the Second Wave entered as political refugees, it is important to review the system of political refugee status in US immigration law.

US immigration policy included a series of enactment legislations to accommodate political refugees. These enactments allowed significant numbers of people to enter the United States between 1946 and 2004, including:

- Hungarian Refugee Act of 1958, allowing in 30,755 people
- Refugee Conditional Entrants Act of 1965, allowing in 142,103 people
- Cuban Adjustment Act of 1966, allowing in 677,611 people
- Indochinese Refugee Act of 1977, allowing in 175,178 people

- Refugee Parolee Act of 1978, allowing in 139,302 people
- Refugee Act of 1980, allowing in 1,960,535[7]

The mass emigration from Vietnam was one of the factors stimulating the Refugee Act of 1980,[8] but the Act later became the framework for entry into the United States for those exiting the Soviet Union in the First Wave and for many in the earlier part of the Second Wave. This Act amended the Immigration and Nationality Act of 1965 by defining a refugee as any person who is outside his or her country of residence or nationality, or without nationality, and is unable or unwilling to return to, and is unable or unwilling to avail himself or herself of the protection of that country because of persecution or a well-founded fear of persecution on account of race, religion, nationality, membership in a particular social group, or political opinion. The phrase "well-founded fear of persecution" became the focal point for refugee status for Soviet Jewry.

The Act also created the Office of Refugee Resettlement, which was charged with integrating refugees into American life through employment training and placement, English language instruction, and cash assistance for living expenses. These programs were funded and administered through block grants awarded to states accepting the refugees.[9] These benefits became a support system for many of the Jewish immigrants in Waves One and Two who came as refugees.

The annual admission of refugees was set at a 50,000 cap per fiscal year, but, in an emergency, the President could change the number for a period of twelve months. The Attorney General is also granted power to admit additional refugees and grant asylum to current aliens, but all admissions must be reported to Congress and be limited to 5,000 people.[10] The annual cap on the number of refugees admitted was a

---

[7] US State Department, "Table 20. Refugees and Asylees Granted Lawful Permanent Resident Status by Enactment: Fiscal Years 1946–2004," *Yearbook of Immigration Statistics: 2004* (Washington, DC: US Department of Homeland Security, Office of Immigration Statistics, 2006): 65. https://www.dhs.gov/sites/default/files/publications/Yearbook_Immigration_Statistics_2004.pdf.

[8] "The 1965 Immigration Act: Asian-Nation: Asian American History, Demographics, and Issues," Asian Nation website, http://www.asian-nation.org/1965-immigration-act.shtml.

[9] "Refugee Act," Wikipedia, accessed August 18, 2016, https://en.wikipedia.org/wiki/Refugee_Act.

[10] Ibid.

key factor in limiting the number of Jewish immigrants after 1989, as we shall see later in this chapter.

Wave One 1972–1986: Soviet Jewish Exit

From 1972 to 1986, emigration from the Soviet Union to the United States was made up almost entirely of Jews, with a smaller number of Armenians also taking part.

Soviet Legal Framework for Exit

The determining factor for the number of people coming to the United States in the First Wave years of 1972 to 1986 was the number allowed to emigrate by the Soviet authorities. Under Soviet law, citizens of the USSR could apply to emigrate only under very specific and narrow grounds of either family reunification or national reconciliation. The number of those approved had to do with broader Soviet foreign policy considerations. Within the limited legal parameters allowed, the Soviet government did not easily grant any right to its citizens to emigrate, and all travel abroad was tightly controlled and subject to official approval.[11] From 1948 to 1970, total emigration from the Soviet Union to all foreign countries was only 60,000 people.[12] Under these criteria, the main groups of immigrants were a limited number of Jews, Germans, and Armenians. Until the mid-1980s, the vast majority of Jews went to Israel, and only a small number came directly to the United States. Germans typically went to West Germany and Armenians went to France and the United States.[13] Between 1948 and 1986, the proportions of emigrants remained relatively stable, with Jews accounting for 60 percent, Germans 30 percent, and Armenians 10 percent.[14]

Most of the Jewish emigrants from the USSR in this period were from western border republics of the Soviet Union and left for

[11] Steven J. Gold, "Soviet Jews in the United States," *American Jewish Yearbook* 1994: 7.
[12] Victor Rosenberg, "Refugee Status for Soviet Jewish Immigrants to the United States," *Touro Law Review* (winter/spring, 2003): 428.
[13] Sidney Heitman, "Soviet Emigration Under Gorbachev," *Soviet Jewish Affairs* Volume 19, Number 2 (1989): 15. http://www.tandfonline.com/doi/abs/10.10 80/13501678908577633?journalCode=feej19.
[14] Ibid., 16.

religious or nationalistic reasons, and thus most went directly to Israel. According to one researcher who compiled statistics from a number of Soviet and Israeli sources, 291,000 Soviet Jews and their relatives left the Soviet Union on Israeli visas from 1970 and 1988, with approximately 165,000 ending up in Israel.[15] The largest number of Jews emigrating came from Ukraine with 106,700, followed by the Russian Federation with 50,400. The remainder was split between Belarus with 13,800, the Baltic republics with 27,300, Moldova with 29,400, the Caucasus area with 41,500, and Central Asia with 21,700.[16] During this period, only a small number of Soviet people emigrated directly to the United States, with the total number from 1980 to 1989 being 33,311.[17]

US Legal Framework

The First Wave of immigrants from the Soviet Union to the United States from 1972 to 1986 came primarily as political refugees, with some lesser number coming under family reunification. Until the Refugee Act of 1980, Soviet Jews seeking entry to the United States were evaluated using the standards of the Immigration Act of 1965 and were rarely denied refugee status.

The Cold War political competition between the United States and the USSR played a role in the number of Soviet Jews allowed to leave the USSR. In October 1972, the United States and the Soviet Union were in the midst of negotiating a major trade agreement as the Trade Act of 1972. Senator Henry Jackson of the state of Washington and Representative Charles Vanik of Ohio introduced an amendment to this legislation that would prevent any communist country that restricted emigration from receiving the "most favored nation" status required for the benefits of the Trade Act. Their amendment was

[15] Mark Tolts, "Post-Soviet Aliyah and Jewish Demographic Transformation," Research Paper presented to the 15th World Congress of Jewish Studies (August 2–6, 2009): 2.

[16] Ibid., 5.

[17] US Department of Homeland Security, "Table 2. Persons Obtaining Lawful Permanent Resident Status by Region and Selected Country of Last Residence: Fiscal Years 1820 to 2013," *Yearbook of Immigration Statistics: 2013* (Washington, DC: Department of Homeland Security Office of Immigration Statistics, 2014): 8. https://www.dhs.gov/sites/default/files/publications/Yearbook_Immigration_Statistics_2013_0.pdf.

adopted on September 26, 1973. It allowed for annual waivers by the President but proved a major obstacle, and the Soviet Union responded by refusing to implement the Trade Act of 1972. In response to the Soviet invasion of Afghanistan, US–Soviet relations again took a turn for the worse, reflected in renewed restrictions on emigration. In 1986, total emigration from the Soviet Union was barely more than 2,000 people, but this began to change dramatically in 1987.[18] The way the process worked in the early 1970s was that invitations were sent to Soviet Jews from relatives or supposed relatives in Israel that allowed them to apply for permission to emigrate for reasons of family reunification. The assumption was that all of the emigrating Soviet Jews would go to Israel. And, in fact, all of them in this period departed Russia on Israeli visas.

Tracking Wave One Numbers

A brief thaw in relations between the United States and the Soviet Union occurred in the mid-1970s that resulted in an uptick of immigration to the United States and then much smaller numbers due to renewed restrictions from political tensions in the 1980s. The number of people emigrating peaked in the 1970s and then began to drop in the early 1980s. In 1981, 9,447 Soviet Jews emigrated; in 1982, 2,688; in 1983, 1,314; and in 1985, 893.[19] The choice of destination became very important again in the late 1980s, when Soviet officials loosened emigration restrictions.

From 1981 to 1986, only 16,900 Soviet Jews were allowed to emigrate. Of those 48.5 percent, or 8,200 people, went to Israel and 46.2 percent, or 7,800, went to the United States.[20] The trend toward coming to the United States as opposed to Israel during this period became a more significant political issue in Wave Two. This expression of desired place of destination by the emigrants themselves did not correspond with the intentions of Soviet policymakers who expected to channel all Soviet Jewish emigrants to Israel. Since the numbers were small, what was a concern in Wave One became a problem only during Wave Two.

[18] Heitman, "Soviet Emigration Under Gorbachev," 15.
[19] Rosenberg, "Refugee Status for Soviet Jewish Immigrants to the United States," 422.
[20] Heitman, "Soviet Emigration Under Gorbachev," 17.

Table 4.1 *Cumulative Summary of Refugee Admissions to the United States 1975–2003*[21]

Year	Former Soviet Union – total
1975	6,211
1976	7,450
1977	8,191
1978	10,688
1979	24,449
1980	28,444
1981	13,444
1982	2,760
1983	1,342
1984	721
1985	623
1986	799
1987	3,699
1988	20,411
1989	39,602
1990	50,628
1991	39,226
1992	61,397
1993	48,773
1994	43,854
1995	35,951
1996	29,816
1997	27,331
1998	23,557
1999	17,410
2000	15,103
2001	15,978
2002	9,969
2003	8,744

Table 4.1 is presented as an indicator of the size and the ebb and flow of Soviet emigrants to the United States in Wave One and beyond. Since almost all of the immigrants in Wave One from 1972 to 1986 were

[21] US Department of State, Bureau of Population, Refugees, and Migration Office of Admissions, Refugee Processing Center, Summary of Refugee Admissions as

Jews who came as refugees, the focus on refugee numbers is an approximate indicator of the levels of Soviet immigration during these years.

Focusing on 1975 to 1986, it is possible to track the ups and downs of the number of refugee admissions relative to the state of political relations between the United States and the Soviet Union. Once again, it is important to keep in mind that this flow was determined primarily by Soviet authorities and the levels of emigration that they allowed.

Wave One 1972–1986: Refugees

Slava Epstein, Distinguished Professor of Biology at Northeastern University who came from Moscow, described the emigration process during this period: "There was no way to immigrate directly to the US. Well, there was a way, if you had sponsors. So some people immigrated by having real or fake sponsors. Israel was ready to supply you with an invitation, called a window envelope. So you see that in the mail, someone sent you an invitation. Person such and such invites you to reunite. Now when you go to the KGB to present that, you need to come up with a story about who that person is. So I go to the synagogue, somehow it was allowed to exist. And that's where everyone went to consult a wise person through connections. So I go to a wise person, give my envelope and ask what I should tell the KGB. And he says, 'Well, tell me, in your family was anyone killed during the Second World War?' I said, 'No, but my grandfather was missing in action.' He says, 'Ha, that's even better. He wasn't missing in action, he wasn't killed, and he didn't die. He ended up in a concentration camp, was rescued by the Allies, and moved to Israel after that. He changed his name, now he's 100 years old and cannot wait to see his family. Yeah. And then you go to the KGB and present the story.' And the KGB officer knows very well that it's all a lie, but the KGB officer does what his superior tells him. And what his superior tells him depends on how much Russia needs to buy grain from Canada.

"So anyway, we were stripped of our Russian citizenship, we didn't even get a permit to leave, but since there were no diplomatic relations with Israel at the time, no direct connections by air, water, whatever,

of December 31, 2015, https://2009–2017.state.gov/j/prm/releases/statistics/251288.htm.

there was no way to get to Israel from Russia other than through a third country. That third country happened to be Austria. Everyone would go to Austria, to Vienna, and people who actually wanted to go to Israel would fly to Israel from there. But if you didn't, that was the time to tell the authorities greeting you there that Israel had never been on your mind, that you actually wanted to go to America. And then paperwork would start. That somehow involved moving you from Austria to Italy for a period of time. It was only in Italy that we would actually be able to apply for American status. You would wait in Italy for the American decision. So it was Russia, Vienna, Rome, New York."

Consistent with Epstein's experience, after receiving permission from Soviet authorities to emigrate, most families would apply for an Israeli visa issued by the Dutch Embassy in Moscow, which represented Israel in the Soviet Union since diplomatic relations were severed by the USSR after the 1967 Six-Day War. After obtaining an Israeli visa, they would fly to Vienna because there were no direct flights from Russia to Israel.[22] In Vienna, they were cared for by the Jewish Agency, a quasi-government Israeli agency, and flown to Israel.

The policy of the Austrian government was that even though Soviet Jews left the Soviet Union on an Israeli visa, they were free to continue on to whatever country would legally admit them. This was not an important issue initially because, between 1968 and 1973, almost all of them went to Israel. This was in accord with the Israeli government's interest in increasing the population of Jews and in bringing in educated, skilled human resources to the country.[23] In 1950, the Law of Return in Israel stated that every Jew was entitled to come to Israel. In 1970, those who were non-Jewish spouses and descendants to the third generation were also allowed to enter the country on the Law of Return.[24] This was important in view of the large number of mixed marriages in the Soviet Union.

---

[22] Ann Cooper, "US Abruptly Closes a Route for Soviet Jews," *New York Times*, November 6, 1989. http://www.nytimes.com/1989/11/06/world/us-abruptly-closes-a-route-for-soviet-jews.html?mcubz=3.

[23] Fred Lazin, "Refugee Resettlement and 'Freedom of Choice': The Case of Soviet Jewry," *Center for Immigration Studies* (July 2005): 1.

[24] Eunice Kim, "Deconstructing Essentialist Identities: Reimagining the Russian Diaspora of the Third Period" (PhD dissertation, Department of Slavic and Eurasian Studies, Duke University, 2013), 20.

Those Soviet Jews who did not travel on to Israel were considered "dropouts," and they were referred to the American Joint Distribution Committee (JDC) and the Hebrew Immigrant Aid Society (HIAS). These agencies then moved those people to Rome for assistance in getting visas for countries other than Israel.[25] Very few such people in this period came to the United States. In 1971, for instance, only 250 Soviet Jews came to the United States.[26] However, as the number of dropouts grew, this created some ambiguity in what was intended to be a pipeline of people to Israel. For those seeking entry into the United States, many of them had the right to enter under the category of family reunion or by claiming political refugee status under the Immigration Act of 1965.[27] The issue caused some dissent in the Jewish relief agency community, along with concerns of Israeli officials.

Kira Makagon, Executive Vice President of Innovation at RingCentral, came from Odessa, Ukraine, with her family at age thirteen: "My family was part of the Jewish emigration, so my uncle, my grandparents, and my parents and I emigrated. We left Ukraine, the Soviet Union at that time, in 1976 and got here in 1977. We went through Europe, by the traditional route, which is that you arrived in Austria, and from there you moved to Italy, where you waited for months to get your visa to wherever you wanted to go. I think all of this was sponsored by HIAS and other Jewish organizations."

After passage of the Refugee Act of 1980, Soviet émigrés applying for US visas in Rome as refugees had to prove to an immigration officer that they had a well-founded fear of persecution in the Soviet Union. As a practical matter up to the late 1980s, almost all of their verbal representations of potential persecution were accepted.[28] Immigrants to the United States came either as political refugees or through family reunification.

Michail Pankratov, President and CEO of MMP Medical Associates who came from Russia at age twenty-six, described the general process: "I was always under pressure but at the end it turned out to be very easy. We applied, and three months later we got a call, 'You have one week to leave the country.' I had my wife, and we had our son. Here in Boston, there was someone we knew, and the Jewish community

---

[25] Lazin, "Refugee Resettlement and 'Freedom of Choice,'" 1.
[26] Rosenberg, "Refugee Status for Soviet Jewish Immigrants to the United States," 422.
[27] Ibid.
[28] Lazin, "Refugee Resettlement and 'Freedom of Choice,'" 1.

accepted us. My wife is Jewish, so we were accepted by the Jewish community and almost immediately, in two weeks, I went to work as a laborer." His entry into the workforce at this level was only a way station on his successful entry into the biotech sector.

Individuals from this First Wave show a mixed motivation that was definitely a reaction to anti-Semitism in the USSR, but also a desire for freedom and self-realization that was expressed through their success in professional and business life in the United States. However, only Jews were allowed to leave.

Michael Schwartzman, founder of ValueSearch Capital Management, came in 1974 at age twenty-nine from Novosibirsk. He explained how he and his friends were thinking of ways to escape the Soviet Union together. His personal reason was not so much because of anti-Semitism, but because he never liked socialism. He believed in freedom of choice, freedom of oppression from the government, and freedom to pursue one's interests. As he grew older and had a family, it became harder to scheme about how to leave, as he was responsible for his family. He described the procedure for getting permission to emigrate: "In 1973, when I was applying for an emigration visa while at Akademgorodok in Siberia, the only route that had any chances of success was to apply for emigration to Israel by claiming, in my case correctly, that 'I'm Jewish and I would like to move to my ancestral land. I have nothing against the Soviet Union.' The last sentence wasn't quite my thoughts, but declaring the deep dislike I had for socialism would be equivalent to applying for a prison sentence. Even this partially accepted reason by the authorities could not be used by active and retired officers of the Soviet armed forces, or by anyone with current or expired security clearances.

"When I applied, I was twenty-seven, and since graduation I had not worked on anything requiring a security clearance. Israeli authorities, once notified covertly of the desire of a Soviet citizen to emigrate, would arrange an invitation – another roadblock thrown in by the Soviets – from a real Israeli who would state, for example, that he was an uncle of the emigrant. I think the Soviets knew about the ruse but chose to ignore it. Life was complex, the Soviets needed to have an emigration scheme that was restrictive enough to keep the population bottled up but workable enough to satisfy the pressures they were experiencing from outside. Remember that two years later, in 1974, the Jackson-Vanik amendment came to life. Once we crossed the Soviet border, the first stop was Vienna, from where those immigrating to Israel were departing after a day or two of rest and paperwork conducted by Israeli authorities. Those who wished to immigrate to the United States, Canada, or Australia continued on to Rome, where several charities were processing the immigrants and where American intelligence services kept an eye on those immigrants who might have been in possession of valuable information, and also those who might have been recruited by the KGB to be planted in the United States."

This worsening of overall relations led to a drop in the number of Soviet Jews authorized to leave the country. Jewish emigration fell from 35,000 in 1973 to 21,000 in 1974 and 13,000 in 1975.[29] Only 28,132 immigrants came from the Soviet Union to the United States in the entire decade of 1970 to 1979.[30]

Timur Shtatland, Development Scientist II at New England Biolabs, described how his family attempted to emigrate from Kiev, Ukraine, at the very end of the First Wave period, but the Soviet authorities had shut off permission to emigrate in the early 1980s. He was not able to emigrate until the Second Wave began under *perestroika*, and he ended up in the United States at age twenty: "In 1972, my mother's brother immigrated to Israel, and ever since then we were considering emigrating, and then in 1979 we finally made up our minds. That year, just incidentally, happened to be almost the last year of major emigration out of the Soviet Union. They basically closed the gates. We applied for an exit visa in '81 and were denied, of course, predictably, on purely formal grounds."

The numbers were limited to a trickle by the 1980s and began to increase only at the end of the 1980s as Gorbachev repaired relations with Israel and began liberalization as part of de-escalating tensions with the West through a series of summits between Reagan and Gorbachev.

Wave Two 1987–1999: Brain Drain

The people who emigrated from the former USSR during Wave Two did so for a wide range of reasons. They are the largest group of post-Soviet

---

[29] Rosenberg, "Refugee Status for Soviet Jewish Immigrants to the United States," 421.

[30] US Department of Homeland Security, "Table 2. Persons Obtaining Lawful Permanent Resident Status by Region and Selected Country of Last Residence: Fiscal Years 1820 to 2013," 8.

immigrants and have made a correspondingly large impact on the US tech sector. As opposed to those who came in the First Wave, the main push for emigration in the Second Wave was primarily social or economic, rather than the result of anti-Semitism.

When Mikhail Gorbachev came to power in 1985, he began his attempts to reform the Soviet Union. In the midst of the social turmoil unleashed by Gorbachev's reforms and ultimately with the dissolution of the Soviet Union in December 1991, the pressure for broader emigration began to build, particularly among the educated intelligentsia. What began as a trickle of Jewish emigrants turned into a river of Jewish and non-Jewish emigrants from the Former Soviet Union in the 1990s, causing a substantial brain drain, a significant portion of whom came to the United States.

Soviet and Post-Soviet Legal Framework

Gorbachev's initiatives to de-escalate Cold War tensions included both a series of Summit meetings with US President Reagan and redeveloped ties with Israel. Over the course of four Summit meetings, Reagan continued to push human rights and particularly the right to emigrate for Soviet Jews as a necessary sign of good faith. Gorbachev initially resisted these demands as an intrusion on internal Soviet matters not subject to foreign intervention. He also feared that unrestricted emigration would lead to serious losses of human capital. However, he began to waver, and, in 1986 at the Reykjavik Summit allowed human rights onto the agenda and then made a significant relaxation of emigration requirements in 1987.[31]

As the 1980s drew to a close and the economic and political problems of the Soviet Union continued to deepen, a wave of economic anxiety led to increasing emigration not only by Jews, but also by other ethnic groups, and Soviet permission to emigrate increased dramatically. From 1987 to 1989, some 450,000 people left the country, with another 450,000 from 1990 to 1991.[32] The period from 1985 to 1990

---

[31] Fred Lazin, "The Role of Ethnic Politics in US Immigration and Refugee Policy: The Case of Soviet Jewry," The Center for Comparative Immigration Studies, Working Paper 175 (San Diego: University of California, February 2009): 12–15. https://ccis.ucsd.edu/_files/wp175.pdf

[32] Lilia Shevtsova, "Post-Soviet Emigration Today and Tomorrow," *The International Migration Review* Volume 26, Number 2, Special Issue: *The New Europe and International Migration* (summer 1992): 241.

brought a significant qualitative change in the stimulus for immigration that continued to accelerate during the 1990s. Soviet social scientists noted that while earlier emigrants had left the country for political reasons or as a result of discrimination against them as Jews, the new emigration had an increasingly economic stimulus as people began to worry about jobs, social stability, benefits, and future prospects.[33]

The ability of the full range of Soviet citizens to travel abroad and possibly to emigrate moved rapidly away from Soviet exit restrictions when, in May 1991, the Soviet Parliament passed a law abolishing all former restrictions that had stood in the way of Soviet citizens wishing to emigrate. Implementation was postponed until 1993. In the meantime, the Soviet Union was dissolved officially on December 26, 1991. The Russian Federation as a successor state implemented the bill, giving citizens the right to travel abroad. This law required issuance of a new external passport, with restrictions on issuance of these passports that limited the ability of scientists or military people who had special military-related experience and special scientific/military skills to emigrate. The key point is that the right to leave was expanded to the non-Jewish sections of the population.

Similar laws were adopted in other newly independent countries of the former Soviet Union. For instance, in January 1994, the Ukrainian Parliament adopted the law "On the Procedure of Exit from and Entry into Ukraine by Citizens of Ukraine" that guaranteed the right to leave and return without restriction and that was strengthened and guaranteed in the Constitution of 1996.[34]

In the midst of the early 1990s, socioeconomic crisis projections were made of the widespread emigration expected from the former Soviet Union. A poll taken in 1991 indicated that 50 percent of the respondents were ready to emigrate.[35] At roughly the same time, experts at the Russian Academy of Sciences projected that emigration would reach from 1.5 to 2 million people per year and peak in 1993.[36] Among some Europeans, there was a fear of being overrun by immigrants that they could not absorb.

[33] Ibid., 242.
[34] Olena Malynovska, "Caught Between East and West, Ukraine Struggles with Its Migration Policy," Migration Policy Institute, Online Journal (January 1, 2006). http://www.migrationpolicy.org/article/caught-between-east-and-west-ukraine-struggles-its-migration-policy/.
[35] Shevtsova, "Post Soviet Emigration – Today and Tomorrow," 244.   [36] Ibid.

There is a wide range of estimates as to how many scientists and researchers left the Soviet Union during the period of *perestroika* and later in the 1990s. Soviet sources estimate that as many as 150,000 members of the intelligentsia, mostly scientific and technical, left between 1987 and 1991.[37] The ethnic composition of emigration also began to change, with increasing proportions of emigrants being non-Jewish Russians and Russians of German background.[38] The major recipients of immigrants between 1987 and 1994 were Israel, the United States, and Germany.[39] Graham and Dezhina estimate that anywhere between 7,000 and 40,000 researchers left Russia between 1993 and 1996. In 1994, Russian researchers reported, "More than 30,000 Russian scientists are now working in the USA and in Israel. There are more than 4,000 in Germany, 600 in France and 95 in Korea."[40] According to US census numbers, more than 10,000 mathematicians, engineers, and scientists born in Russia or the Soviet Union immigrated to the United States in the 1990s.[41]

US Legal Framework

As we have noted, the initial surge in immigrants caught Western officials by surprise, leading to changes in US immigration procedure and efforts to channel Jews to Israel. US officials fully expected the vast majority of Soviet émigrés to go to Israel. The United States actually provided substantial financial support for the move to Israel, with $405 million given to Israel between 1973 and 1993 to resettle Soviet Jews.[42]

As a result of political negotiations between President Reagan and General Secretary Gorbachev, Soviet officials relaxed permission for Jewish emigration. As the numbers emigrating continued to escalate in 1987, 1988, and 1989, US officials were overwhelmed with requests for refugee status. The determining factor on the volume of immigrants to the United States became US immigration policy rather than Soviet regulatory control. The category of political refugee became the most contentious point in the US immigration process. As the numbers of Jewish immigrants swelled, resistance began on the US side to accepting all of those who wanted to enter. Still, those "dropouts" who applied

---

[37] I. Orlova, Y. Streltsova, and E. Skvortsova, "Contemporary Migration Processes in Russia," *Refuge* Volume 14, Number 2 (May 1994), 13.
[38] Ibid.   [39] Ibid.   [40] Ibid.   [41] Ganguli, "Immigration and Ideas," 258.
[42] Lazin, "The Role of Ethnic Politics in US Immigration and Refugee Policy," 6.

for political refugee status from the US consular officials in Vienna or Rome had a very high rate of success. Achieving this status was important not only for purposes of legal entry, but also for receipt of social support benefits upon arrival. The US State Department announced that it would no longer accept applications for refugee status for those who left the Soviet Union on Israeli visas after November 5, 1989. Up to then, almost all Jewish emigrants left on Israeli visas. After leaving the Soviet Union, some "dropped out" at the transit point in Rome, requesting US refugee status. After November 5, 1989, applicants for US refugee or parole entry status had to be made through the US Embassy in Moscow, and the standards for refugee status became much stricter.[43] Applicants were more closely scrutinized for actual "well-founded threat of persecution" to them personally.

In response, Senator Frank Lautenburg of New Jersey sponsored new legislation to establish a "presumption of refugee status" for any Russian Jew or Evangelical Christian. The Lautenberg Amendment, Public Law 101–167, enacted on November 21, 1989, established that presumption without showing individual discrimination. This presumption allowed relatively easy issuance of a US visa for anyone coming under its terms. Those covered as refugees by the Lautenberg Amendment also became eligible for Special Cash Assistance and for Federal Public Assistance Programs including, but not limited to, Social Security, Medicaid, Food Stamps, and Temporary Assistance for Needy Families.[44] Despite the legal presumption established by the Lautenberg Amendment, the actual flow of Russian and Ukrainian Jews in this period was primarily to Israel. This was due to the impact of the overall cap on refugees who could enter the United States on an annual basis through that status.

While the focus of our inquiry into the US visa status for immigrants in the First Wave and those in the first half of the Second Wave has been on refugee status, other visa routes were opened in the 1990s. The Immigration Act of 1990 amended the visa preference system by adding categories for employment-based immigration and investor-based immigration. It placed a cap on total immigration, broadened the definitions of immediate relatives, and established the lottery system for Green Cards for permanent residence status. In addition to the cap

---

[43] Cooper, "US Abruptly Closes a Route for Soviet Jews."
[44] Jeffrey B. Perry, "Two-Tier Immigration: The Lautenberg Amendment Legacy," *Black Agenda Report*, June 11, 2013. http://www.blackagendareport.com/content/two-tier-us-immigration-lautenberg-amendment-legacy.

Table 4.2 *Refugees and Asylees Granted Lawful US Permanent Residence Status, 1991–2000*[45]

Country	1991–2000
Armenia	2,161
Azerbaijan	12,072
Belarus	24,581
Georgia	2,593
Kazakhstan	4,269
Kyrgyzstan	1,248
Russia	60,404
Ukraine	109,739
Uzbekistan	19,539

on H-1B and H-2B nonimmigrant business visas, the Act established new categories of O and P for athletes and entertainers, Q visas for international exchange, and R for religious workers and other defense and specialist education programs. There were also J-1 student visas and L-1 visas for intra-firm transfers. Many of these other visa categories become critical to understanding the legal status of Third Wave immigrants as they expressed it in their interviews.

Tracking Wave Two Numbers

During the Second Wave, Jews continued to come to the United States as refugees, but decreasing numbers were coming under that status as the overall number of immigrants rose dramatically. The majority of Jewish and non-Jewish immigrants in this period, and even more so in the Third Wave, came to the United States under family unification, student visas, green cards, and various employment-based visas.

During these years, US officials began to differentiate between the newly independent countries of the former Soviet Union in keeping some, but not all, immigration statistics. Purely as an indicator, we created Table 4.2, which is focused only on refugees but does differentiate by country of origin and thus gives some idea of the composition

---

[45] Refugee and asylees granted lawful permanent resident status by region and selected country of birth: Fiscal years 1946–2004, US State Department.

of the flow into the United States. Table 4.2 indicates the relative predominance of immigrants from Ukraine, Russia, and Belarus.

Ukrainian government statistics indicate that, between 1991 and 2004, 2,537,400 individuals emigrated, with 1,897,500 moving to other post-Soviet states and 639,900 moving to other, mainly Western, states.[46] By the end of the 1990s, the scale of emigration decreased dramatically. In 1991, 310,200 individuals emigrated, while in 2004, 46,200 emigrated (28,900 to former Soviet countries and 17,300 to other countries), almost five times less in total.[47] The decline was primarily the result of inability to continue to obtain refugee status in the West, particularly in the United States and Germany.[48] This is a reflection of the change from restricted exit regulations in the Soviet period to limits set by US immigration policy for the number of refugees the government was willing to accept.

Interviewee Experiences of Wave Two

Eugene Boguslavsky, Release Engineer at Facebook, came at age fifteen from Minsk, Belarus: "We left in 1989. It took us about a year to arrive in the United States because that's when Gorbachev basically opened up the borders for Jews to emigrate. Since there was such a huge number of people leaving, and everybody wanted to go to the United States, the United States basically needed time to process everybody. There was a temporary place set up in Italy and in Austria, where basically we were on hold, waiting for our paperwork to get processed to get refugee status. Then, once you got a green light, you got on a plane and arrived in the United States. We stayed in Vienna for about a month, and then we moved to Rome and stayed there for another eight or nine months. And from there we flew to Brooklyn, New York."

Alexander Vybornov, Product Line Manager of Medical Products at IPG Photonics, came in 1993 at age eighteen from St. Petersburg. He reflected upon whether the driving factor for emigration was prejudice or economic opportunity: "I think it's a combination of cultural and economic backgrounds. My family was doing pretty well in Russia, but I think it's an inherent call of insecurity, if you will, of being an immigrant and knowing that you can do better. I think that's also

---

[46] Malynovska, "Caught Between East and West."   [47] Ibid.   [48] Ibid.

creating the drive. So it's the mentality. I think it's not the sole contributor, but it's one that drives you. These factors are very strong drivers, because growing up in the 1990s in Russia, I first recognized how bad things could get. And my family, of course, bore the brunt of that. But I also kind of witnessed it, seeing my grandmother or my parents standing in line for basic necessities like soap, or dealing with crime or poverty. You always know you can never rest on your laurels no matter what country you live in. I think this view is very common with many immigrants, not only Russians. Definitely the negative aspect of this is a very strong stick, if you will, and it's a stick and carrot type of situation. You know definitely the opportunity for higher quality of life in general, whether it's in the form of a day-to-day level or of advancement and acceptance in society. But I think it's also the ability to reach your potential. So that's sort of the carrot."

Some people came seeking personal safety. **Alex Bushoy**, CEO of New Concept Group, who came at age thirty, described the conditions in Moldova that led him to emigrate: "I don't think I wanted to emigrate, I probably didn't want to emigrate. I was forced to move out because it was very dangerous. We decided that we needed to move, it was literally a time of war in Moldova, and it was very unsafe. My wife was staying home with the children because they were very young. I was making money, trying to provide for the family, and I remember vividly that I had a gun at my house, should my wife have to use it just in case something happened when I wasn't home. It was expected that at any time you could be robbed, or if you didn't speak the Moldavian language, you could be in trouble. People could stop you in the street, and if you were not native or you didn't speak Moldavian, you could really be beaten up. It was a very dangerous time."

Similarly, **David Gukasian**, Senior Systems Manager at Battery Ventures, who emigrated at age seventeen, found his homeland of Armenia turned into a war zone. He explained: "After the breakup of the Soviet Union things really collapsed. I'm not sure if you're aware, but Armenia was a war zone at one point. After 1991, when Armenia reclaimed independence, there was the Azerbaijan military conflict tension. So it was a complicated situation. That's kind of what drove the decision to move to another country. My mom's father and her sister had already been here for five years, so they were able to procure us the appropriate paper work, visas, and all that business. That process took around five years to get here."

Some came to the United States and were essentially marooned here as their homeland dissolved. **Alexandra Johnson,** Managing Director of DFJ VTB Aurora, was originally from Vladivostok in the Russian Far East and came for graduate school at age twenty: "So technically, I never immigrated. It's just I love the country that doesn't exist anymore. And, by the time I was done with school, the Soviet Union fell apart and I remember visiting what became Russia in the early '90s and that was quite disturbing."

Some interviewees came to take advantage of opportunities in the tech sector as early as 1990. **Alexis Sukharev,** Founder and CEO of Auriga, came from Moscow at age forty-four, just before the dissolution of the Soviet Union: "The first time I was invited to visit California was in early 1990, and the company was formed in December of 1990. Before that I didn't have a company, I just had this idea. The Contract Programming Division of Hewlett-Packard invited me to California. At that time, they didn't have their own company units in India. They used Indians. I know for sure Chinese and Hungarians, maybe somebody else for programming." This led to his founding of the first US–Russian joint venture for software outsourcing, which became a major player in that sector.

Felix Feldchstein, a Medical Devices Consultant, came from Dzerzhinsk, Russia, at age thirty, with the intention of starting his business in the United States: "As I moved, I immediately had to make not only the transition from Russia, Soviet Union, to the United States, but also a transition from the academic world to the industrial and entrepreneurial world, which is as sharp as a change of country and language. We did have some initial success and created manufacturing technology. We started some sales, but then our investors pushed us into several mistakes. You know, pushing sales too early without good clinical validation and without creating a good marketing story of why people should buy the product at this price and how they would pay for it. So finally, my first company didn't make it. Later, we got involved in the development of optical imaging technology. In addition to scientific activity, we created a company that was incorporated by a group of American lawyers working on the commercialization of Soviet technologies and trying to move them to the US market. So they encouraged us first to file several patents, and then they found some resources to bring us here to create some connections, trying to raise funds, trying to do business in United States. So we created a commercial company in

1996, and it was growing and growing, and, by 1999, it was clear that somebody needed to be here to nurse this company. I came here for three months on a business trip. I sent all my team back to Russia and, you know, tightened my belt and decided to stay until we got money or something else happened. It was the winter of 2000. So we created a business plan, and we were sure we would have crowds of investors wanting to fund it. We were ready to present our business plan to the broader investor community around March, April, and May of 2000. But it took much longer than we thought. So I had to postpone my return ticket several times. Then finally we raised funds. I brought my family here and filed for permanent residency. So I never came here intentionally; I came here on a business trip with a return ticket and am still here."

As the private technology business sector began to develop in Russia, a flow of people came to the United States for purely business reasons, particularly in recognition of market opportunities. **Eugene Shablygin**, CEO of WWPASS Corporation, came from Moscow at age thirty-two, after establishing Jet Info as a successful computer systems integrator in Russia in the early days of the transition to a market economy. He founded a US company as well to straddle both markets: "Since I was the CEO of the Russian company and the Chairman of the Board and the President of the US company, and since both companies, US and Russian, had overlapping ownership, there was a rare legal way to immigrate to the US through an executive green card. So that was my path. I was working with an attorney, and then, in 1995, I got a green card. I got it primarily because I thought that since my life was in computers, and because computers were coming from America, I needed to be closer to the origin of this technology."

As private business in technology and particularly in certain areas of software began to emerge in Russia in the early 1990s, multinational firms began to develop links with talented Russian researchers and scientists. These links then led to exchanges of personnel and ultimately, in some cases, to immigration to the United States. **Tatiana Kvitka**, CEO of Design by Light, came from Moscow at age twenty-five, as part of her business career: "I was the first one of my family to move from Russia to here, and I never immigrated. I came for work, basically. I worked at the end of the '80s, actually 1989, for the first technological startup in the Soviet Union. It was called ParaGraph International and the company was developing handwriting recognition technology. I came to the US because ParaGraph had signed a contract with Apple Computer for localizing software for the Russian market and for developing handwriting recognition technology that later was used by Apple in their first PDA [personal digital assistant], Newton. ParaGraph signed the contract with Apple, and it was the time of the military *coup d'etat* going on in Russia against Gorbachev. So, at some point, Apple said, 'We don't want to have our key project dependent on your political situation,' so they requested that the core group of people be moved to California. That's how Stepan Pachikov, the founder of ParaGraph, decided to open an office headquarters in Silicon Valley. Basically, he moved part of the company here, ultimately including me."

Different forces, including individual initiative by Russians and others in the former USSR realizing the opportunities that their talents and training offered them in the United States, sparked the movement of talented people in the tech sector. The channels for this movement were in their infancy and were often unconventional. Serial entrepreneur and investor, **Max Skibinsky**, described his path to the United States, coming from Moscow to take advantage of opportunities in the technology sector: "When I was in the middle of my third year at Moscow State University, I had this employer in Seattle who was paying an amazing $50 a month salary. I was writing some 3D surgical modeling code for them. I was connected in this very minimal fashion to the United States software development ecosystem. After a while I started telling people, 'Hey, do you want to bring me over to the United States? I'm ready.' I needed visa sponsorship, and I had no idea what an H1-B visa was. After two years of begging and emailing and doing all of these $50 contracts, I eventually found a company that said 'OK, come over, it would be interesting, let's talk in person.' And that is basically how I got to the United States the first time. I would say it is a little bit of a compliment to call them a company. The company was more or less two Russian guys who had immigrated many years before. And they thought of the idea of inviting people from Russia, getting them H-1B visas, and then marketing their services on a local contract and keeping the difference. It was kind of a Russian play: they invented kind of a model of an Indian body shop for outsourcing software. But, then again, it was extremely early, so it wasn't any sort of industrial trend at that time; it was more of just people experimenting."

Skibinsky described the process: "There are two parts to this: The first part is legal and the second part is more personal, so I will chat about the legal first. The immigration system back then was completely broken, and then it got slightly fixed. Fundamentally, if you were an entrepreneur and wanted to do something yourself, you really needed to twist yourself into these crazy knots to do something. Literally, at my first job in Silicon Valley, I had to educate my employer. I told them: 'I know you want to employ me directly but you cannot. You need to pay these strange, weird guys in Chicago. Actually, what you pay them is not going to be my salary. They're going to take their cut, and basically you can't compensate me directly for the work.' And the employers were like, 'OK, weird, strange, but we want your skills so much that we won't resist.' They had interviewed twenty people, and I was the first interviewee who they really wanted to work for them."

For some individuals, the process of immigration was the result of talent searches by US companies, sometimes ones founded by Soviet immigrants. **Dmitry Skavish**, Cofounder and CEO of Animatron who came from Yelets, Russia, at age twenty-nine, explained: "I just posted my resume somewhere, then somebody called me from Parametric Technology (PTC). It's a big company in Needham [Massachusetts]. It was actually founded by a Russian guy. They did an interview on the phone. Then they actually flew me to Boston for an interview on site. They gave me an offer, and I moved in 1999 under an H-1B working visa."

As international companies began to enter the Russian market, they hired and trained Russian staff who later became part of their international human resources, some of whom ended up in the United States. A senior executive in investment banking who traveled this route came to Silicon Valley in 1993 at age twenty-one to work at Goldman Sachs. She noted: "It was very easy. For some reason, I didn't have any problems, maybe because I had been trained in Citibank in Russia, so I knew how to work with the high-profile customers." The Second Wave was clearly rich in very skilled and talented entrepreneurs, researchers, software engineers, and professional service personnel who contributed world-class capabilities to the surging growth of the US tech sector in the 1990s.

Another significant path to permanent status in the United States came from student and professional exchange programs. As with other areas, exact numbers are difficult to pin down. The Institute of International Education in the United States estimated that the number of Russian students in US universities grew from around 1,000 in the mid-1990s to around 8,000 in 2004. In seeming contrast, the Russian Ministry of Education estimated that, in the mid-1990s, the number of Russian students in foreign countries, including short-term youth work programs, was 20,000, growing to 50,000 in 2004.[49]

Sten Tamkivi, Cofounder and CEO of Teleport, had his first experience in the United States as an exchange student from Estonia, and he later became an active participant in the Silicon Valley technology ecosystem as an executive with Skype: "I was an exchange student in 1994–1995. I spent my junior year in high school here, three years after Estonian independence. There was a Rotary Club exchange program that I applied for. I was living here hosted by a family in Cupertino [California] who were Rotary members. I went to Monte Vista High, a public school that had just been included as a pilot in Al Gore's Superhighway of the Future schools. There were probably 1,400 students and 800 computers, and fiber optic networks. I was using Yahoo on the domain yahoo.cs.stanford.edu. This experience didn't make me a geek. I was a geek before, but it was very, very nice to experience."

Marat Alimzhanov, Lead Staff Scientist specializing in oncology at Acceleron Pharma, explained the complicated process he endured in obtaining a visa to come, as part of his doctoral training, to the National Cancer Institute for a few months in 1994. Born in Russia but raised in Tselinograd, Kazakhstan, he found himself caught between those two countries: "In Soviet times, when I was about sixteen years old, I had to decide about my nationality to put on my passport. I actually had lived in Kazakhstan for almost ten years and my whole family was there, so I wrote 'Kazakh.' But, in 1994, a couple of years after the Soviet Union collapsed, I had a dilemma because I needed a passport to travel abroad. I was in Moscow and yet I was not a citizen of Russia. I went to some Russian authorities and they told me, 'Oh, you have to apply for Russian citizenship and it will

[49]   Irina Ivakhnyuk, "Brain Drain from Russia: In Search for a Solution," *Center for International Relations, Reports and Analysis* Volume 15, Number 6 (undated): 3. http://pdc.ceu.hu/archive/00004817/01/rap_i_an_1506a.pdf.

take so much time.' And I said, 'Well, I need it now. I was born in Russia, my mother is Russian, and I have the right to be a citizen of Russia.' I just filled out the application and they said, 'No, we cannot help you.' So I called my father and I said, 'Can you do something about it?' And he said, 'Yeah, just come to Kazakhstan and we'll make a Kazakh passport for you and then you can apply for the visa.' It was funny, actually: When I went to Kazakhstan, they gave me an old Soviet-style passport, but then there was just a stamp on it, you know, 'Kazakhstan.' At that time, two years after the Soviet Union collapsed, they didn't even have their own passports. It was a big mess. It was chaotic. People really didn't know what was happening at that time. But, basically, it worked out. I went with this makeshift Kazakh passport to the American Embassy in Moscow, and they gave me a visitor visa."

For those who had been in Europe or the United States at the time of the dissolution of the USSR and the accompanying dramatic drop in science spending, there was no lab to go home to and no way to continue doing research in the country. There was also concern in nonproliferation and security circles in the United States and Europe about unemployed Soviet scientists assisting "rogue" countries, namely Libya and North Korea, with nuclear weapons programs. There is no evidence that anything like this took place, but a number of Western-financed programs were put in place to provide employment for high-level researchers in Russia and other former Soviet republics. These included the International Science and Technology Center in Moscow, the International Science Foundation funded by investor and philanthropist George Soros, the US Civilian Research and Development Foundation for the Independent States of the Former Soviet Union, the NATO Science Program, the American–Russian Biomedical Research Foundation, the Association of Scientific Societies of Russia, the US Industrial Partnering Program, US lab-to-lab activities, and others.[50] Along the same lines, the United States enacted the Soviet Scientists Immigration Act of 1992, which was in effect for four years and provided employment-based exceptions for immigration to the United States for up to 750 beneficiaries.

---

[50] R. Adam Moody, "Reexamining Brain Drain from the Former Soviet Union," *The Nonproliferation Review* (spring-summer 1996), 93.

Wave Three 2000–2015: Integration into the Global High-Technology Sector

Wave Three was characterized by the integration of the countries of the former Soviet Union into the global high-technology sector and by an increasing level of economic and political diversity among them. Fundamental to this period were rising oil and commodity prices that lifted the economies of Russia, Kazakhstan, and Azerbaijan, while economies of the other successor countries remained stagnant. The energy-producing countries depended heavily on taxes and fees on energy to fund their state budgets. For example, the budget of the Russian Federation depended on the energy sector for 60 percent of its revenue. Oil prices in November 1998 stood at $16.44 a barrel.[51] By November 2000, the price of oil had risen to $45.78 a barrel,[52] and by June 2008, it had risen to $151.72 a barrel.[53] Largely due to these rising oil prices, the Russian economy not only stabilized, but also grew dramatically. In the later years of Wave Three, after the economic crisis of 2008, Russian economic growth declined to 1 percent annually from 2009 to 2013.[54] However, for the early years of the Wave Three period, rising government revenues in Russia and Kazakhstan in particular led to renewed support of science and opportunities for innovative researchers, as described in Chapter 3.

As the largest of the post-Soviet countries, Russia set the stage for increasing openness and integration into international economic, scientific, and political institutions. The push for restoration of Russia's status as a leader in science and technology began under the presidency of Vladimir Putin. This early initiative was amplified by his successor, Dmitry Medvedev, into a call for transition to an innovative, knowledge-based economy. As will be described in Chapter 7, ties grew with Silicon Valley, fostered by Wave Two émigrés and organizations like AmBAR and forums like Silicon Valley Open Doors and the Russian Technology Symposium, as well as with the innovation ecosystem in Boston through university ties with Harvard, MIT, Northeastern, Babson, and others. During that same period, a number of

---

[51] *Macrotrends*, http://www.macrotrends.net/1369/crude-oil-price-history-chart.
[52] Ibid.   [53] Ibid.
[54] Susanne Oxenstierna, "Russia's Defense Spending and the Economic Decline," *Journal of Eurasian Studies* Volume 7, Issue 1 (January 2016). http://www.sciencedirect.com/science/article/pii/S1879366515000287.

multinational companies, including EMC, Intel, Boeing, Microsoft, and others, established development centers in Russia which further absorbed intellectual human capital and helped keep people in place there.

As a result of these factors, the brain drain of the scientific and technical intelligentsia from the former USSR to the United States declined markedly. Students and entrepreneurs were still able to travel to the United States to pursue their scientific and business interests, but without having to emigrate. Within these circles, there was a feeling that while they had a lot to learn in the United States, their opportunities were becoming much better in their home countries. Many began to think of themselves as citizens of the world, but with a Russian, Ukrainian, or other national cultural background.

Meanwhile, as indicated in Chapter 3, government spending on education and science in Ukraine, Georgia, Armenia, and Belarus continued to decline, which continued the pressure to find employment abroad for members of their technical intelligentsia. Talented individuals from the stagnating economy of Ukraine or from those countries with limited domestic markets like Belarus, Georgia, Armenia, and Moldova continued to seek opportunities in the United States and Europe.

Special notice should be taken of the role of the software and IT industries in the countries of the former Soviet Union. The industry was founded primarily by academics who were impoverished by the collapse of the Soviet Union and who went into private business. Only a few actually emigrated to the United States, but a number of them established representative offices and even headquarters in the United States. Most importantly, the dramatic growth of their companies has absorbed large numbers of scientific and technical workers who did not have to emigrate in order to make a living wage. For example, EPAM, founded in Minsk, Belarus, and now headquartered in the United States, grew from 50 programmers in 2001 to 1,700 employees by 2006 in Russia, Belarus, and Ukraine, and is listed on the New York Stock Exchange. The second largest software outsourcing company, Luxoft, a Russian firm, grew from 100 employees in 2001 to 1,600 in 2006.[55] A number of Russian software product companies, like

---

[55] Stan Gibson, "Russian Software Makers Search for Growth," *eWeek*, January 2, 2006. http://russoft.org/docs/?doc=1169.

ABBYY, Acronis, and Parallels, that have found foreign market success have also increased opportunities available without requiring emigration. The industry, including many other significant companies not mentioned, rapidly spread throughout the former Soviet Union, absorbing the intellectual scientific intelligentsia throughout the region and, in the process, mitigating pressures to emigrate.

Post-Soviet Legal Framework

In the early years of the Wave Three period, there were no significant changes in the legal framework for travel or emigration from the countries of the former Soviet Union that had been established at the beginning of Wave Two in the 1990s. The formal regulation of emigration remained the same, while concern about preserving national intellectual capital began to grow by the end of Wave Three, particularly in Russia. What did appear were incentives to stay in Russia as well as some incentives for former Soviet scientists and researchers to return.

Almost every Russian ministry or administrative body at all levels had some element of an innovation program. Russia established new government-funded bodies including Rusnano, the Russian Venture Company, and the Skolkovo Foundation, to carry out this transformation. These were established along with major reforms of the Russian university system and the Russian Academy of Sciences, as discussed in Chapter 3. The improvement in budget support reduced the economic incentive for emigration and provided a means for those committed to scientific research to continue working in the country. This did not completely resolve the crisis of education and research, but it did mitigate the direct pressure to leave. Recently, Russia has tried to target support for technology innovation and support of scientists on a policy level through programs like its National Technology Initiative started in 2014, and a May 2016 decree by Prime Minister Medvedev to set up a national Agency for Technology Transfer.[56] The relationship between the Russian political elite and the emerging technical elite came into conflict around the contested 2011 State Duma elections

---

[56] Ksenia Zubacheva, "Russia Direct Report: 'Imagining the Future': Russia's Plan to Boost Technology by 2035," *Russia Direct*, May 30, 2016. http://www.russ ia-direct.org/archive/russia-direct-report-imagining-future-russias-plan-boost-t echnology-2035.

and the popular protests against corruption and vote rigging that followed. Shortly thereafter, President Putin's reelection and tightening controls over dissent led to renewed emigration, particularly by what was called "the creative class."

Russia's relationships with the United States and the European Union were already under stress by the time of the upheaval in Ukraine. Russia's ongoing support of the self-proclaimed Republic of Donetsk and Republic of Lugansk in Russian-speaking eastern Ukraine, along with recognition of the Crimean referendum to join the Russian Federation, led not only to sanctions, but also to an upsurge in Russian nationalism. Accordingly, most forms of collaboration in business and technology have been shut down by the United States, the EU, and Russian governments, and other forms of private collaboration became much more difficult due to public opinion.

The growing opposition to the effects of integration of Russian science into the global technology sector was clearly expressed by President Putin when, in June of 2015, he criticized foreign organizations that were "working like a vacuum cleaner"[57] to find Russian talent wanting to emigrate. In response to ongoing international economic sanctions, Russia has increasingly supported policies for domestic production of technology and software to replace foreign products. Technology globalization is increasingly replaced by technological autarchy, reminiscent of the Soviet era.

US Legal Framework

As of early 2018, the United States was divided in its attitude regarding immigration policy and the continuing benefits of recruiting foreign talent to come to the United States. Prior to the November 2016 US presidential election, there continued to be programs designed to attract entrepreneurial talent. On August 26, 2016, the US Department of Citizenship and Immigration Services announced a new International Entrepreneur Rule that would allow foreign entrepreneurs to be legally resident in the United States if they satisfy certain conditions. Implementation of this measure was delayed to March 14, 2018, amidst speculation that it might be

---

[57] Irina Reznik, Ksenia Galouchko, and Ilya Arkhipov, "Putin Faces Growing Exodus as Russia's Banking, Tech Pros Flee," *Bloomberg*, September 20, 2015. https://www.bloomberg.com/news/articles/2015–09–21/putin-faces-growing-exodus-as-russia-s-banking-tech-pros-flee?cmpid=yhoo.

cancelled entirely. Whatever desire there may be from the technical intelligentsia of the countries of the former Soviet Union to emigrate to the United States, the primary determining factor will be the extent to which the United States will continue to admit them. And any assumption that US immigration policy will continue to be friendly to attracting talented foreign individuals appears problematic for the foreseeable future.

Tracking Wave Three Numbers

As was shown in Table 4.1, the Third Wave shows a gradual decline in refugee admissions between 2000 and 2003. After 2003, the State Department reported no refugee admissions from the former Soviet Union. From this point, all immigrants from the former Soviet Union were to be admitted according to the standard immigration system, including employment-based, student, and special talent visas.

Beginning in 2000, the statistics for permanent residence status were recorded separately for each country of the former Soviet Union. Those numbers are summarized in Table 4.3 for the period from 2000 to 2013. Some databases continue to include a category for the former Soviet Union, creating ambiguity for country of origin. Given the difficulty of reconciling databases, this table should also be seen as indicative only. The table shows the predominance of immigrants from Russia and Ukraine.

Although the absolute number of immigrants in Wave Three is larger than Wave Two, it is also a longer period of time by three additional years. Wave Three is also less dominated by the pressures of anti-Semitism and economic collapse that led to the brain drain of Wave Two and is more a reflection of the perceived opportunities in the United States.

Interviewee Experiences of Wave Three

Many of the Third Wave immigrants had already been successful in their home countries and were drawn by challenges and opportunities in the United States, about which they had a fairly sophisticated understanding. One such individual is **Evgeny Medvednikov**, Serial Entrepreneur and Angel Investor from Arkhangelsk, Russia, who came at age thirty-four: "I decided to change something in my life. I

Table 4.3 *US Permanent Resident Status by Country 2000–2013*[58]

Country	2000–2013
Armenia	39,637
Azerbaijan	13,290
Belarus	32,796
Georgia	17,016
Kazakhstan	20,757
Kyrgyzstan	7,747
Russia	188,572
Soviet Union (former)	36,005
Ukraine	119,682
Uzbekistan	44,338
Total:	519,840

was not very happy with St. Petersburg, the weather, the political and economic situation. Also, my first daughter was six and had to go to school. We decided to choose another school where she would have more options in her future life. For example, if she speaks both English and Russian, she has more options to choose where to get higher education, where to work, and so on. Also, I had sold my company in Russia so I wasn't connected to St. Petersburg and now had an option to change location.

"I met with Rafael Soultanov [another interviewee] who helped me get a long-term L-1 visa. He wanted me to help his company, and I had no option but to choose Silicon Valley. To get an L-1 visa, your sponsoring company must have subsidiaries in Russia and you must be transferred to the subsidiary in the US. The advantage of an L-1 is that it is not like the lottery for an H-1 visa. You just need to conform to the checklist of requirements. For example, if you apply for an H-1, you have only a 33 percent chance that you will get that visa. But to get an

[58] US Department of Homeland Security, "Table 2. Persons Obtaining Lawful Permanent Resident Status by Region and Selected Country of Last Residence: Fiscal Years 1820 to 2013," and "Table 3. Persons Obtaining Lawful Permanent Resident Status by Region and Country of Birth: Fiscal Years 2004 to 2013," *2013 Yearbook of Immigration Statistics* (Washington, DC: US Department of Homeland Security, Office of Immigration, August 2014): 6, 12.

L-1 visa, you are not connected to any time frame. An H-1 visa is only within a specific time frame, while an L-1 visa is more convenient for a family. For example, all my family can work and study here. For an H-1, the spouse can't work."

On a different note, some interviewees understood the limitations of their home markets for their business and career ambitions. **Davit Baghdasaryan**, Senior Security Engineer at Twilio, came at age twenty-four, from Armenia: "I'm a software engineer and the opportunities for working on exciting projects are much less in Armenia compared to here. And also having goals, it's incomparable. Silicon Valley is the heart of IT, high-tech. We were working for a company that was a partner of a US-based company; it was involved in outsourcing. We were working in Yerevan. Our team was pretty important, but then the US company needed someone who would oversee their software security architecture. They offered me the opportunity to come here and work for them on an H-1 visa, and I was excited. I said I would come here and take the job."

Mikita Mikado Teploukho, Founder and CEO of Quote Roller, subsequently renamed PandaDoc, came at age twenty, from Belarus. He described his startup: "Quote Roller is a sales proposals automation tool. We have thousands of clients from all around the world. From day one, we decided to incorporate in the US. Belarus doesn't have decent laws about intellectual property and stock options, micro-transaction processing services, or other important things needed for a startup of our kind. We bootstrapped the company and, after breaking even, I decided to apply for an L-1 visa to open a physical office in the US. Belarus is not great when it comes to hiring sales and marketing talent. I was denied this visa, which slowed us down and forced us to hustle. Next, I had to apply for an O-1 visa, which took over seven months and $15,000 in expenses. The process was a nightmare in terms of the amount of work I had to put into it. Right now Quote Roller gives jobs to eight US citizens, some as contractors and some as full-time employees. I came to the US to hire people and to build an international business. It wasn't easy at all." His company subsequently raised $15 million Series B funding from Microsoft, HubSpot, and others, and grew to sixty employees in the United States.

Moldovan entrepreneur, **Grigore Raileanu**, Founder and CEO of Noction, came at twenty-nine, also on an L-1 visa, which he said was "not because I am an investor, but because we already have a business in Moldova and the US company was a sister company. So it was an intercompany management transfer."

The process of immigration, even for economic opportunity, was not necessarily as smooth or straightforward for many people as it might seem, as noted earlier by Mikita Mikado Teploukho. **Sergey Kononov**, Senior Delivery Manager at EPAM Systems, was originally from Kharkov, Ukraine, and came at age twenty-nine through his previous employers, Luxoft and Deutsche Bank: "The first time I came to the US it was 2010, probably. My employer, John, was from Deutsche management, but formerly I was working for an outsourcing company. He brought me here and said that I needed to stay for two months to work on the project. I had just gotten married, and I was missing my wife so much after a month. I told him I needed to go back. Next time, again, they said you have to come to the US for one or two months. I said that it was too long. So they told me, 'You can take your wife with you. We'll pay for the tickets. We'll make all the arrangements and give you more money so she can go shopping. So with this huge addition to my salary, it was enough so that she could entertain the idea of coming to the US. So we came. It was a Saturday. It was in Secaucus, New Jersey, and it was a very dirty hotel. I didn't like it but she was like, 'Wow.' So we relocated here."

For many interviewees from the Third Wave, Silicon Valley and California more broadly were huge lures. **David Yang**, Founder and Chair of the Board of ABBYY, who was originally from Yerevan, Armenia, came to the United States at age forty-three: "My wife and I wanted to spend some time outside of Russia. We thought about China, maybe somewhere else. We just wanted to make our kids more open to international life and to learn different languages while they were small. ABBYY has sixteen offices now in thirteen countries, so we could choose. We are an international company and we do linguistic software that actually makes people closer. Our model, or our idea, is to help people better understand each other. So we thought we should live internationally, but we never thought of the US for some reason, I don't know why. But then I broke my hand badly and I had my first surgery in Moscow. They did fix it mostly, but then they told me that it wasn't fixed completely. Other people said that at Stanford they have microsurgery in orthopedic surgery, very high level, and so I moved here for the surgery. And then I spent almost a year just recovering from the hand surgery. I called my wife and said, 'Beautiful weather, beautiful people, the center of technologies, our office here in Milpitas is a thirty-minute drive. Why shouldn't we think about spending some time here?' She came, and in a week we bought land in Portola Valley and we decided to build a home, so now we are finishing the papers to build a home."

Tatul Ajamyan, Founder of Wakie, originally from Spitak, Russia, came at age thirty-two to expand his company internationally: "Well, our business was growing in Russia, and we understood that it was becoming a global product. We wanted to launch it in other languages and expand globally. So there is no better place than Silicon Valley to expand your product globally. We launched the English version here, and it started spreading. When some media from Silicon Valley gives you notice, then the Brazilian, Indian, Chinese, and Russian media also get this piece of material and start to use it. So it's really easier to expand from here. And there are so many clever and experienced people here who can help you with your business, to make your business perform better, and a lot of investors, of course. So it's a culture that is very, very healthy for growing a startup company like ours. We came on multiple entry visas. B-1, B-2, and now we are applying for an O-1 visa, which is for extraordinary success in some field. If you have some interesting success, you can apply for that kind of visa."

Roman Kostochka, Cofounder and CEO of Coursmos from Tolyatti, Russia, came at thirty-seven to explore bringing his company to San Francisco. "I came on a tourist visa and after that I obtained an H-1B business visa." He explained why he decided to settle on the West Coast: "Because it's very good weather. It's very nice people, smiling, I don't know, it's the ocean. I was in many places in the world, and in California, in Silicon Valley, and all this coast, it's the best. The West Coast, it's the best nature in the world, I think. And if we talk about business, of course, it's Silicon Valley. It's the best place to begin doing some work with a startup. It's a great environment for bringing up kids. For example, my partner's daughter is going to school with children whose parents work at Google and Facebook; it's networking in school."

A significant trend in the Third Wave was the increased recruitment by multinational corporations of talented people from the former Soviet Union that began during the Second Wave. This often led to transfers to the United States as part of a company's overall human resources deployment. These professionals became part of a global movement of intellectual capital. **Dmitry Kuzmenko**, Senior Software Engineer at Google and who had previously worked at Microsoft, came from Latvia at age twenty-three: "I was choosing between Canada, New Zealand, the UK, and the United States. All of these countries at the time had professional immigration programs, but the Microsoft opportunity came kind of unexpectedly, so I jumped on it because the US looked more appealing." **Gregory Tseytin**, Freelance Consultant in R&D in Computing/Language Processing, came at age forty-four from an analogous situation in St. Petersburg on a J-1 visa to work at IBM in San Jose. **Dmitry Kovalev**, Consulting Systems Engineer at Cisco and originally from Moscow, came at age twenty-four, through a transfer from Cisco in Amsterdam to San Jose: "Cisco sponsored me with the move. It was an intercompany transfer on an L-1 visa." **Andrey Doronichev**, Senior Product Manager at Google who came at age thirty from Moscow, explained his own path: "Part of the reason why I moved out of Russia was because unlike most of my friends, I had never moved before. Like most Russians, they at some point have to take a step and move to Moscow. And I didn't even have to do that, and I felt an itch to make my step and go somewhere. Well, I didn't come to the US straight away. In fact, I left Russia in 2006 and I moved to Prague in the Czech Republic. Then I moved to Dublin, Ireland, then to London, and finally I moved here. I only came to the US three years ago."

Part of this Third Wave was the ongoing recruitment of very talented people to come to the United States under various visa categories that encouraged and allowed them to enter. **Andrey Klen**, Cofounder and COO of Petcube who was born in Egvekinot, in the Russian Far East, and raised in Ukraine, came on an O visa: "It is for people with an exceptional talent. There are a few ways to show this exceptional talent. They include media publications and achievements in my area of expertise, and I provided all of that. Normally, an O visa is for some sort of artist, athlete, and all that stuff, but I'm guessing IT people and entrepreneurs are a huge part of that nowadays."

Vlad Pavlov, Founder and CEO of rollApp, came at age thirty-six, from Dnepropetrovsk, Ukraine, also on an O visa, among others: "I had different types of visas. While in Silicon Valley I had three types of visas. I had an E-2, which is an investor visa; I had O-1, which is a 'smart guy' visa. I often joke that I have an official certificate from the American government that I am smart. And now I have an EB-1, which is another way of saying that I am smart. The EB-1 is a green card. It's actually on the way. I am going through a process called Adjustment of Status. It is already approved, but I still haven't received it."

Shamil Sunyaev, Professor of Bioinformatics and Medicine at Harvard Medical School, came from Moscow at age thirty-one, after receiving his PhD in molecular biophysics from the Moscow Institute of Physics and Technology. His move was part of the same kind of attraction of talented, trained people to come to the United States during this period: "I wasn't quite sure where to go, and initially I wanted to stay in Europe since I thought that was a more fun place and a better place to live and all that. But somehow I finally decided to come to Boston. I came in August of 2002 on an H-1B visa. There was a discussion about what kind of visa to use, but I think the idea was, as a junior faculty, the O type of visa would be very difficult to get. So I came to this position with Brigham and Women's Hospital and with my faculty appointment through Harvard Medical School."

Even as the Third Wave period came to a close due to external political conditions, individuals continue to come to the United States for the business opportunities available. Serial Entrepreneur **Dimitri Popov**, originally from Zelenograd, Russia, came at age forty. He expressed the feeling that the Third Wave might be ending and speculated about what might follow: "And for the last year in Russia, I was just absorbing the total deterioration of the business environment. Everything was becoming so bad. You know what I mean. So I just decided to buy a ticket and fly to the US. I worked for one of the venture funds last summer in Kazan. We worked together with their CEO and someone from here, from Silicon Valley, who speaks good Russian and also worked for them. And I think he heard something about me from those guys in Kazan. When I met him finally this year in January, he invited me to work for him. They applied for my work visa, for an H-1B, and it looks like I got into the quota. This is why I am here, because I believe that they will approve my visa by October 1. Then I will have a chance to stay here and work and be more successful here than in Putin's Russia for the next maybe three to five years. I'm not looking for a green card or any other citizenship. I'm not sure I really want to be an American citizen because of taxes. Everything else excites me, but taxes are an issue."

The future of globalized science and innovation as represented in the experience of the interviewees in this book is uncertain. The general atmosphere for the integration of Russia into a global innovation ecosystem has diminished substantially, bringing to a virtual halt this period of transfer of human capital from Russia that we have characterized as the Third Wave. However, depending upon US openness to receive them, there remains the possibility that immigrants from Ukraine and other countries of the former Soviet Union might continue providing a supply of scientists, researchers, and entrepreneurs fitting the Third Wave profile, or alternatively, becoming the genesis for a new Fourth Wave. As the Russian saying goes, "We will live and see what happens."

PART II

The Immigrants' Experiences, Integration, and Contributions

5 | Entrepreneurial Spirit, Creativity, and Innovativeness

Startups in the United States

While numerous interviewees had participated in startups in the former USSR, mostly as founders or cofounders, far more did so after their arrival in the United States, and some had founded startups in both locations. That so many were able to start companies in less than ideal circumstances before they emigrated is a testimony not only to their entrepreneurial spirit and creativity, but also to their willingness to take risk and accept failure. Those traits were exhibited by them and many other interviewees who started firms in the United States, a far more hospitable institutional environment than in their home countries. It is common knowledge that entrepreneurs willing to start new ventures constitute a small percentage of any country's population because of the personal characteristics such initiatives require. Those characteristics are the central topics of this chapter.

We begin with the experiences of interviewees regarding their startup ventures in Silicon Valley or the Boston-Cambridge area, organized by the three waves of migration. Within each wave, interviewees' stories are organized into three broad categories. The first group is made up of those whose startups are involved in biotech, pharma, and medical products. Those in the second group are involved in software, the Internet, communications, and IT. We identify the third group as other industry sectors since those businesses either did not fit into the other two categories or the entrepreneurs' multiple startups crossed over into different industries, preventing a simple categorization. We note that the vast majority of entrepreneurs in all three waves founded companies in the software, Internet, communications, and IT sectors, reflecting the prevalence of their technical education in physics, math, and computer science, and the fact that software startups usually are less capital intensive than those in sectors like biotech, pharma, and medical products. Not surprisingly, no Wave One interviewees had started businesses in biotech, pharma, or medical products, thus we begin Wave One with entrepreneurs in the second group.

Wave One 1972–1986: Software, Internet, Communications, and IT

One of the few interviewees who came during Wave One with a background to become involved in technical businesses was **Igor Razboff**, Founder and CEO of Scoros International. He grew up in St. Petersburg and graduated with a PhD in mathematics from St. Petersburg State University. He elaborated: "My background is math, but I didn't really want to be in science anymore. I wanted to be in something more practical. I felt that the most appropriate area for me to enter would be CAD-CAM, computer-aided design and computer-aided manufacturing. So I was in software development for several years. Not long after, I came to Boston and worked for AT&T for about three years. And then I went back to computer design and computer manufacturing, and in 1991, I started my own business in multimedia called Animation Magic. When I started that business, Russia opened up and my partner, who was also from Massachusetts but originally from California, suggested that we should probably hire some animators from Russia because that would make us more competitive in the marketplace. We would have better animation and therefore more success, so I went to Russia and we opened a business there in 1992. Animation Magic was later acquired by Capital Multimedia, a public company out of Bethesda, Maryland. I became VP of R&D and later CEO of that company from 1994 to 1997.

"I later sold the assets of that company and continued to work there for about three years as Vice President of Cendant Software from 1997 through 2000. Then I went back on my own, starting my own business, again in software development. I continued to do that, and also for a few years I did innovation, broader than just software development. While I was working with Russia, one of the things I was coming up with was how to get access for American companies to some of the innovations that were developed in the Soviet Union during the Cold War. That's when I expanded my business into broader innovations, beyond just software. Numerous companies were very interested in learning more about it, so I was in the area of general innovation for some time. We brought people from US companies to Russia, and arranged meetings with scientists and researchers in hopes of finding something that was interesting for American businesses." With his varied business activities, Razboff had certainly fulfilled his objective of building on his scientific background in practical ways.

Leonid Raiz, Cofounder of Rize, had lived in Moscow and St. Petersburg and spent eleven years in the Russian Far East near Vladivostok. After attending night school at Leningrad State University, he transferred to the regular day program, majoring in mathematics with an emphasis on applied math, and graduated first in his class. Yet he ran into a roadblock: "I couldn't find a job. I graduated magna cum laude, but couldn't find a job in the USSR for various reasons." Although he spoke little English, he found a job within six months at a Massachusetts company, Computervision, doing work that he described as being "very close to what I was doing back in Russia. I was quite successful back in Russia from the beginning. I was noticed, and they promoted me, but I was always on the engineering side rather than on the managerial side." He subsequently joined PTC, a highly successful Massachusetts company started by Russian immigrant Sam Geisberg and an American cofounder: "I joined the group even before it became PTC, before it got financed. So that was my first taste of a very early startup. I was employee number 5, even before the eventual CEO came on board and before a venture capitalist put money into the company. That's when I started. And it was fun, and it was interesting. I think the company had only one round of financing and never went for a second round. Eventually, I was put in charge of the entire development organization, both here in the US and for the two development shops PTC had set up in Israel. In my case, I'm not really a people person, but more of an engineer, and my style was hands on. But, by the middle of the 1990s, PTC started becoming too big for me, the company outgrew me, and I decided to leave in 1997.

"By that time, I already had an idea of what I wanted to do, which was to develop a product with automated design for a very much underserved market segment – architects, structural engineers, HVAC engineers, and so on. So I left PTC with an idea to do something in this market with not necessarily the same set of tools, but the same idea of automating design for building professionals. I started a company, which at that time was called Charles River Software, in October 1997, and a couple of months later was joined by a former colleague from PTC, Irwin Jungreis, who had a Harvard PhD in math, a very smart guy. His idea was also to address this market, so we started working on it, interviewing architects and finding out how different they were from mechanical engineers. We got venture capitalists to finance us, including Atlas Venture and Rich D'Amore at North Bridge Venture Partners. We ended up selling the company to Autodesk in 2000."

Autodesk is reported to have paid $133 million for the company which had been renamed Revit. Raiz continued: "If you drive by the Autodesk building on Route 128, that's my baby. Autodesk is extremely successful with the product, and I understand that they are now selling $700 million of our product every year. I stayed with Autodesk for two more years, making sure they wouldn't destroy my baby.

"About two years ago, I joined another group of guys who are developing a new product in 3D printing that is very different from everything else in the field. So we'll have to wait another six to twelve months to see if we make a big splash. I am in charge of the software piece of phase development. So that's what I'm doing right now." That company is called Rize, which emanated from a predecessor company called File2D, founded in 2011 and specializing in 3D printing, of which Raiz was one of two cofounders. The company was backed by venture capitalists. Raiz's talents have made him a valuable member of founding teams in companies that have added much to the Massachusetts innovation economy.

A highly successful serial entrepreneur, innovator, and thought leader in Silicon Valley whose activities and innovations have extended throughout the country and globally, is **Kira Makagon**, Executive Vice President of Innovation at RingCentral. Coming from Odessa, Ukraine, as a refugee with her parents, she credited them with setting her on a path to success. Highly accomplished academically, she earned a BS in computer science as well as an MBA from the University of California, Berkeley. She spent her early career as a software engineer, and, by the early 1990s was vice president of Development at Scopus Technology that was later acquired by Siebel Systems. After five years, Makagon became a cofounder and senior vice president of Octane Software, a pioneer in multichannel web-based applications for sales and service. Within two years, the company was acquired by a leading analytics company, E.piphany for $3.2 billion. In 2006, Makagon cofounded and became President and CEO of NebuAd, a company that aimed to help advertisers gain visibility with their audiences across all media. In 2008, in another serial entrepreneurial venture, she became Founder, CEO, and President of Red Aril, a company that provided audience and media optimization solutions for digital marketing to provide advertisers a SaaS – software as a solution – platform. Red Aril was acquired in 2011, by Hearst Corporation. Makagon then joined RingCentral as Executive Vice President of Innovation. RingCentral provides cloud business communications and collaboration solutions. In responding to the recognition she has received for the scope and impact of her accomplishments, Makagon said: "As a serial entrepreneur and business and technology leader, I've dedicated my career to breaking new ground both by creating cutting edge software products and by speaking up as a veteran 'only woman in the room' in Silicon Valley. I enjoy giving back by mentoring young entrepreneurs and serving as an advisor to early-stage companies and work with early-stage investors such as Sierra Ventures and Illuminate Ventures." In recognition of her outstanding contributions in business and technology, Makagon was named among the 2015 Most Influential Women in Business by the *San Francisco Business Times*, was one of the 2015 Women of Influence awarded by the *Silicon Valley Business Journal*, and was also recognized by the 2015 Golden Bridge Business and Innovation Awards.

Wave One 1972–1986: Other Industries

George Gamota, Founder and President of Science and Technology Associates Management, originally from Lvov, Ukraine, had a story to tell about his innovation activities in the United States, as well as his activities as a consultant for the US government, primarily in Ukraine. Much of his career after obtaining his PhD was spent at Thermo Electron, where, beginning in the late 1980s, he was the head of R&D for that highly diversified group of high-technology companies. The firm was later called Thermo Fisher Scientific after merging with Fisher Scientific. During that period, he was also involved in startups: "During my time at Thermo Electron, I started a couple of companies, laser companies and other technology-oriented companies. I provided seed money and was on the board, but I didn't get involved with the actual management because I had my management job at Thermo Electron."

Perhaps Gamota's most interesting display of creativity was in trying to bring entrepreneurship to Ukraine on behalf of the US government.

He noted that the billionaire investor and philanthropist George Soros had provided funding to the National Science Foundation that lasted a couple of years, about which Gamota shared an important insight: "I really started to understand what the needs are and assessed that one of the biggest problems that Ukraine had, and the same thing as in Russia and other former Soviet countries, was that while they were doing research, it was almost anathema to apply the research to something useful. Software development or IT activities were easy to start in Ukraine, in part because there was no one to oversee you. There was no fire inspector coming to check, and so a lot of the issues that would be involved in manufacturing did not exist. Many actually did software development in their homes. The most successful businesses, even today, twenty-five years after independence, are doing software development, and I would say that Ukraine is right up there below India. They're doing very well with software development, especially in the capital and the western part of the country. They're very competitive, highly educated, and speak English. The education system, technical education, is excellent. Math and science are very strong." Gamota's experience reflects how he successfully applied the business expertise he had gained in a large American high-tech company to startups in the United States as well as to fostering entrepreneurship in his native Ukraine that, as he recounted in Chapter 2, he had fled as a child during World War II.

Wave Two 1987–1999: Biotech, Pharma, and Medical Products

While no interviewees had founded firms in biotech, pharma, or medical products in Wave One, there were a substantial number in Wave Two. We begin with **Alexander Vybornov**, who founded his startup, Dental Photonics, while a student at Babson College. Vybornov explained the company's beginnings and its relatively long road to fruition: "So halfway through Babson College, I started thinking about ideas and opportunities in the medical field, and mostly technical ideas my father had at the time. We started working on this as a kind of a project. Essentially, I went through Babson's Entrepreneurship Intensity Program with Professors Timmons and Spinelli, as well as Professor Bygrave who was my advisor, and then participated in business plan competitions. In fact, I think we placed second or third in competitions, like at UC San Francisco in 2005. When I graduated in 2005, Dental Photonics did not yet exist. We had significant patents for laser technology that were invented by my father and which he could actually commercialize on his own time, independently from his company because of special provisions in his contract. Between 2005 and 2007, this was more of a project, although we did raise our first seed round from an angel investor who we were very lucky to bring aboard at the time. In 2007, that's when the project evolved from more of a concept project to the first round of investment in Dental Photonics. Specifically, our mission was to create a product that would be a next-generation system for surgical, mainly dental, applications. We later sold the business to IPG Photonics, and now we're essentially a part of IPG Photonics Medical Division, and we're taking it to the next level." Alexander Vybornov's startup no longer exists as a stand-alone entity but has integrated into a larger US company with prospects for strengthening that larger organization. Vybornov has stayed on as Product Line Manager of Medical Products at IPG Photonics.

Our next Wave Two entrepreneur, **Alex Nivorozhkin**, brings us to a third industry under the overall umbrella of biotech, pharma, and medical products. He ultimately cofounded three companies in the Boston area after building a career in academe and industry. Originally from Rostov on Don in southern Russia, Nivorozhkin earned his PhD in 1985 from Rostov State University, majoring in physical organic chemistry. The choice of fields was perhaps not surprising since both his parents had degrees in chemistry and were working in that field. He himself worked as a chemistry scientist at the university. He said: "I never thought about leaving. I was young and I was a rising star. I loved generating new ideas and working on them. I was curious and always wanted to grow out of there and try to learn something new, maybe with foreign travel."

Nivorozhkin's life changed when he was offered a postdoctoral position in Denmark in the lab of a prestigious scientist who became his mentor. In spite of what he called, "in every respect, a kind of cultural shock," he added, "My work in the lab all fit together with theirs. I never felt anything surprising and appreciated the world-class experience I had gained back at my alma mater. But going back to Russia after *perestroika* was professionally suicidal: the economic regress destroyed academia. I got an invitation to work in Paris, in the lab of someone my mentor knew. It was a terrific societal experience in Denmark and

France. I think in Paris I first realized how applicable to the real world my science knowledge and skills were, but I knew the future was in the US. The US is a real opportunity, and you hear about a lot of people going and staying there. So I came to Harvard in 1996 as a visiting scientist in the chemistry department, in a really big lab with the best people from around the world." After several years at Harvard, Nivorozhkin spent about five years in industry as a Research Scientist at Epix Medical and as Head of Chemistry at Inotek Pharmaceuticals, both Boston-area companies. In 2004, he moved to the Massachusetts General Hospital as a Senior Program Manager at the Center for Integration of Medicine and Innovative Technologies. Around that time Nivorozhkin cofounded Neo-Advent Technologies, a privately held company headquartered in Marlborough, Massachusetts. The company is a preclinical contract research organization offering capabilities needed for the advancement of client drug development programs.

Nivorozhkin explained the transition to being an entrepreneur from his role as a scientist: "It was a process that involved exposure to the entrepreneurial culture and hearing people saying, 'Let's consider business,' or something like that. You get used to it, and you start trying to comprehend it. First you start dreading it, and then later the American success story starts getting into your head. You say to yourself: 'People are doing this, and why?' And you think: 'I am filled with enthusiasm in meeting new challenges. Maybe I should start being my own boss and implement my own ideas and help other people by benefiting myself.' But it was a slow process." He then said to his potential American business partner: "'Maybe we need to start something.' So we started to think of creating something, but only on the back burner, slowly. And since pharma was growing, I switched to pharma. My partner was, in fact, a person I had hired, and he was telling me compelling stories about people from MIT who had started companies in the late 1950s. I had a vision that he understood. We decided, 'Why don't we form a company and, maybe for starters, write a couple of grants?' So that's how we started, really by cross-fertilizing our thinking. And we are still together. Without him, I wouldn't have been able to build the specialized brick-and-mortar labs we needed for work on all our projects with over $1 million in equipment. We have complementary skills. He's more like an operating person while I'm more the scientist. And we've built from scratch several labs for different effective business opportunities, like materials science for the US Department of Defense, the Department of Agriculture, and the Department of Energy. Then we realized that it had been a good move to reorient ourselves back to pharma." Nivorozhkin then reverted back to the situation in Russia: "I see the best scientists from Russia are here in the US. Being in the US requires being an innovator and seeing the world like an entrepreneur. You can't be sitting in a tree and all of a sudden get to where you might want to be."

True to his quest for learning and challenging himself by becoming involved in new things, he became a serial entrepreneur when, in 2007, he cofounded his second startup, Boston Biocom, a biomedical technology company that received seed funding from Pfizer Corporation. Boston Biocom develops and commercializes life science technologies in oncology, infectious diseases, immune system disorders, and central nervous system diseases. In 2013, Nivorozhkin cofounded a third startup, Amorsa Therapeutics, an early-stage neuro-pharmaceutical company focused on improving the health and well-being of patients affected by nervous system diseases. The company develops and commercializes treatments of nervous system disorders including treatment-resistant depression and pain. In the process, Nivorozhkin has coauthored more than sixty publications and holds twenty patents. His career is testimony of the value that scientific professionals from the former USSR have brought to the US innovation economy.

Another successful Wave Two entrepreneur was **Kate Torchilin,** CEO of Novaseek Research, who came to the United States after receiving undergraduate and master's degrees in chemistry from Moscow State University. Along with her husband, she emigrated to Boston, as her parents had done a few years earlier. Her interest in science, as she notes later, had been strongly supported by her father, Vladimir Torchilin, an eminent pharmaceutical scientist, also an interviewee. After earning her PhD in biochemistry at Tufts Medical School, followed by an MBA at Harvard Business School, she joined Thermo Electron as director of Global, Clinical, and Toxicology Markets, where she stayed for five years. She left to become cofounder and CEO of Enspire-Medica. She explained: "So I left a larger employer to start my own company. I had an idea that seemed very cool at the time, so I jumped from the big corporate world, and, for about two years, I worked on one idea that ultimately didn't work out. Then I started another company that was actually going fairly well.

I helped researchers from Harvard Medical School who were spinning out their technologies, doing technology transfer from the other side, from what I had been doing at the larger company. Those researchers wanted to start their own companies based on their discoveries, and I helped them do that. We very quickly sold that company to a much larger one so that the assets could be utilized worldwide."

Torchilin went on to describe her situation after selling that company and starting her next venture in 2014, NovaSeek Research, a company involved in clinical data analytics and biospecimen sourcing to accelerate research. She explained: "What we've built is a very novel way to give pharma and academic researchers access to clinical data or biological specimens for patients with really specific disease profiles. There is more and more personalized research going on, and researchers need access to patients with very specific clinical profiles. So we built a platform that could bridge the hospital electronic data system with the researchers to find the right patients and either get the data or the biological specimens for research. Largely speaking, all my ventures have been in the life sciences." Torchilin demonstrated how her ability to apply her knowledge and skills gained in a major high-technology firm, together with her superlative scientific and managerial education, facilitated her path to become a serial entrepreneur.

Another scientist turned entrepreneur who also followed in his father's footsteps is **Alex Polinsky**, CEO of OncoTartis and Everon Biosciences. He came in 1988, after having earned his PhD in chemistry at Moscow State University, where he initially stayed to teach and do research. After a few years, he was invited to the University of California, San Diego, to work with a famous scientist who, since 1958, had known his mentor at Moscow State University. Polinsky recounted: "I cannot overestimate the importance of those two and a half years at UCSD for me, in spite of the minimal $19,000 salary, since I really learned how science operated here, what drove it, and what could be the driver for success. I was thirty-two at the time. And I saw how important it is in science here to be connected, to get grants. You have to be a good salesman of yourself and of your science. Nothing negative, it's just a fact. And I decided that that's not me. Later on, I realized that within this large group of people calling themselves scientists, all with PhD degrees, there are actually two categories: scientists who seek to understand how things work in nature, and engineers who take things that are known and combine them, one way or another, to produce something that was never done before, new products, for example. I just had a feeling that I'm an engineer and not a scientist. That is when I left UCSD and started my first company with a medical school professor who became my partner and friend for many years. We found an angel investor who provided seed money. We promised him that we just needed $500,000 and six months to become self-sufficient. I did get a patent for what I invented, but a year passed and we still needed financing, and by that time we were into probably $1 million already. Our angel gave us a deadline, and, right before it hit, Amgen liked what we were doing, invested in the company, and decided to support our research. And that's when we stopped receiving money from our investor. We became independent. We now had at least fifteen people, with only four of them being Russians. That was in 1993. We parted ways with Amgen amicably in 1996. I have to say that Amgen was very gracious in terms of our parting conditions, which allowed us to survive. In 1997, our company was bought by another biotech, and we became part of a company called Agouron Pharmaceuticals. Before long, that company was acquired by Warner Lambert, so at that point we became part of a big company, although we were still very separate. And then, in 2000, Pfizer acquired Warner Lambert to get their hands on the Lipitor drug, and so we became part of Pfizer. With my background, I was sort of the entrepreneur in charge because everybody else was traditional pharma people, and we came up with the idea of organizing a Pfizer incubator. As time went on, Pfizer got new managers, and, in 2008, my boss left and I left, so I found myself looking for what to do next. I did some soul searching, and I was able to get back to doing interesting innovative projects on my own again."

Polinsky then described his current activities: "Right now, I am developing two projects. I'm CEO of two biotech companies. I cofounded this company with a friend, a relatively recent friend. I didn't know him in Russia, I met him here. I think he is probably the most accomplished biologist who came from Russia. He works at the Roswell Park Cancer Institute in Buffalo, New York. You start the company where the brains are, so the companies are in Buffalo. I am playing a managerial role and also contribute my share of scientific experience. One company is called OncoTartis, and the second company is called Everon Biosciences. The first company is more a conventional biotech company with a drug. We are at the stage of clinical trials. It's a really good drug against blood cancers like leukemia and multiple myeloma. The second company is, for me, the most exciting because we are trying to tackle the problem of aging as a disease, and as a disease it can be treated. Both companies came from basically one little idea. It was in the middle of the financial crisis in 2008, 2009, so we couldn't raise money here in the States. We went back to Russia, and one of the advantages of our age was that a lot of people with whom we drank and played had become directors of institutions or venture capitalists. We found a venture fund whose director was a biochemist, who got excited about this idea that, at that time, was really more like an 'on a napkin' kind of idea. He gave us seed money, and we started working on these two companies." Polinsky provides insights into various transitions, including changing from being a pure scientist to a business person, and going from a small company to large companies and then back to startups. His experiences can be instructive not only for those in biotech, but also those in other areas of science who might face similar dilemmas of pursuing science or starting companies or considering involvement in larger corporations.

Wave Two 1987–1999: Software, Internet, Communications, and IT

Although few interviewees from Wave One worked in these industry sectors, there were many from Wave Two. We begin with the experiences of **Sergei Burkov**, Founder and CEO of Alterra, who became a serial entrepreneur, founding several companies after spending some years as a research scientist in the United States. After receiving his PhD from Phystech in Moscow, he was invited as a visiting research scientist at Brookhaven National Laboratory, later moving on to Cornell University, the University of Wisconsin, and McMaster University in Canada as a researcher in theoretical physics. On his transition to industry, he noted: "I cofounded a startup with another Phystech graduate, Alex Freed, here in the US, and we started a company called Bilbo Innovations. We manufactured foot pedals for personal computers, believe it or not. It was for people with disabilities and for games. We didn't sell it, we didn't go public. It became kind of a cash cow for a while, and then we phased it out." As for his second startup, he explained: "Then, with the same partner we started another company in software so that was a better idea for a startup, and it involved security. We were kind of hipsters, we did security and encryption before it was cool. That was in 1996. That company was acquired by VASCO Data Security, which is a small public Belgian-American company in the security space.

"I worked for a while for that company and then started company number three, Dulance, which involved product search, comparison shopping with emphasis on rare items and small stores. All comparison shopping sites, or almost all, rely on structured data feeds. So, essentially, it's a price list. It's kind of data extraction from unstructured resources, and by doing this we extended the number of available online stores from roughly 100,000 by the competition, to 1 million in North America only. Then Google bought us in 2006." It was described at that time as "the first of a new breed of shopping search engine based on crawling technology." Google announced the acquisition as part of a broader plan, as follows: "Google announced today that it is to open a research and development center in Russia later this year as part of its ongoing investment in Europe. The center will be based in Moscow and run by Sergei Burkov, PhD," noting that he had founded two other companies in addition to Dulance. After leaving Google some time later, Burkov went on to create his fourth startup, Alterra, a virtual, artificial intelligence travel agent: "I left Google because I'm an entrepreneur and Google is a big company. How I got the idea of Alterra is a long and convoluted story. Frankly, I don't consider it the next big thing. It's more of a hobby than a business just for now. But we'll see."

Stas Khirman, Managing Partner at TEC Ventures, was originally from Odessa, Ukraine, and earned a master's in mathematics at the University of Kiev. He and his family emigrated to Israel in 1991, prior to coming to the United States. He explained: "I was twenty-five. Gorbachev opened the border, and it was time to leave." When asked whether he had intended early in life to leave the Soviet Union, he responded: "Every day. It was funny because when I was a student, I considered mathematics as a profession and software development as a hobby. But when I came to Israel, I realized that being a software developer was actually a good profession. So I forced myself to switch professions." In Israel, Khirman initially worked for a large military company called Israeli Airforce Industries, and also with a small company called VDOnet, where he was deeply involved in developing products much like Skype's. He explained that he was a founder and was very engaged in developing the company's products, videostream and video-phone. The company was eventually divided into two entities, one of which was sold to Microsoft and the second to Citrix. In the process, Khirman and a business partner moved to New York for a couple of months in 1997: "Here it became apparent that if you're doing a startup, you have to move to Silicon Valley. It's where the money is and where the people are. Our product was pretty revolutionary, and there wasn't even a name for this technology, but it's really DPI, deep packet inspections. So we essentially reconstructed Internet application activities by observing sequences of individual packets." His company that developed this technology was called Narus, which was later acquired by Boeing. In more recent times, Khirman was a cofounder of Skyrider, a company involved in developing a technology to organize and monetize peer-to-peer networks. But as Khirman explained, the company ran into a financial crisis and was forced to close. He added that he and his associates had maintained good relationships with the investment community, and their failure was well accepted since it was at the forefront of a revolutionary technology with a relatively good chance of succeeding. It took little time for Khirman and his associates to rebound with the Israeli–American startup ChooChee, which developed cloud-based communication systems. The company was sold in 2013, to Deutsche Telekom, for a reported $100 million.

Meanwhile, in 2012, Khirman and **Anna Dvornikova**, another interviewee, created TEC Ventures. The fund invests in very-early-stage startups located in Silicon Valley that include entrepreneurs of various nationalities. Dvornikova had come from Moscow to study political science at Stanford University and later earned a law degree at Berkeley. In recognition of her accomplishments, Dvornikova was ranked by the *Silicon Valley Business Journal* as one of the 100 most influential women of 2014 in the Women Venture Capitalists category. Khirman and Dvornikova also applied their entrepreneurial talents to be among the cofounders of AmBAR, the American Business Association of Russian-speaking Professionals, and Silicon Valley Open Doors, both of which will be discussed in Chapter 7.

Serial entrepreneur and investor, **Max Skibinsky**, came to the United States after graduating in physics and math from Moscow State University. His experience provides an opportunity to combine important topics of this book, including starting businesses in the United States; managing mixed teams of Americans, Russians, and other nationalities; timing the sale of a company; and working as a venture capitalist. Skibinsky's story is rather typical of the experiences of other successful Silicon Valley entrepreneurs. His first business was a US-based outsourcing company that utilized Russian software developers in that country to do most of the technical work for clients, and he explained some of the difficulties in managing that business: "Overall there was a big disconnect in lots of situations, so I was serving as a bridge. I was already starting to understand local American clients much better, and trying to educate my team. The dot-com crash killed my business eventually. So I felt that it would be much more interesting instead of me being this outsourcing startup, I thought, 'What if I build my own company, not just outsourcing but a product company? A Silicon Valley startup.' During that time, I started prototyping what would eventually become my company in 2004, and in 2005 it got funded. So that's the company I ran for the next five years. The company eventually became involved in entertainment, but initially it wasn't an entertainment company but it had a lot of gaming influence."

Skibinsky went on to describe the process of starting and funding the company: "When I fundraised back then, our key advantage was considered to be our cutting edge, new technology that is now commonplace, what was called AJAX. In 2004 and 2005, it was considered the 'rocket science' of web development. What we were showing to investors was, 'Look, you can make a completely interactive environment out of web pages. You can have a web page where other people will come in, they will see each other, they will get avatars like in a virtual world, they can chat, they can watch video clips together, and they can play chess. So you can convert the web into a totally immersive interactive environment. And the key is you don't need to download any software for it.' It was a virtual world built totally with web technologies. But we came to realize, 'Oh my God, Facebook is getting the same number of customers we get in a year over just a few days. So why don't we go there and try to do it on Facebook, and make it a little bit more like traditional games that we know people enjoy?' We thought, 'Why not just use our own technology and use it to build this new breed of social games?' And that, as they say, is history. We started, after the first title, growing 1 million users per month, and then we started monetizing the company we called Hive 7. I recall our first month's revenue was around $150,000 without really trying. Having a product company is a totally different universe than just being a contracting service provider with an outsourcing shop. And people were lapping it up, and they absolutely loved that experience. The games were free, and the add-ons made the money. Actually, it was like Farmville that became popular years later. That's the nice part about startups: Nobody really knew the answers, and we had to figure them out as we went along."

Skibinsky went on to explain the next developments: "Later, we were acquired by another company, and that company was in turn in the middle of acquisition talks with Disney. So, after the acquisition, we were all a part of Disney. My company was here in Palo Alto, and it was around twenty people." Skibinsky explained that after the acquisition, he was free to explore other career directions including becoming a venture capitalist: "Instead of being the one that founded the startups, I could now invest in startups. So I started dabbling in small-scale angel investing at that time. A year or two later, I'd become a mentor at an early-stage incubator called 500 Startups, which was an interesting formative experience. If you're doing angel investing, you do a few companies a year. It's a very small sample size. When you're mentoring for incubators, you see dozens. Building on that, in 2013, I became one of the investment partners at Andreessen Horowitz, which is a top-tier venture fund on Sand Hill Road here in The Valley. It was at this point that I was thinking, 'OK, you've come a really long way from the $50 per month salary in a Moscow apartment.' The whole investment thing is a totally different ball game than being a startup founder. And I would say that, to a degree, it was a huge privilege that they picked me from all of the candidates, since the demand for positions in that fund is overwhelming. That exposed me to the top of the top-tier investment thinking and investment rationale behind one of the best funds in the country right now. Fundamentally, I looked at whatever interesting deals required a strong scientific background, like physics and math, to evaluate them. You basically had to look at it and make decisions and present your findings to other partners. Initially, when I started, I thought we were going to do a lot of bitcoin deals, but that wasn't what actually happened. After all this, I funded only one bitcoin startup.

"After a couple of years, I left to do my own cryptography and cryptocurrency startup, so that's what I'm doing right now, because again we had lots of exposure and brainstorming about the bitcoin space. It was a hard choice. And I have a personal attraction to mathematics and the certain elegance and potential of bitcoin because really, from a pure mathematical perspective, as Marc Andreessen likes to repeat, that's one of the greatest breakthroughs in computer science in the past twenty years, and once you dig into the details, you understand he's completely correct on this. So we feel that cryptocurrencies is in a super early stage where people are literally learning how to send the first packets around, and any companies on this track can have a very disproportionate impact on a planetary scale. Of course, that's a vision at this point. It might not work out exactly the same way, but I felt extremely tempted to do it. And we've already closed our first round of funding and hired the first employees, and so that train is leaving the station." Skibinsky continuously displayed the creativity and entrepreneurial spirit that is a fundamental theme of this book. In the process, he has created employment for others, including bringing to the United States talented professionals from the former USSR. As such, he typifies the type of highly educated technical professional that was so prevalent among our interviewees on both coasts of the United States.

In contrast to the previous entrepreneurs who created their startups in the United States or the former USSR, serial entrepreneur **Mike Sandler** developed his first startup elsewhere. After receiving his PhD in physics from the Moscow Institute of Physics and Technology, Sandler came as a refugee to Ottawa, Canada, because it was easier for him to immigrate there than to the United States. After a series of odd jobs, he worked as a contractor for Nortel, a major Canadian telecommunications company, but soon realized: "I could do additional work on the side, and I never say, 'no', to any opportunity. I had a few people who wanted to hire me as a subcontractor to do some programming, and I started trying to get more contracts. I managed to run many contracts at the same time, working fourteen hours a day off site or in my basement." By late 1999, Sandler had founded IPMeeting.com, a provider of Java-based web collaboration solutions for self-help, an application that allowed sharing and real-time live discussions. Within a year, the company was acquired by DWL, Inc., where he became director of R&D, and that firm was later acquired by IBM.

In 2003, Sandler founded Epiphan Systems in Ottawa. A decade later, his motivation for the crucial decision of opening a Silicon Valley office in the heart of Palo Alto, where the company employs around a dozen people, was to be close to a large cluster of his major customers. He continues as CEO of the privately held company, which is a leader in high-resolution video and audio video capture, multimedia display broadcast streaming systems, and multimedia display recording systems and accessories. The company was named one of Canada's fastest growing technology companies in the fifteenth annual Deloitte Technology Fast 50 Awards. Sandler elaborated: "Our units have already been in space twice, and one unit is working right now in the International Space Station. One member of our team is a former astronaut and cosmonaut and has been in space four times – three times for the US on shuttles, and once he spent half a year as commander of the International Space Station." In 2013, Sandler and Epiphan Systems were inducted into the Space Foundation's technology hall of fame in recognition of the company's innovative technology product developments. Along the way, in 2006, Sandler had founded Mediphan as a subsidiary of Epiphan Systems, and established an operation in New Jersey. Mediphan is a remote medical diagnostics technology company that develops audiovisual capture, streaming, recording, and replay products and technologies for telemedicine and remote guidance applications. Sandler and his companies are clearly becoming more deeply involved in the US innovation economy with their sophisticated audiovisual products.

Another serial entrepreneur with a master's degree in physics from the Moscow Institute of Physics and Technology, **Alex Miroshnichenko**, had a very successful career in both startups and larger companies. He described his career trajectory: "I was really lucky because I got hired into a startup. I was employee number four and the first engineer doing real kinds of computer science, file systems for QStar Technologies in the Washington, DC, area. That was perfect, because in a startup you do everything, and you get exposed to this business culture and everything else. I was going to The Valley every six weeks or so, and to New York and Boston, and a bunch of other places. My former boss in Moscow had moved to the US in 1991 as a Jewish refugee, so he looked around for a few weeks and was looking to fill jobs. He called me and said: 'Come on over for an interview.' And then I became his boss in Maryland, but he has actually been in The Valley since the late 1990s, and he is one of the cofounders of another high-flying startup."

After three years as the director of engineering at QStar Technologies, Miroshnichenko moved to Veritas Software as senior director of the advanced technology group and developed many advanced technical and business skills over a twelve-year period. He followed up that experience with time at two other software companies, PowerFile and Acronis. In mid-2007, he became the cofounder and CTO of VIRSTO Software, a software provider that optimizes storage performance and utilization in virtual environments. The company was reported to have raised more than $24 million in venture capital from major venture capitalists. VIRSTO was sold to VMware in 2013, for a reported $184.5 million. Miroshnichenko then became chief architect at VMware for two years, before striking out on another startup in February 2015. He reflected: "The fact that I ended up in situations as I did is really something, and I had no connections to Russia whatsoever. It's all in American culture, it's all kind of this business. You get thrown in, and you do your stuff. You say things because they make sense, and you have the skills, and it's just perfect. I didn't know much back then. I got extremely lucky in more senses than one."

Another highly accomplished interviewee started his business while a professor of operations research and numerical analysis at Moscow State University. **Alexis Sukharev,** who was born and raised in Grozny, in the Chechen Republic of southern Russia, started his Moscow-based company, Infort, in 1990. He realized the wealth of mathematical talent in the Soviet Union that could be valuable to Western and other computer companies and described early negotiations with Hewlett-Packard: "I ran a project that we started in 1991, and I sent my people to California. We set up a system for managing their resources for contract programming worldwide, such as in India, and it was a successful project." In 1993, Sukharev partnered with a colleague in New Mexico and the company's name was changed to Auriga. Eventually Auriga became the leading US-based company outsourcing software development projects in Russia and Lithuania. Sukharev noted: "Auriga has been recognized for many years as one of Central and Eastern Europe's top service providers." In 2011, the company was ranked number one as the Engineering Services Outsourcing Vendor worldwide by the industry's leading ranking agency. Sukharev has published widely in academic outlets as well as in the press, including the *New York Times,* the *Washington Post*, and *BusinessWeek*. And, in 2004, Sukharev was affectionately named "the Godfather of Russian Outsourcing" by the *New York Times.*

Greg Rublev attributes his entrepreneurial spirit to selling newspapers in the Moscow subway as a young teenager. A serial entrepreneur and COO of CompleteCase.com, he described starting his first company in the United States as being the result of losing his job at a firm where he had been doing part-time web programming: "'I'll do this on my own,' I thought. So I just got a bunch of clients. I developed some sort of software. I just put it back out there saying, 'Hey, there's a place you can put these kind of things.' And a bunch of people started emailing me asking if I could do something for them. I began to get questions like, 'Can you build a system that will automate divorce documents?,' which is where we're at right now. So all of that was going on during the last twelve months when I was finishing my High-Tech MBA degree at Northeastern University. Meanwhile, this whole time I was working on AsisChem as a side project that I started in 2005. I had a cofounder, and he moved to Russia. We started with a couple of chemists, and within two years we actually acquired a lab in St. Petersburg. So I said, 'All right, I'm going to do this chemistry thing full time.' So that was my first startup since the newspaper thing." Describing his next startup, LeanWagon, founded in 2011, he said: "So yeah, I got this idea to do a Facebook page for weight loss. I was working with two other guys when we started LeanWagon, and that went on for almost three years. We raised a little money, had some customers, but just didn't raise enough money to continue. I thought, rather than continue, just give the investors the money that's left and take the opportunity to exit in late 2013." He then described his current startup, CompleteCase, that was founded in January 2014: "Think of it like TurboTax for divorce. You answer questions and the forms are prepared for you." He described ongoing growth directions: "We're about to start testing products that'll help manage their parenting time calendars and also support payments. So with MediationMate, you can look on your phone and say, 'Oh, I have to go pick up my kids at 3 PM today, and there's a note saying Russian homework is due so don't forget to do it.' And we just rolled out wills, so you can get a will now. And we're just testing a mobile app right now." Speaking of future plans, Rublev stated: "I think, with MediationMate, by the end of next year, we want to have maybe fifty employees, half in the Boston area and half in Ukraine, and maybe $5 million in revenues."

Like Rublev, **Alexander Kesler,** President of InSegment, showed his entrepreneurial spirit with small startups in Russia at a very early age.

Kesler and his family came to Boston as refugees, and he immediately sought to continue his education, initially at Newbury College and later transferring to Babson College, where he studied entrepreneurship. He described his time there: "So Babson was an amazing experience. I loved it. I did well at Babson, which I am proud to say." Kesler, however, wasn't satisfied with just being a student and at both institutions was engaged in businesses such as repairing computers and computer consulting. Displaying his creativity, he said: "So when I came to this country, the first thing I did was have business cards printed where I called myself an independent computer consultant, and I put up some ads in the Russian community about setting up your computer, teaching you about it, and so on. Later, I became a salesman for cell phones at schools in the area, and eventually set up a distribution network, and all of this was very successful."

After graduating, Kesler had an idea that turned into his first major startup in the United States, zTrace, which was a software application that could locate lost computers with an invisibly installed app. Kesler went on: "And so I called my best friend from high school in Russia, who went to Moscow State University as a computer science and math major, and said: 'Let's work on this idea together.' He was still in Moscow, and that's where we had our development team. I ended up running that venture for six years. We sold our app in over sixty countries, having good distributors in Europe, Japan, and the US, with the US being our number one market. We ended up selling the company after six years in 2006, to an Internet security company that was a direct competitor and had a patent, while I had only a patent pending. I had a very interesting experience with that company, being the founder and president, and going from just me as an employee to twenty-five people at one point. The positives of the experience are unbelievable. I got to travel the world and was exposed to various cultures from Belgium to Germany, the UK, Japan, and Korea. It was a good experience developing software. I love software and Internet security. I love technology. I bounce back because the positive approach has to remain and persist. So that was my zTrace experience."

Kesler's major startup is inSegment, which he founded in 2007. The company specializes in digital marketing that spans B2B lead generation, website development, and search campaign management, mobile marketing, programmatic media buying, and social media monitoring and outreach. Kesler explained: "The Boston headquarters has about twenty-five employees, and the Romanian office also has about twenty-five people. The company also opened an office in Kharkov, Ukraine, with six people. Most of the leadership roles are located in the Boston office, while foreign offices are dedicated to back office operations." An indication of the company's success is that inSegment was recognized as a 2016 Future 50 award winner by SmartCEO, an organization that honors the fifty fastest growing mid-sized companies in the Greater Boston Area. Also, the *Boston Business Journal* named inSegment as one of the fastest growing private companies in Massachusetts having more than $1 million in revenue in 2015. Additionally, inSegment was recognized in 2015 as one of the top twenty-five most influential brands in B2B marketing worldwide by the London analytics firm, Onalytica, and Kesler was named one of the top ten most influential B2B marketers in the United States.

Another serial entrepreneur is **Oleg Rogynskyy**, Founder and CEO of Semantria, who came to continue his education six years after Kesler. Rogynskyy, who was born and raised in Dnepropetrovsk, Ukraine, described the fate of his early startup that allowed customers to share photos on Facebook: "When Facebook did its own thing, ours just evaporated overnight. So that was the first real venture I tried building, and I met my really close group of friends at that time, and most of us are still in the software business. I then started working remotely with a company in Boston called Lexalytics. They are a small lifestyle business that was doing very smart algorithms for the time, but they were very engineering focused, with not much business going on." Rogynskyy helped the company grow dramatically, utilizing its platform to enter a cloud-based version of their business that began to grow faster than the original model.

Around that time, he became interested in starting his own company, which became Semantria, a firm specializing in text analytics and sentiment analysis technology that operates in the Cloud. Its attractiveness was its lower cost and greater ease of use, allowing small and medium-sized businesses to access the technology. Sentiment analysis analyzes text, providing the capability to understand contextual relationships, even in different languages. It analyzes who is talking, what they are saying, and, most importantly, emotional content, to show how the person feels about what they're saying. Rogynskyy explained how he started Semantria: "I randomly met the head of one of the leads of Lexalytics who was also a Ukrainian. He asked, 'Hey look, are you working on something on your own?' And I said, 'No, but I would love to.' He said, 'I'm looking for something to invest in, so if you have any ideas, you seem like a promising person.' So I shared with him my idea for Semantria." Rogynskyy was able to grow the company significantly, based on Lexalytics's platform. After protracted negotiations lasting more than a year, Semantria was acquired by Lexalytics in 2014. Rogynskyy stayed with the company and served as president of Lexalytics for a year, still focused on growing Semantria, which had become the growth engine of Lexalytics. After leaving Lexalytics, Rogynskyy spent time mentoring a number of Stanford startups in the accelerator, StartX, and, in early 2016, he initiated another venture of his own in Palo Alto, People.ai.

Tocobox Founder and CEO **Dennis Bolgov**, who was born in Astrakhan in southern Russia, gained substantial experience in US technology companies and became a serial entrepreneur, selling one of his companies and moving on to start his current venture. Bolgov earned a master's degree in theoretical nuclear physics from the Moscow Engineering Physics Institute and a PhD in theoretical physics from the Institute of Chemical Physics at the Russian Academy of Sciences in 1998. In describing his early career as a software developer, he noted how working in high-tech companies around 2000 was an excellent experience for developing his technical capabilities. He described his early situation at Lobby7, a wireless consulting startup in the newly developing mobile industry named after a lobby at MIT: "It was a lot of fun because it was a VC-funded startup, and a lot of young kids just out of college had stellar careers after Lobby7. It was just like a launch pad for many good ideas and people. It was acquired by Nuance in 2003, and I was one of the four people who got acquired. And then a week later, they bought Speechworks and all of our technologies came together. Before we were acquired, we were getting a lot of help from my guys in Russia, and Nuance was very welcoming to our Russian team since we were very helpful to them. By the time I left Nuance, I was promoted to technological lead of a multimodal group, which was a combination of interfaces like speech, visual, and touch. The knowledge we created actually became part of Nuance's mobile speech platform, and later it most likely became part of the Siri interface.

"Around that time, I incorporated Warelex to effectively utilize my team in Russia, and my company continued to work with Nuance, but I had other clients as well. We had many big-name companies like Nokia, Sony, Intel, and Verizon, that were doing mobile technology projects with us. And as for Nuance, at some point we were involved in pretty much most of their mobile technology projects, and sometimes they were checking with me which technologies they had in their different offices around the world. In 2008, Warelex got acquired by a German company, Shape, which was called ShapeServices back then. They acquired us because we had a great team of forty developers with very solid experience working with many big companies. We had development processes in place and some mobile technologies that were interesting to them. Also, we had our own suite of applications where you could use your mobile phone as a webcam or headset or second display for your PC. I worked for Shape for five years, which was part of the acquisition deal. I wanted to be careful with any new venture so as not to compete with Shape since I have very good relations with everyone there and had many friends there. We started thinking about making a simple messenger-like interface for an email client, and we developed an application called Mail Ninja. But, by the time we launched, it was too late since we launched two months after the Mailbox app came out. Mail Ninja is still in the app store, but we do not support it any more because there is no real business there. The app is actually good, it is fun to use, and we still have some loyal users who say that our email app is the best. When I realized it was not going to fly, I came up with the idea of email for kids since I have two kids. I was thinking about an application for my seven-year-old and I wasn't able to find any. So it seemed like a good application to work on."

In January 2013, Bolgov and a cofounder started Tocobox and raised $500,000 in a seed round basically from friends and family. Speaking of Tocobox, he said: "I was rushing the app to release, which was a clear mistake. It actually became usable only two months after the release. We received a good user retention rate and positive feedback from our users. We also had some user acquisition problems because we have two users – the child and the parent – and the child who is using the app is not the one who creates the account. This additional step creates some extra friction. After that, we received some press and became visible as a search engine. We are now number one on Google search for email for kids. At the moment we have 50,000 users. Right now Tocobox is a freemium service that has very good functionality, with only minor limitations that you can turn off with a paid subscription. We have users paying for these premium features, so we're optimistic. The next step for us is creating additional revenues by convincing more users to pay since they want safety for their kids, and showing ads to children is not an option for us. We are now considering turning Tocomail into a paid service." Bolgov found creative and profitable ways of using his technological expertise, first serving the needs of multinational corporations and then turning to his personal interest in serving his young family's and other families' online activities.

Technology executive and entrepreneur, **Igor Balk,** came from Moscow, where he earned a master's degree from the Moscow Institute of Physics and Technology, followed by doctoral work in electrical engineering and computer science at MIT. An accomplished researcher and scientist in information technology and communications, Balk holds a half dozen patents issued in the United States. Before entering MIT, he cofounded VISO, a startup that he described as being "the first online supermarket. It was right on the MIT campus and did campus delivery, and we extended it to take on online orders, and then it became a supermarket." He eventually left the PhD program "to join another MIT startup, IntelliSense, and, in 2002, it was acquired by Corning for $750 million." That year, Balk started TDC, a company providing technology and professional services, that existed until the end of 2015. He was traveling back and forth during that period between the United States and Russia and realized that Russia at that time had excellent technologists whom he described as: "a competitive advantage. You want to utilize it and not throw it away, and I know how to deal with Russians. Especially at that time, that talent was untapped and very cheap, so my company could do better stuff and cheaper. The quality of these Russian people at that time was much higher than others from India or China." The flagship of the company was called TaskPoint, easy to use online project management and collaboration software that introduced a patent-pending natural communication concept to task management. It was named among the top ten online project management tools, was recognized as one of the world's top 100 most promising startups by *Innovate* in 2010, was a finalist in the 2010 MassChallenge startup accelerator competition, and was named one of the top ten IT startups in the MIT $100K startup competition.

In November 2006, Balk became involved with another startup, cofounding Unison Technologies and serving as CTO until 2009. The company is described as "the world's first fully unified communications system combining telephony, email, instant messaging, and collaboration in a single Linux server, creating a single platform." Unison was acquired in 2015 by RepEquity, a leader in digital brand marketing and reputation. Unison was acquired to gain access to its mobile technology and experience design. The company name was retained after the acquisition, and the entire Unison staff was reported to have become part of the RepEquity team. Balk was still running TDC as well as a relatively new startup which he had founded. He explained: "Global Innovation Labs does two things. First and foremost, it helps entrepreneurs from different countries to see that the world is small and that they can sell their products all around the world. We also do a lot of stuff with big data analytics. We do some consulting work, and we also have a project in the bioinformatics space. All of our activities are either big data or education oriented." Balk leveraged his superlative technical prowess to contribute to several US companies and to cofound several startups.

Polina Raygorodskaya, Cofounder and CEO of Wanderu.com, is an example of interviewees who, while not technical professionals, started technology-based companies by partnering with other immigrants from the former Soviet Union. She explained: "I had no clue that I wanted to start a company, but when I got to Babson College, everybody was doing that, and so I felt like I needed to do it, too." She went on to describe how she started her first company, Polina Fashion, a fashion PR and event company. While not in the high-technology space, the experience whetted her appetite for entrepreneurship: "I didn't have any funding. I was asked to produce a fashion show, and they paid me for doing that. Building a business is very similar to moving to a new country and starting from scratch, and building a family and a career. So it's having exposure to that, and being able to see that people can live the American Dream and have anything that they want to have if they work hard for it and figure it out. Both my parents influenced me with their emigration experience. That's what I've been doing. Polina Fashion expanded over time from Boston to New York and Phoenix, but I closed the business several years ago because I wanted to concentrate on my new startup."

Raygorodskaya described Wanderu, which she started with a cofounder in 2011: "Wanderu enables travelers to find and book ground travel in North America, similar to Kayak's service for air travel. In planning a trip with friends to US national parks, we decided to rideshare cross-country, partially to get more people to pay attention to what we were doing, but also because it's a sustainable form of transportation. Well, we ended up getting stranded in the Jefferson Forest in Virginia because one of our rideshares had cancelled. So we ended up having to rent a car, and that's sort of where the idea for Wanderu came to us. We thought: 'There must be a better way to find and book bus and train travel.' And so we just got really passionate and obsessed with solving this problem." She described her cofounder, who provided the technical expertise for the new venture: "Igor was one of our friends on the trip, and we had met at Russian math school in the Boston area when we were both young teenagers. He was born in Moscow and had come to the US at age nine. He went to school for electrical engineering as an undergrad at McGill, and then studied patent law at Boston University Law School. We brought on a third cofounder, and we started building relationships with bus companies, trying to convince them they should do this, and then started building out the technology." She went on to describe the early days: "It was just a whole lot of pounding on doors and trying to convince people that they needed to believe in us when we had nothing. It was a lot of work, months and months of trying to get through. But it's also being strategic, right? I realized we needed an advisor who knew the industry well, from a big bus company.

"We came across a former Greyhound CEO who was very forward thinking and led a lot of innovation and change at Greyhound. I thought that he would be a perfect kind of advisor for us in the beginning. And he realized that all of the things we were saying were things he had wanted to do in the past. After some convincing, he agreed to join as an advisor, and then he became one of our earliest investors, and now he's on our board of directors. At that stage, we had just started building the product, but we knew what we wanted to build. We're still building it up, the technology especially, and it has been four years. Two of our lead investors are Alta Ventures and Metamorphic Ventures who participated in our first two rounds. We're not raising funds right now, and I can say that fundraising is very challenging, I mean, very challenging. Running a company is also very challenging. Every day is a new challenge, so I go about it like I go about everything else: Just do it." In the process, Raygorodskaya and her cofounders built a company with thirty employees, mostly technical professionals. In recognition of her achievements, Raygorodskaya was included among the *Business Week* Top 25 Entrepreneurs Under 25, *Inc.* magazine's 30 Under 30, and *Fortune*'s 10 Female CEOs to Watch.

Wave Two 1987–1999: Other Industries

Serguei Beloussov, Founder and CEO of Acronis, who was originally from St. Petersburg, is a highly successful serial entrepreneur who has been involved in many different industries, and thus his career defies a simple categorization. Like Igor Balk and a substantial number of other interviewees, he graduated from the Moscow Institute of Physics and Technology, where he earned a master's degree in physics and electrical engineering with high honors and a PhD in computer science. Beloussov has been a very successful entrepreneur not only in the United States, but also in Singapore and Russia. He is still involved in a number of companies, both as an operating head and as a partner in investment funds that finance numerous startups as well as growth companies. From 1995 to 2003, Beloussov was CEO at Rolsen, a company he founded in Russia that offered its own brand of TVs and DVD players and that worked with LG Electronics of Korea. By 1999, Rolsen was the largest consumer electronics manufacturing company in Russia and, by 2015, had annual revenues of approximately $500 million and employed thousands of people. In 1996, Beloussov founded Solomon Software in Singapore, a distributor of US mid-market enterprise resource planning (ERP) systems throughout southeast Asia. The firm was sold to Great Plains three years later, which in turn was acquired by Microsoft for approximately $1.1 billion. Beloussov then founded SWsoft/Parallels in Singapore in 2000, a company involved in desktop visualization for the Mac and automation for cloud services providers. By 2001, the company had moved all of its R&D to Russia to leverage engineering talent from Phystech. That firm is still operating, with Beloussov as chairman. In the early 2000s, Beloussov founded the SWsoft/Parallels spinoff, Acronis. More will be said later about Acronis since this is Beloussov's current major activity.

Showing his ability to initiate startups while still running major businesses, in 2008, Beloussov founded Acumatica, another of his startups where he continues to serve as chairman. That company develops web-based ERP software to improve the productivity of mid-sized operations by giving customers access to their business applications and documents from anywhere using a web browser. Displaying versatility, Beloussov has also been a cofounder in two investment firms with somewhat different objectives in which he is still active as a partner. In 2010, he became the senior founding partner of Runa Capital, which is associated with the much larger investment firm, Almaz Capital Partners. Runa's more than $30 million fund is focused on under $3 million-round investments in Internet, software, and mobile companies with Russian technology teams. Similarly, in 2012, he became a cofounder of Qwave Capital, a $100 million venture fund headquartered in Cambridge, Massachusetts, focused on scientific technology and next-generation technologies, primarily in materials science. Since 2009, he has also been a venture partner and advisor at Almaz.

Beloussov's primary activity is CEO of Acronis, the company he founded as a spinoff from SWsoft/Parallels in the early 2000s. Acronis provides cloud data protection through backup, disaster recovery, and secure file sync and share solutions. The company has US offices in Burlington, Massachusetts, its current US headquarters; San Francisco, its initial US headquarters; and Arlington, Virginia, as well as in Singapore and Moscow. Acronis has more than 650 employees worldwide with more than $150 million in annual revenues, and serves 5 million consumers and 500,000 businesses in more than 180 countries and in 14 languages. The company has more than 100 patents, and some of its products have been named best product of the year. Beloussov has more than eighty patents issued and numerous others pending. He has been running the company since May 2013, and, as he noted, from May 1995 onward he had been spending a lot of time in the United States. He added: "But I've only been at Acronis for two years and I'm spending 95 percent of my time here, and I do have business partners. So basically, I have Acronis which I run, and then there is a portfolio which is run by others, but of course, I am participating. In that portfolio, between Parallels and Runa, we have fifty companies, so we are really different, and it's complicated. It's three or four venture funds in which half of the upside is between my partner and me, but maybe one-third of the money comes from the two of us. And like any venture fund, you have a lot of companies, but also many other partners." Speaking further of companies in the fund, he said: "So I start a business and build it to a state where it has a mature business model. It can still be growing very fast while having a mature business model. Now that I think about it, I hadn't thought about business in a conscious way, not until maybe seven or eight years ago. Before that, it was just happening."

Beloussov provided other interesting comments about himself and his attitude toward business. Once, in responding to a business partner, he answered: "That's it. I cannot help you anymore. I'm not a business person. I hate business. I want to be a scientist as I had planned." His ambition toward science is not surprising since he had gone to schools specializing in math and physics and had participated in and won numerous olympiads in the Soviet Union, including regional olympiads in physics. While he realized the financial attractions of business, he regarded it with some trepidation: "When I was in the Soviet Union, I built computers but I was still sure that I was going to be a scientist because doing business is a bad idea. We were taught it's not really capitalism, it's speculation, and in Russia you could go to jail for that. But it wasn't just about jail, it was just about unfair capitalists exploiting poor workers and peasants. They make money and those workers and peasants live very bad lives. I believed it was much better to be a scientist, and if you need money, you can always make it with your own hands." But displaying more insight about his transition to business, he added: "Still, I guess I was already a little bit entrepreneurial unconsciously, even from a young age." Serguei Beloussov was certainly one of the most successful serial entrepreneurs among our interviewees. Wanting to be a scientist and never really intending to go into business, he actually showed entrepreneurial spirit and engaged in entrepreneurial activities at a fairly early age. He did so in spite of the fact that some of his negative comments about business showed signs of anticapitalist imprinting that was a primary goal of the communist regime.

Wave Three 2000–2015: Biotech, Pharma, and Medical Products

Wave Three interviewees represent a highly diverse geographic sample since, in addition to those who came from major Russian or Ukrainian cities, many hailed from other regions in those countries as well as from

Armenia, Estonia, and Uzbekistan. Although the vast majority of Wave Three entrepreneurs were involved in software and related sciences, one dramatic exception is **Evgeny Zaytsev**, Managing Partner of RMI Partners and General Partner at Helix Ventures. His industry experience is different from other Wave Three interviewees, as is his success as an entrepreneur and venture capitalist. For those reasons, we present his story in considerable detail. Originally from Barnaul in the Altai region of Siberia, he came to the United States to attend Stanford Business School and ultimately became a successful venture capitalist in Silicon Valley, investing in innovative therapies and medical devices in the life sciences sector. Zaytsev graduated from medical school in his home region and founded a medical student society that connected Soviet medical students to counterparts in Europe, traveling himself on exchanges to Spain and Italy. He explained the evolution of his career trajectory: "As time went on, I realized I didn't have the practice of medicine as a calling and was more interested in pure science. In the 1990s, I became involved in a research group at the Institute of Medico-Ecological Problems in Barnaul that investigated the high rate of carcinogenic and environmental damage to the Siberian population that had been exposed to nuclear tests. Although the Soviet Union had collapsed, the new Russian government allocated millions of dollars to that research project. I was honored to have worked with the top Soviet scientists in that field and was in awe of their brilliance and accomplishments." Zaytsev went on to become chief scientific officer of the research group, and, during his time there, it became an internationally recognized research center. Zaytsev went on: "After a while, I developed an interest in the business side of science. It was the chaotic years of the 1990s, and scientists and academics, among countless others, had jobs in name only, with salaries unpaid or in arrears. So I began researching MBA programs in the US and applied to Stanford, Harvard, and Wharton because they were top schools and offered scholarships to international students." He was accepted to Stanford, where he became enamored of the Silicon Valley business culture, climate, and lifestyle.

In the summer of 2001, Silicon Valley was hit hard by the dot-com bubble and the ensuing recession, a time during which Zaytsev had to find his own internship during his MBA program. He applied to venture capital firms, highlighting his medical and business backgrounds as well as his ability to analyze the science behind proposed ventures in the life sciences. He elaborated: "After two months, I received a call from Dr. Graham Crooke at Asset Management Company who invited me to be his business partner, and we are partners to this day. Graham also introduced me to the company's legendary Pitch Johnson, an early and highly successful Silicon Valley VC who had invested early in companies like Amgen and who also taught as an adjunct faculty member at Stanford. While I was an intern, I analyzed the scientific and economic merits of life sciences investments and was offered a full-time position after I graduated in 2002, and I eventually became a partner." He noted that the life sciences side of the company was so successful that it was spun off as Helix Ventures, with himself and Crooke being the managing partners along with Philip Sawyer. The firm focused on investments in therapeutic innovations in the biopharmaceutical and medical devices sectors. By that time, Zaytsev had been involved in more than thirty investments in biomedical ventures that included numerous successful exits, primarily as his firms were being acquired by major companies including Sanofi, Boston Scientific, and Baxter. That success allowed Zaytsev to start a second venture capital firm by becoming managing partner of RMI Partners, a global life sciences venture capital firm that managed investments for RusnanoMedInvest, a subsidiary of the Russian government-owned Rusnano. The company had $500 million under management to implement innovative projects in medicine and pharmaceuticals. This new role required him to travel to Moscow monthly in order to manage the business. RMI partners invested in US opportunities and facilitated technology transfer from its portfolio companies to Russia. Zaytsev noted: "I decreased my involvement in RMI because they reduced their capital allocation to the venture portfolio, but my other activities are actually still going on. In my entrepreneurial role, one of my US portfolio companies, InteKrin Therapeutics, successfully accomplished a phase two trial in Russia for multiple sclerosis, and now I am building an antiviral company also on the basis of US–Russia collaboration."

Evgeny Zaytsev's success story is a great example of how his early medical education in Russia as well as his research experiences there provided the basic capabilities that allowed him to succeed as a Silicon Valley entrepreneur and venture capitalist in the life sciences. Coupled with his Stanford MBA education that provided much of the business acumen necessary for that success, the scientific training from the former USSR allowed him to specialize in niches that were fast growing in the United States, specifically biopharmaceuticals and medical devices. Zaytsev utilized his education acquired in both countries to fulfill his personal career objectives of being deeply involved in life sciences, but from the business and investment side, as he had wished to do after deciding that being a medical doctor was not his calling.

Wave Three 2000–2015: Software, Internet, Communications, IT

David Yang, a serial entrepreneur who founded nine companies including the multinational software company, ABBYY, has been recognized by the World Economic Forum as a technology pioneer. Yang grew up in Yerevan, Armenia, the son of an Armenian mother and a Chinese father. A classmate of Serguei Beloussov, presented earlier, he earned two master's degrees, one in computer science and one in applied solid state physics from the Moscow Institute of Physics and Technology, equivalent to a PhD. Many years later, Yang came to the San Francisco area for a medical procedure and stayed after coming to appreciate the pleasant weather and the exciting and supportive business climate. Since Yang had an office in Silicon Valley, he was able to remain in the United States on an employee transfer visa. He had previously spent considerable time in New York and Chicago in the late 1990s and early 2000s, developing marketing and sales for another one of his startups, Cybiko, while still involved with his earlier startup, ABBYY. In describing his startups that had been primarily in Russia before bringing ABBYY to the United States, Yang recounted: "I started with my friend a company with an electronic dictionary called Lingvo. I was learning French and wanted an electronic dictionary that would replace a paper one. I never thought I would give up physics at that time. But that was the time of *perestroika*, and all our energy was aimed at changing so many things, potentially at least, things we perceived should be changed. My product was the genesis of ABBYY, but I left the company temporarily in 1998 to start another called Cybiko. That company's product was a handheld computer that was similar to the more well-known Palm Pilot handheld device. The times were too difficult for a small startup, and we had no chance to actually go strongly after the US market because of the 1998 financial crisis. So we finally had to sell the company in pieces to different buyers in

2003. And during all this time, I was still acting as chairman of the board of ABBYY. But I had switched 100 percent of my time to Cybiko."

He continued by describing his next startup: "After that, we opened FAQ Café in Russia and started several restaurants around that time." Yang and his partners parlayed that experience into a startup called Iiko. Its product was an ERP system for restaurant management. He explained: "I'm not really active in that company now, but I am chairman of the board. I think I spend about two hours a month, but it's profitable, growing well, and has a strong team with about 100 people." Yang went on to explain his next startup, a spinoff from Iiko: "Plazius/Platius is a mobile payment and loyalty platform for retail and restaurants. The system allows people in a restaurant to pay their check by phone without waiting for the server. It's like an Uber for restaurants. In Russia, we have 3 million users, and it's growing. We sold 51 percent of the company to Sberbank, which is Russia's number one bank." With his large minority position, Yang serves as a member of the board of Plazius/Platius. He continued: "Now we are thinking of bringing it here to the US because the current US systems are much more complicated to implement because it's far more effort to teach personnel. Our system is integrated into the business process, so you don't need to change to utilize this royalty and payment system."

Yang's primary activities currently focus on his role as founder and chair of the board of directors of ABBYY, his original startup founded in 1989, but a company in which he had spent little time in actual management because of his numerous other startups. ABBYY has four headquarters, one in Milpitas in the San Francisco Bay Area, and the others in Munich, Kiev, and Moscow. Most R&D projects are conducted in Moscow. The company has sixteen offices around the world. More than 40 million people from over 200 countries use ABBYY's products, including Samsung, Dell/EMC, and Panasonic. The US market is a major priority because it is more developed than most in terms of technology and accounts for around 40 percent of ABBYY's sales, while Europe and Russia each account for 20 percent. Yang explained: "We have about 1,300 people, most of them involved in R&D as programmers, engineers, and linguists. Around fifty of them have been working with ABBYY for around twenty-five years. In one instance, we found really good people who were doing translation things, and we acquired them many years ago. Now they have grown so fast and they're doing beautiful international business. And we will probably spin them off, too, so they will become independent as a very fast-growing company. They were originally Russia-based, but now they, too, have an office here in the US. They provide language services of high human quality as well as other language services from semi-human and semi-machine to pure machine translation. The company has two faces, one to the consumer as a B2C [business to consumer], while the other is to businesses as a B2B [business to business]. So the project has taken them eight years, but they are profitable and growing in the linguistics services industry, which today is a $20 to $40 billion market. Nobody has done this scalable, supply–demand marketplace for linguistic services. This team is brilliant."

Following his entrepreneurial instincts, Yang cofounded yet another startup in 2014, called Findo, in Menlo Park, California, to enable people to access any of their information in their email, the Cloud, or their computer files. Customers can search from one place using descriptions and can collaborate from messengers, share quickly, and search on the go. Yang's cofounder is Gary Fowler, an American who has cofounded seven startups. So, including Yang's nine startups, between the two men, they have cofounded sixteen different companies. Yang again has proved his ability to work with cofounders to start and build successful businesses. As he has progressed in his life as a serial entrepreneur, Yang has gravitated toward focusing on the United States, initially as a customer base, but increasingly as a location of major offices or headquarters for his companies and, most recently, as the site of his latest startup.

Anna Uvarova, Founder and CEO of online photo service 3DBin, came to the United States after earning a master's of public administration from Moscow State University. Originally from Tomsk in Siberia, she enhanced her education by learning programming and taking courses in entrepreneurship and leadership, fields that could provide background for potentially starting a company. She explained: "I actually wanted to get the best education here as well. That's why I came here. So we started the company in 2009, and I went to the Berkeley global faculty colloquium entrepreneurship program founded by Jerry Engle and David Charron. This is a program for startups by entrepreneurs from all over the world. Intel has a collaboration with Berkeley's Haas School of Business, and they find the best companies and they let them compete. Intel sponsored me to attend the colloquium. I was volunteering for AmBAR and for Triple Helix, and my main goal was to meet awesome people and learn from them. This is how I was moving forward to starting my own company. I started it right here in the US with my cofounder in Tomsk, who had been doing research for years on 3D and 360 technology. I was a business cofounder who built the business strategy and sales channels. So we got together and started 3DBin. We have about 40,000 users. We have awesome results. People love us. We are the most efficient and affordable service out there. And just after we launched the app, we were named one of the best iOS applications in July 2014 by the NextWeb. We were right there with Facebook Messenger. We also won third place in the Funders and Founders startup competition. Still, we're upgrading all the time, changing and adding new features. Right now we are getting more money and new investors, and we're going to be doing some awesome things. We don't have any major clients yet, but we're just starting a big sales process and looking for a head of sales." Uvarova's success reflects her entrepreneurial spirit and creativity and her ability to work remotely with her cofounder in Siberia.

Rafael Soultanov, Founder and CEO of iBuildApp, came to Silicon Valley from Ufa, an oil-producing city in the Russian Ural Mountains region. After his highly profitable music distribution and web-based shopping businesses in Russia were, like so many others, crushed by the Russian 1998 financial crash, he decided to come to the United States to study business to build on his computer science degree from Ufa State University. After starting in a small college in San Francisco, he transferred to Golden Gate University to improve his English. Like Anna Uvarova, he soon became more involved in business pursuits utilizing his computer science background, and he began his first US startup that was a competitor to Craigslist. That venture received money from an angel investor, a friend who owned a transportation company. Soultanov described how he gave up his equity to the angel investor: "He refused to invest any more because he felt that the company would grow on its own, but six months later, the company closed down. He was Russian because I only knew Russian people. I did have American and Indian friends, but it was easier to get to and talk to Russians and get cash faster. Plus, in 2002, there wasn't much startup investment here in The Valley, some VC but not angel. So I didn't sell the company. I just gave it to him and started my second company. It was consulting and basically getting orders for website development and web systems software development. It was called SolveItLabs, and it continues to exist. So I decided to disengage even though we were doing around $2 million in revenue in 2008. We had around fifty engineers in Russia and five managers in our Menlo Park, California, office with clients including Nielsen Media, Samsung, and Dell." In spite of the scale of that operation, he explained that the company was now in limbo, basically because of his new venture: "I have a startup called iBuildApp, and you cannot sit on two chairs. I'm not that skillful. We brought some of our programmers in Russia into our iBuildApp Russian office and let others go. So we've moved on to iBuildApp. We do iPhone development very well, mobile apps, and business people were asking for an iPhone app. This is how iBuildApp was born. We decided, 'Let's see if we can automate the system where its Internet goes to a box called iBuildApp, and the app is coming out as a mobile app.' It's not completely free. We are responding to business needs, not consumers, actually small business needs. We target small businesses that want to create their own apps on an iPhone, Android, or iPad. We price it according to usage, by customer downloads. We're staying competitive by marketing and also trying to differentiate ourselves, trying to get customers to understand why iBuildApp is so different. It's not just because over the course of four years we built patented technology to automatically build apps. We built a whole ecosystem around market-place developers, a whole community of designers. So we're trying to differentiate ourselves from an app store."

Regarding his investors, Soultanov explained: "So far it's mostly angel investors, mostly Russians here in The Valley and in Moscow. One of them is COO of a Russian payment system like PayPal. In Russia, it's mostly business people, while in the US, for some reason, it's technical people. Some of them are friends of mine. In Russia for some reason, I was able to raise money from COOs, while in the US it was mostly CEOs of technical companies. I don't know why. It's interesting. That's just the way it played out. I was able to show them I had clients like Southern Methodist University and others including churches, music and entertainment companies, and ecommerce. I tried getting to American investors and talked to quite a few of them. First, it's an angel investment, and it's more intimate. They want to know you, and they want to trust you. But when I talk with Russian guys, they see that I'm able to navigate, that I know sales and technical stuff. I know management, and I'm able to operate it, not just that I have an idea and plan to hire a manager. American angel investors feel more comfortable investing in people they know. We raised only $1.5 million so far, while we have $2 million in revenue after four years. VCs usually prefer financing those they know who can build an executive team and who come to them and ask for help to build it. They like you to build an American executive team because you want to be an American company. Usually, we see it's all about numbers. In Silicon Valley, you really have to prove to VCs that you are actually going after big markets, so you have to talk about opportunity. And our app fits into the growing market of today's world. We are actually growing and still innovating to keep our product competitive with new additions and programming for our app to continually update it. We'll try to grow as much as possible unless we are contacted by Cisco, for example. You never know. We've had lots of exciting talks. We're probably going to be a very good company, but let's see where the market goes." Soultanov's experience as a serial entrepreneur is a vivid example of moving ahead into ever more challenging but potentially lucrative startups. His story typifies the journey of a technically trained individual who, while successful as an entrepreneur in his home country, had to retool and retrain himself in the Silicon Valley technological environment.

Another serial entrepreneur who began his activities in the former USSR is **Vlad Pavlov**, who came from Dnepropetrovsk, Ukraine, in 2010 at age thirty-six with a master's degree in computer science from Dnepropetrovsk National University. He is the founder and CEO of rollApp, a Palo Alto-based online virtualization platform. The platform runs Windows and Linux applications on any device by virtualizing apps on its servers so that users can access them without any special browser features. RollApp's platform is made up of free and open source software with the objective to bring PC apps to browsers by hosting apps in the Cloud. The company is reported to have raised $1 million in 2013, after an initial $350,000 in earlier seed funding from TMT Investments.

Pavlov's startup experience in the United States follows on his substantial background with other startups in the former USSR, as well as his experience with larger companies like Microsoft and Intel. Pavlov noted: "At some point, I quit a larger company to start my own outsourcing company. In our best times, we had around fifty people, and we had pretty good clients like Microsoft and Vodafone. We were able to develop algorithms that were ten to fifty times more efficient than those of our competitors." Some time later, Pavlov was headhunted by Microsoft Russia: "But I was working there like a bureaucrat, even though Microsoft still had at that time a kind of startup spirit. Anyway, I felt like I could do more and started INTSPEI, that would basically monetize the results of work I had been doing. I incorporated it in the US, but hired a CEO and spent most of my time in Ukraine running the engineering team to develop a project. The next step would be to go to the States and start selling it. But we ran out of money just before the 2008 financial crash happened, just before we were going to raise our next round." Like others, Pavlov's experience back in the former USSR provided the impetus and experience for him to develop his Silicon Valley startup.

Another serial entrepreneur whose first startup was in the former USSR is **Sten Tamkivi**, Cofounder and CEO of Teleport, a software startup that includes numerous former Skype personnel. Teleport has developed an algorithm to assist companies in selecting the best location worldwide for their operations. Originally from Tartu, Estonia, Tamkivi first came in 1994, as a high school exchange student. Returning to Estonia, he later became the first General Manager of Skype, which was founded there. He explained that he took on that position after selling his first startup, Halo: "I sold Halo to DDB Worldwide, which was an international chain. Then, after I joined Skype, eBay acquired Skype in 2005, and we were suddenly working with people in San Jose, and I started coming more regularly to Silicon Valley. I would go to a Facebook meeting and have a Skype business card. It was a business trip, but you get into things, into the Silicon Valley part of life much more. And now I'm here studying at Stanford Business School and now staying here for a while, and that's my third reason for being here."

Serial entrepreneur **Mark Kofman** was born in Tallinn, Estonia, and he came to Silicon Valley to found his fourth software startup, Import2, which, by 2015, was operating profitably with revenues of more than $1 million. He received a degree in computer science from the University of Tartu, followed by a master's in information systems from the Royal Institute of Technology in Stockholm. He worked for several years as a programmer in major companies in the former Soviet Union and Europe, and he had been a serial entrepreneur in Estonia with the same cofounder. Kofman noted that when starting their fourth venture, their experience in earlier startups led them to believe "that if you wanted to do the business that we were doing, then most of the customers will be in the US, as will most of your potential partners. So it made a lot of sense to be here. We got into 500 Startups, which is a Silicon Valley seed fund and accelerator founded by PayPal and Google alums. We raised $100,000 from them and other individual investors, and that helped us get some kind of a base here for the company. And then we moved our families here as well. I had been with my cofounder, Anton Litvinenko, since we were together at university, and have worked with him on basically everything since that time, such as the three companies we have started together so far. He is Russian, but he spent most of his life in Estonia, and I think he may have been born there." Regarding investors, he recalled: "They helped us with cash flow in the early days, and now we're at a stage where the company is profitable. We pay our own salaries now, but for the first years the investment supported us for sure. We have investors locally here in the US, we have investors with a Skype background, we have investors who are from the UK, and we have one from Russia. My partner and I have done three startups together, which is a rare combination where people build something and have known each other for fifteen years. We fight like cat and dog, but we can still work peaceably together. I think investors saw a team that worked well together and was able to deliver."

Discussing their earlier companies, Kofman noted: "We have most of our engineers in St. Petersburg, so the fact that we know the language helped a lot because we fly in there and can talk to them personally. There is a big pool of software talent in Russia, and they have good experience. Our whole team today is eight people. We have my partner and me, three other people here in The Valley, and three in St. Petersburg." Kofman went on to describe his latest startup, Import2, a software company specializing in data migration: "The main idea of what the company does is to help you move the data you already have into a new product. With Import2, you're importing into something new, something better. So the mission of this company is, 'We are allowing you to use the best software out there, so you're not locked in with a legacy system provider you may have been using for ten years just because of the data. So you can start using the new system, the best one, as simply as with one click.' Because of our one-click software, the data migration can be accomplished not only very easily but also at about one-fifth to one-tenth the cost of traditional migration. We've helped about 10,000 businesses so far. It's hard to say who might be our biggest because we have served multi-billion-dollar companies like Salesforce, Zendesk, and Zillow mostly, but also thousands of very small companies."

In discussing competition and the company's future, Kofman explained: "So far we've been super focused on one thing, which is data migration, and there is no company out there focused on the same thing. That gives us an advantage because we don't have to do everything. Most of our competitors are doing consulting, which is about all kinds of information services, so they have to know their way around all things. Given our experience with 10,000 customers, that is probably about 100 times more than any competitor has, and we keep investing into our software. But honestly, while developing the last product, we were trying to automate everything to zero click, but we found that's actually not what people want. People still want to have some control. They want to feel like, before something happens, here's the explicit action we need to take, instead of the computer deciding for itself." As to his future plans, he added: "We are working on a new product that is complementary to our original idea. It's partially like making the data migration idea obsolete. But if you want to launch another product on top of that business, we will have to be careful and try not to kill the original idea and the good aspects of the company. More investment might become a question some time, but not at the moment."

Umida Stelovska (born Gaimova), Founder and CEO of parWinr, had far different origins, having grown up on a cotton-producing collective farm near Samarkand, Uzbekistan. She received her bachelor's degree in German from Termez State University. Her love of the German language was a window on a much bigger world that she strived to be part of: "I remember my father always telling me to dream big and study well at school. When I was in the fifth grade, my father took me to my German teacher and told him that he wanted me to become a great teacher like him. From that day on, I was given extra homework every day, and when I was in the seventh grade, I was assigned to teach German in the lower classes. When I was in the ninth grade, my teacher asked me to teach the eleventh grade. My first day was so embarrassing for me, and I was shaking because they were older than me and now I was to assign them their homework.

I remember how my classmates would laugh at me when I told them that I would definitely go to Germany one day. Maybe reading stories about the outside world in our German textbooks made me become a dreamer of the impossible, and my passion for traveling the world, learning about different cultures, languages, and people, and doing the big things in life grew bigger and bigger. I knew that the outside world was different, and life outside of our country was exciting to me. Today, I proudly can say that I have lived in five countries and traveled to more than twenty."

Stelovska poignantly described the difficulties of building a startup, even while having a patented technology: "I left Uzbekistan for the Czech Republic in 2006. Just before I left, my father passed away and I became the only hope for my mom and my siblings. I badly wanted to start working so that I could send money to my family. But the only jobs I was offered were floor cleaning jobs. That's when I said to myself, 'If I can't get a decent job, then I should create those jobs for others like me,' and I started PrAgEmInt employment agency to find legal jobs for illegal immigrants. That was the beginning of my entrepreneurial adventure. Then I met the best Thai chef and started the Thajemny Svet [Magical World] restaurant chain. Later I got my visa and came to the United States in 2009, where my passion for technology grew bigger and bigger. After I understood what a patent meant and the value of my husband's patents, I had a vision for a new technology that I believed would soon disrupt many industries. So I created parWinr. When I convinced my husband to start the company in April 2011, I barely spoke English and I had really no idea about doing business in America. My idea was to offer YouTube users the ability to convert their videos into trivia-poll games and reward people for the engagement. During the process of creating the company and hiring the first employees, many things sounded simple, but I didn't know that doing business needed more than just building the product you believe that people would buy.

"Today, I'm even more convinced that the technology I envisioned six years ago has great potential, and living in Silicon Valley has taught me a lot. As of this year, I have a clear roadmap and strategy to launch the platform where businesses and consumers can monetize their creativity. I know it is never easy for anyone to start a company and get funded fast. Especially, if you're a foreigner, a woman, or have no diploma from Stanford or MIT, or no experiences working at Google or Facebook, the chances of your company getting funded are minimal. But what I believe is that if you are able to dream, and if you're capable of following your dreams, then you will find ways to succeed. You are already different than others because you didn't stay back, or you didn't give up and quit when you faced tremendous challenges every day. It may take ten years to experience 'overnight' success. Years of trial and error in launching your product, testing it in different markets, understanding the problems you are solving for your customer, and coming up with innovative solutions, all that takes time." Stelovska provides an insightful reflection on her entrepreneurial journey to Silicon Valley and her in-depth understanding of what it will take to bring her venture to the next level.

Andrey Klen came to San Francisco as a cofounder of Petcube, an interactive remote pet monitoring system that lets pet owners watch, talk to, and play laser games with their dogs or cats from any location using a smartphone. They can also share access to their pet video camera with friends, family, or anyone on Petcube Network. The product is an interactive wifi camera sold usually at major retailers like Best Buy, Nordstrom, and Amazon. The private company is headquartered in San Francisco, with additional offices in Kiev, Ukraine, and Shenzhen, China. Born of Ukrainian parents in Egvekinot on the Bering Sea in the Russian Far East, Klen was raised in Smila near Kiev. Unlike most of our high-tech entrepreneurs, Klen graduated with a degree in journalism from Shevchenko State University in Kiev. Although he settled in the United States in 2014, he had earlier visited Seattle, where a girlfriend had moved to, and he sought to explore a new Western type of life. He got into design during that time, although he had previous experience in writing for art and culture magazines in Ukraine. When he returned home from the United States, he got together with a friend, Jaroslav, now the CEO of Petcube, to create the first social media agency in Ukraine. After two years, their venture was acquired by the Ukrainian company, Prodigi. They later met the former CEO of Prodigi: "That partner, Sasha, eventually had the idea that evolved into Petcube. I designed interfaces for the camera and Sasha asked me to pursue the project with him, and then Jaroslav joined to handle the business operations. Our venture was actually called Petcube while we were in Ukraine."

Klen described the turning point in the venture: "Then we got an offer from Hax Accelerator, and it was at this point we understood that everything was getting serious and we needed to focus on Petcube. So we decided to quit our jobs and go to China and figure out how we could make this hardware a reality. We had been there for four months and then launched a crowdfunding campaign in China with Kickstarter that went pretty well. We became the most crowdfunded pet product ever. We decided to have our headquarters in the US, but we stayed in China for a while talking to factories, and our team grew to ten people including new hardware engineers. Right now, in 2015, we are almost thirty people. Hax Accelerator funded us at first, and we also got some angel investment from Semyon Dukach, the head of Boston Tech Stars. He thought our idea was good and became an investor almost immediately. So we had money to sustain our growth. This was really our first real startup. And the idea is great. It's really viral. Everybody loves pets. It's really fun and cute, and people really love it. I think we were lucky enough to have this idea and have the ability to develop it into this sort of product. We've sold more than 5,000, I believe, and are now producing another batch of 3,000. Our customers are mostly in the US now, because the whole culture of being pet owners is higher here than anywhere else. And the whole industry is really huge. It is $56 billion yearly for pet-related supplies, so the market is here. We knew that we needed to come to the US, so very early we registered our company in the US, and we have patents pending on our whole product, both the camera and laser." Petcube's initial success led to an impressive funding round in early 2016, when the company raised $2.6 million from a consortium led by Almaz Capital as well as AVentures, and Y Combinator.

Andrey Klen is one of the few of our entrepreneurs who is not highly trained technically and is not involved in the business side of the venture he cofounded, since his role focuses on his major strength, design. Klen's story and experiences of bringing a technologically based product to the US consumer market, outside of gaming, stands in contrast to the more common business-to-business technology products and services described by other interviewees.

Wave Three 2000–2015: Other Industries

We conclude this section with the relatively long and informative story of **Max Polyakov**, Founder and Chairman of Noosphere and the venture capital firm, Noosphere Ventures. He came to Silicon Valley already having been a highly successful serial entrepreneur in the former USSR and elsewhere, and who, from all appearances, is continuing that trajectory in the United States. Before describing his multitude of startups, approximately half of which have been in the United States, we note his wide range of interests and expertise that include space technologies and predictive analytics, consumer Internet, AdTech, EdTech, FinTech, and enterprise software. Polyakov holds numerous patents related to these fields. Additionally, his educational background is highly eclectic and undoubtedly has been influential in developing his breadth of expertise.

Born and raised in Zaporizhia, Ukraine, Polyakov began his higher education at Zaporizhia State Medical University, where he studied medicine for six years to become a gynecologist. However, he came to realize that he preferred going into business. For the next seven years, he started a number of companies and then studied international economics at Dnepropetrovsk National University, focusing on problems of the world economy. He analyzed the paradigmal model of information development and its implementation, earning his PhD in 2013. During much of that period, he continued to be involved in creating numerous new ventures in Ukraine and the UK. He stated his personal philosophy on LinkedIn: "Humans fundamentally change the world around them through a combination of knowledge and technology. By creating close connections between my employees, partners, communities, and academia, and leveraging this collective knowledge with robust technology platforms, we are trying to drive a real revolution in the private space industry."

Polyakov's experience with startups began in Ukraine, in 2000, when he became a founder and CEO of IT Ukraine, which he described as "one of the largest and first technical outsourcing companies in Ukraine." His quest to expand that business and start new ventures took him to Scotland, where, in 2005, he founded HitDynamics, an on-demand software platform that combined bid management software, search marketing reporting, and conversion tracking in a single platform. After a year, the company was acquired by Hitwise, which in turn was sold in 2007 for $240 million to the Experian Group. This appears to be Polyakov's first major exit through selling a company, although he had remained with Hitwise for more than a year as vice president. He repeated that strategy in 2012, with the sale of his next startup, Maxima Group, to a media group in the Commonwealth of

Independent States (CIS). Maxima had developed one of the largest portfolios of B2C businesses in the CIS. Concurrently, Polyakov started several other companies, including Cupid, plc, which he cofounded in 2005 and served as COO. A leader in the online dating industry, the company launched an IPO on the London Stock Exchange in June 2010, with an initial market capitalization of £45 million. By 2012, Cupid had more than £53 million in revenue, 54 million users, and 500 employees. Another company he cofounded in 2006 was Maxymiser, a data analytics firm with 400 employees and offices in New York, London, San Francisco, Edinburgh, and Dusseldorf. Maxymiser was a leading provider of cloud-based software that enabled marketers to test, target, and personalize what customers see on a webpage or mobile app, and its customers included Lufthansa, HSBC, and Tommy Hilfiger. The VC-backed Maxymiser was sold in 2015 to Oracle. According to *BusinessInsider*, "Oracle paid well for Maxymiser since this was a strategic acquisition of a thriving company – not a struggling company picked up for a bargain."[1]

In 2007, Polyakov founded TrafficDNA, which he described as "a big data and technology expert in traffic management, analytics monetization, and product promotion on web, mobile, and social media platforms, where clients are able to receive a full-service advertising solution." MURKA, founded in 2009, creates games for a gambling vertical and has launched more than ten projects for social networks and mobile platforms.

With his large number of startups and also his frequent exits, Polyakov had strong views of venture capital: "We're very careful with our companies not to take too much VC capital because that means you lose your flexibility. As a founder, you continue to be on the company's board as one of seven people, but you're not CEO anymore. They basically dilute your management expertise. This happens with VCs because now they control your company, and they have different goals. The VCs want to double or triple their money, or even, like, six times. So that's why, very often for them, they look at your decisions and will change your strategy. We still have profitable companies and strong partnerships worldwide. They are really helpful for

[1] Julie Bort, "Oracle Paid Well for Maxymiser, Says Person Close to the Company" *BusinessInsider* August 21, 2015 http://www.businessinsider.com/oracle-paid-well-for-maxymiser-2015-8.

our innovative projects, not only in terms of finance, but also as sources of knowledge and experience."

Since arriving in Silicon Valley in 2012, Polyakov has founded seven additional companies. Three were founded that year, the first being ClickDealer. That company has grown into a global marketing agency offering advertising services in social and mobile verticals. Renatus is in the mobile and social game publishing industry. And Noosphere, focused on space activities and solar panels, was founded with lofty goals. Polyakov elaborated: "Our goal is the fundamental creation of the new information-oriented society where the information system serves the whole of humanity and develops a fertile atmosphere for personal self-fulfillment. That's why we're looking to space activity and solar panel activity, especially here in Silicon Valley. Although we have companies making money, for the venture part, we really want to address the problems that can help humanity live longer or better or safer. Things that we don't see being done, so that's why we value this position." In 2013, Polyakov founded two additional companies, Together Networks and Databrain. The first is a leading social and online dating business, while the second is a machine learning platform for mobile and web applications that combines big data analytics with predictive modeling. In 2014, Polyakov founded Universal Commerce Group, an international e-commerce company that provides electronic payment services, a shopping search engine, and a data-centric cloud computing service. Polyakov's most recent startup is EOS or Earth Observing System, a private space company delivering products based on proprietary satellite data retrieved from an in-house system of satellites that are delivered by privately launched rockets.

Polyakov's lofty goals with EOS and Noosphere seem in contrast to his other businesses, many of which emphasize social networking or IT solutions for consumers and businesses. Yet this juxtaposition seems to fit well with his philosophy toward business, which emphasizes the importance of knowledge: "So what we want to do in the future somehow is to create an instrument so that corporate knowledge exists as a separate ecosystem, like financial knowledge or people-intensive knowledge." When asked about what career he might have had, if not in business, he responded: "Research most likely. Some scientific stuff. Yes, of course. But we should make money, and then after we make money we could have some other interesting ideas beyond making money. Other ecosystems exist like knowledge, so it's good synergy between making money and doing something good." That perspective seems to summarize Polyakov's business philosophy, which embodies the familiar phrase, "doing well and then doing good."

Polyakov's forays into numerous Internet businesses have provided a financial foundation that could allow him to pursue his more lofty goals of harnessing knowledge and combining that ecosystem with launches into space utilizing satellites, with a goal, as he said, "to address problems that can help humanity live longer or better or safer." His philosophy of "do well and then do good" likely stems from his background and experiences starting with his relatively poor childhood, which likely provided an incentive to make money, something he noted several times. His focus on space is understandable since both his parents were rocket scientists. His humanitarian instincts reflect his training as a medical doctor, and his global business view could well have emanated from his education in international economics focusing on problems of the world economy. In many respects, Max Polyakov's business philosophy and his basic identity could stem from the imprinting he received in the former Soviet Union during his childhood that could have continued through early adulthood, a theme to be revisited in Chapter 9 on identity. His story, like that of Evgeny Zaytsev and others, illustrates that there is more than one way to fulfill career objectives without losing the valuable expertise gained from one's education and earlier experiences, a theme that pervades this chapter and the next.

This chapter has focused on the experiences of those of our interviewees who have become entrepreneurs in the US innovation economy. The next chapter presents the experiences of the talented professionals we interviewed who are primarily in roles other than as founders or venture capitalists.

6 | Research, Development, and Applications in Academic and Industry Settings

The previous chapter focused on entrepreneurial spirit and the experiences in the United States of the entrepreneurs among our interviewees, with some references to their entrepreneurial activities in the former USSR. This chapter recognizes contributions of those who were not usually entrepreneurs, although some were involved in startups. Like the entrepreneurs, their involvement in the US innovation economy has also added substantial value. Often, entrepreneurs are the primary if not the sole focus of technological contributions to an innovation economy, but here we recognize the contributions that can be made by various technological professionals in different organizational settings. This chapter is organized by the three immigration waves, and different groups of professionals are featured sequentially in each wave. The first group includes research scientists, primarily academic researchers in universities and hospitals as well as industry, and who are almost uniformly MDs or PhDs or their equivalents in the former USSR. The second group moved beyond pure research roles toward development, applications, and even business functions in industry, with most having science or engineering backgrounds. Within both groups, we have separated the interviewees' stories, as in the previous chapter, into biotech, pharma, and medical products as one category, and a second category that includes software, Internet, communications, and IT. Wave One is an exception since it features only one interviewee, an engineering researcher and inventor. Few interviewees in the biotech, pharma, and medical products group were among the entrepreneurs in Chapter 5, one important reason being the relatively heavy capital-intensive nature of those sectors, which often made it more practical to engage in research, development, and applications in other non-profit and profit-oriented organizations.

Wave One 1972–1986: Research Scientist in Engineering

The sole academic interviewee who arrived in Wave One was **Alexander Gorlov**, Professor Emeritus of Mechanical Engineering at Northeastern University, where he had worked since 1976 after being exiled from the Soviet Union, as he recounted in Chapter 2. He received his PhD in mechanical and structural engineering from the Moscow Institute of Transport Engineers. Gorlov was a world-renowned scientist in the field of renewable energy, including wind and water power. He described one of his inventions: "I developed the helical turbine in 1994 that increased efficiency and power from water and wind compared to other turbines of their size. In 2001, the American Society of Mechanical Engineers honored me with the Thomas A. Edison Patent Award. And one of my turbines was installed in 2015, at the top of the Eiffel Tower in Paris, where it provides power to operate the tower's commercial areas." Gorlov published more than 150 technical papers and books, and obtained twenty-five US and international patents. His inventions and their resulting products built upon his earlier projects in the USSR, which included involvement in the design and construction of Leningrad's Pushkin metro station. He also worked on the Aswan Dam in Egypt, and, while at Northeastern, he did extensive work in Korea and was awarded honorary citizenship for his achievements there. Gorlov's accomplishments are in contrast to that of many scientists who remained in the USSR in that his inventions were made into important commercial applications around the world.

Wave Two 1987–1999: Research Scientists in Biotech, Pharma, and Medical Products

Vladimir Torchilin, originally from Moscow, is a University Distinguished Professor at Northeastern University and Director of the Center for Pharmaceutical Biotechnology and Nanomedicine. A renowned cancer researcher, Torchilin, in collaboration with a physicist at Rice University, developed a treatment that combines chemical and physical modalities to more efficiently destroy a tumor without harming nearby healthy cells. Torchilin also had significant experience as Associate Director of the Center for Imaging and Pharmaceutical Research at Harvard Medical School and Massachusetts General Hospital. He received three degrees from Moscow State University, including a PhD and a Doctor of Science in bioorganic chemistry, and he was awarded the Lenin Prize in Science and Technology, the highest honor in the USSR.

Beyond his distinguished academic career, Torchilin has been involved in several startups. He explained: "The first one came when we were giving a talk in the early 1990s, and right afterwards another professor asked, 'Have you tried to build a company on that?' I said, 'I have no idea what it is or how to do it.' So we met the next day, and he said, 'If you would like to see if somebody would be interested in developing it and be involved, we can do it, but for this you have to give us a substantial part of the company.' It turned out eventually that he had suggested very fair conditions, and he didn't try to take advantage of our complete ignorance. It went very successfully in the sense that we almost immediately got an offer to sell the company, actually within half a year. Unfortunately, we sold the company at that time. Later, we understood that if we had still owned the company for just another half year, we could have gotten five times as much because we were accumulating data very fast, and the more data you have, especially in animal experiments, the higher the price. But what is done, is done. But it was a lesson: never rush, think about it. I still believe that this was probably one of the best ideas we ever had. I'm involved in three or four other ventures, and if they're successful, probably I will get something, but you never know. For instance, we're just developing a very new, very interesting idea with a team from California. I don't bet on these ideas becoming commercial successes, but the research is a new direction and is educational. The biggest challenge is to raise money. Certainly, they expect the return to be almost instant and huge, but you are trying to convince them that there is a great potential, and this is what you need the money for." Torchilin's experiences in becoming involved in startups while not actually becoming an operating entrepreneur are similar to many researchers in universities and research hospitals and other such nonprofit institutions whose work leads to inventions and potential commercial applications. But many, like Torchilin, prefer to remain as researchers and be satisfied with being founders of startups, choosing to let others take over the business aspects.

Another accomplished biotech academic is **Dmitri Petrov**, the Michelle and Kevin Douglas Professor of Biology at Stanford University who came to the United States in 1990 at age twenty-one. With a master's degree in biology from the Moscow Institute of Physics and Technology, he came to study and conduct research at Washington University with Dan Hartl, and he later earned a PhD in biology at Harvard working with Richard Lewontin. Petrov referred to Lewontin as "one of the greats in evolutionary biology in the United States. He actually is a student of Theodosius Dobzhansky, a Russian immigrant who got out of Russia just in time, was not killed by Stalin, and was one of the fathers of genetics and population genetics in the United States. So just talking to Lewontin opened my eyes to this field of evolutionary genetics and population genetics. And I suddenly got a sense that's the way my brain works, and it was a good field for me. It combines the analytical thinking that is common in physics with the variation that I found so appealing about biology. It was a great education, very thorough. What I love about the American system of graduate education is that, at least in evolutionary biology, it's very free and there are very few requirements. You're kind of left to your own devices with some guidance, but you have to figure out what you want to do. I got into the Society of Fellows, which was also an amazing experience, so I stayed at Harvard for three more years, five in all. I was also a Research Fellow in the genetics department at Harvard Medical School before being hired at Stanford as a faculty member. At Stanford, my lab does theoretical, computational, and experimental work to address questions in molecular evolution and molecular population genomics." Seizing an opportunity to see his translational research applied in an innovative commercial environment, Petrov was a consultant to Natera, a global genetic testing and diagnostics company that develops noninvasive methods for analyzing DNA to manage genetic diseases.

Vadim Gladyshev, Professor of Medicine at Harvard Medical School, came as a postdoctoral fellow, after completing his PhD at Moscow State University. He conveyed his early excitement about science when referring to his experience at the US National Institutes of Health: "I finally found myself in an environment where competitive research was done. We were learning about something that nobody had done before. It is an amazing feeling when you find something and realize, 'OK, I now know this, and nobody else in the whole world knows it,' and you want to share this discovery with the world. I didn't know if my experience was sufficient or not, if I was good enough for the job. I just had no idea, but I found that my training was really good, and in fact I was better trained than most other lab members. These days, when you talk to other people, yes, there is a general feeling that the Russians are creative people. They may not be the hardest working scientists, but if you have a difficult problem some say, 'Hire a Russian.' On the other hand, there are Russians and there are Russians – it depends on the individual, like in any nation. Probably the general feeling that the Russian scientists are good is because they come from the best universities, like Moscow State. Even if these people haven't published much before coming to this country, they have potential. And that's why so many of them succeed here in the US.

"Also, what distinguishes American science is the way research is organized here. For example, one of the greatest things about American science is that when you are hired as an assistant professor, you become completely independent. This allows scientists to become independent at a young age, when they are still creative. For example, in Germany or Japan and in Russia as well, science is more hierarch-ical, and this is not good. Often, you're not fully independent there, and you should follow what the big boss says. In the US, it's not like that at all. I was young, only about thirty years old, and I got a faculty position and realized that I could study anything I liked. I had energy, I had drive, I had interesting ideas. This is something that works well in the US, which is not the case in other countries. I brought with me from the Soviet Union my education and training and found oppor-tunities to apply them here in the US. So a very important thing is the way science is organized here."

Gladyshev elaborated on his work: "My speciality is research on aging and other biomedical processes, but I didn't start out there. I initially studied an enzyme, and that enzyme happened to be a selenium-containing protein. So I transitioned to the selenium area and continued this research at the National Cancer Institute. When I moved to Nebraska, I also continued this line of research. Selenium is a trace element that is essential for human health, and there are only a few proteins that are known to contain selenium. I guess our impor-tant contribution to science is that we discovered most of these pro-teins. We developed computational tools, and, in a matter of a few months, discovered twenty-five such proteins in humans. Then we realized that these proteins are mostly redox proteins, and so we expanded into the redox biology area. And then we realized that some of these proteins are involved in the regulation of lifespan or in aging, so we extended our research into the aging field.

"I guess a big transition for me happened in 2010, when NIH [National Institutes of Health] announced a program called Eureka, where you can propose a crazy idea, a high-risk, high-payoff idea. I proposed and wrote: 'I will study and understand the nature of aging.' This is important because nobody knows why organisms age. I got funded by this program, and this support influenced me a lot. I became very excited about aging research and thought I could do something important in this field. We're thinking about some sort of a spinoff and other practical applications. We haven't done it yet, but we always have this in mind, and sooner or later it will happen. Obviously, we would like to apply what we've learned to alter the course of aging. I haven't formed a company yet, but we do have some patents. I'm sure it will happen. At this point, our lab is well funded and we're doing well, at least for the next three years. I'm in a lucky situation in that sense. When I go to bed, I often think about how to solve a certain problem, and it's so exciting. That's why we're in science. I think it's just the best possible profession."

In featuring the interviewees in this section, our objective has been to highlight their scientific contributions, and also to provide deeper insights into their thinking and approaches to science, highlighting the evolutionary process through which most have traveled. Gladyshev's story illustrates these points, including the potential developed in the former USSR and the realization of that potential by building a career in the US innovation economy.

Another accomplished and insightful research scientist is **Slava Epstein,** a Distinguished Professor of Biology at Northeastern University who oversees a prominent laboratory focused on microbial discovery in the environment and in the human microbiome. He received a PhD in microbial biology from the Shirshov Institute of Oceanology of the Academy of Sciences of the USSR after completing two degrees at Moscow State University. He explained that scientists so far have been able to cultivate in the lab only about 1 percent of global microbial diversity: "The remaining 99 percent is the largest source of biological and chemical novelty in the world, and it's unexplored and essentially untouched." As he noted on his faculty website: "We uncover novel microbial life forms by inventing novel cultivation strategies that depart from conventional wisdom and provide access to the greatest part of microbial diversity, unexplored species missed in the past." Epstein has conducted extensive field research in highly remote areas of the globe, including among aboriginal people in Australia and indigenous tribes in the Amazon. He talked about the passion he has for his work: "There is a challenge in science that people have known, for example, that has existed for twenty to fifty years. And people try to resolve that question, problem, or phenomenon. And it's exciting to take that on. That's a challenge." Illustrating his passion and perseverance, he then related his favorite fable about two frogs trapped in a pot of milk: "They keep swimming, but the walls of the pot are vertical, and the level of the milk is such that they can't jump, and they get progressively more and more tired. And one of them asks, 'For how long are we going to continue this? At some point, just die with dignity.' That one committed suicide. But the other just couldn't stop. It kept swimming, kept beating the milk, churning, until it turned into butter. And then it jumped out." He added with a smile: "I am the second frog."

The central focus of Epstein's research is to combat antibiotic-resistant bacteria. His research team has developed a new lab-on-a-chip that permits screening many more pathogens in record time at lower cost, including numerous pathogens that until now have not been available for analysis. Epstein explained: "In the past, researchers had to grow massive amounts of bacteria and extract, purify, fragment, tag, and sort massive amounts of DNA from them, a costly and time-consuming process. Our new system reduces the amount of DNA required one hundred-fold. We can use a tiny number of cells, even 10,000 is enough, and produce superior sequencing analysis." With his Epstein chip, or iChip, a dominosized plastic block, he collaborates with researchers from the Broad Institute at MIT and Harvard. Epstein has also been successful in translating his discoveries into commercial applications. He has teamed with his Northeastern colleague, University Distinguished Professor of Biology Kim Lewis, in cofounding the biotechnology company, NovoBiotics Pharmaceuticals. This pathbreaking research, resulting in the first new class of antibiotics in thirty years, has gained international attention, with teixobactin and the iChip being included among the top 100 discoveries of 2015 by *Discover* magazine and recognized as one of the top ten discoveries by United Press International, as well as one of the top five most important medical developments that year by the BBC. Epstein was named along with Lewis among the top 100 Global Thinkers of 2015 by *Foreign Policy* magazine.

Igor Kramnik is an Associate Professor and Biomedical Scientist at Boston University's National Emerging Infectious Disease Laboratory, with his expertise being in tuberculosis research. He had been at the Harvard School of Public Health for ten years after spending five years at the Albert Einstein College of Medicine in New York. He earned his MD degree at Samara State Medical University in his home city, as well as a PhD at the Central Tuberculosis Research Institute at the Russian Academy of Medical Sciences. Kramnik initially expressed some disappointment with his achievements, stating about Russian immigrants in the United States: "I still think that there are some migrant workers and there are elites. I didn't make it into the elites, and I think I'm still perceived as a migrant worker to some degree. That's a very sad statement based on some of my recent experiences." However, he went on with a more optimistic view: "I'm really at a crossroads in my career at this moment. I have several ideas that can be commercially viable. I have a number of things that I've achieved in science, but it's unrelated to how successful my career has been. I've done something important in recent years, and I'm very happy about this. But I really need resources to take it further, so I would say that I'm looking for opportunities where I can be more connected to industry. I'm looking for opportunities for how I can translate what I learned and start doing more applied research. So maybe I might join a company or even start a company. This is something I'm thinking about."

Anastasia Khvorova is a Professor at the University of Massachusetts Medical School and director of the RNA Therapeutics Institute, after having worked in the pharmaceutical and biotech industries for most of her career. She received her PhD in biochemistry from Moscow State University. She came to do a postdoc at Temple University and then became a research scientist at Amgen, an experience that changed the direction of her scientific life. She then worked as Chief Scientific Officer of Dharmacon Products, a division of Thermo Fisher Scientific. The focus of her lab at UMass is enabling therapeutic oligonucleotide delivery to tissues other than liver through chemical engineering. The lab works on identifying chemical and biological properties that drive small RNA tissue distribution, retention, cellular uptake, and biological availability. RNA interference provides researchers with a simple and effective tool to inhibit the function of human genes. The lab has identified novel chemical modalities that support delivery of RNAs to the heart, kidneys, muscle, placenta, vasculature, and brain, tissues previously untargetable by RNAs. The sequence specificity and lasting activity of small RNAs makes them ideal drugs that are expected to transform drug development and the approach to human health. Like others, Khvorova was able to translate her research from the laboratory to practice in that her scientific work came to form the basis of Advira, a startup where she served as Chief Scientific Officer and that was later acquired by Rxi Pharmaceuticals. And, again like other interviewees doing research in the pharmaceutical and biotech fields, she continues to support the improvement of human health care.

Another interviewee who has continued to pursue science, **Anton Manuilov**, is Senior Scientist II at the multinational biopharmaceutical company, AbbVie. He came from St. Petersburg and earned three degrees from the University of Massachusetts, Amherst, including a PhD in molecular and cellular biology. He was then hired as a scientist at the startup, Intelligent Bio-Systems, where he worked for a year before joining Abbott Bioresearch Center, now AbbVie, where he has been rising through the scientific ranks since 2008. He described his experience working there: "It was very different in a good way. Working for a bigger company, you have more resources where you can plan your experiments based on science versus the amount of money you have to buy reagents and equipment. When you have more tools at your disposal, you can get a lot of things done, and I have learned a lot here. I work in a lab conducting experiments, and I really enjoy it. In the pharmaceutical industry, you will always be on a team because it's impossible to accomplish things by yourself, it's impossible to know everything. When I got to Abbott, I started learning from my peers and my colleagues, and I was fortunate to be working with very experienced people in the field. My boss was a professor from Tufts who decided to go into industry. I worked with him and I had an idea, and he helped me to grow that idea. I just got one patent issued and another one is pending, and I have another provisional one as well.

"My goal has been to contribute here, and it's starting to be realized, but it took a long time and we're slowly getting there. If you have a great idea that the company is currently not focused on, you can propose it in a forum where people can listen to it, and they will vote if the idea is worth pursuing or not. You have to write a proposal, and my proposal was just accepted. I collaborated with someone from another department to nourish that idea. This is almost like a startup environment – high stress, long hours, and limited resources. You have a few months to deliver and to make sure that your proposed project gets the next round of funding. It's very stressful. I would like to continue with what I have right now and stay on a scientific track for as long as I can, before they force me into management." Manuilov is not only content but passionate about staying on the scientific side of the business, but he realizes that eventually successful scientists will often be required to move toward more managerial responsibilities, which is common in the pharma/biotech industry. In either case, such people make important contributions to that industry, and, in doing so, to the innovation economy.

Wave Two 1987–1999: Industry Professionals in Biotech, Pharma, and Medical Products

The following interviewees, many of whom hold PhDs in science, have moved away from basic research and into managerial and executive positions, with a number having pursued further education in business and management. **Bella Gorbatcheva** is Oncology Translational Project Manager at Novartis Institutes for Biomedical Research. She might be seen as a transition interviewee in the sense that she follows the group still deeply involved in scientific research but precedes another relatively large group who have used their backgrounds in science and technology to transition to the commercial development side in the pharma/biotech sector. Gorbatcheva came from Moscow in 1993, to become a research technician in the department of immunology at the Cleveland Clinic Foundation, and later became a senior research technologist there in the department of cell biology. She earned a master's degree in engineering and technology, majoring in biotechnology, at the Moscow State University of Food Production, and she worked for five years as a researcher in the Research Institute for Vaccines and SERA of the Russian Ministry of Public Health. She moved from the Cleveland Clinic to become a senior research associate in oncology genomics at Millennium Pharmaceuticals in Cambridge.

After five years at Millennium, Gorbatcheva joined Novartis Pharmaceuticals as a scientist in the oncology area. In 2014, she was promoted to Translational Product Manager for Oncology, where her role was to support the clinical research teams related to biomarker strategies in oncology clinical trials. She explained: "I started out in academics and then I moved to preclinical development and clinical development after completing my High Technology MBA at Northeastern University. My mentor at Novartis helped me get an internship with the Novartis marketing department in New Jersey, so I spent summers there to get exposed to strategic marketing that is personally very interesting to me because you work with scientists and clinicians as opposed to strictly with salespeople and numbers. I think it's very important to get marketers into the development early, who also understand the science, and that's the group they call Oncology, New Product Development, and Portfolio Strategy. And so I spent summers with them, which was amazing. The strategic side of marketing is very interesting." She continued: "At Novartis, it's different. I work at Novartis Institutes for BioMedical Research, which is separated from commercial influence. So it's not easy for me at Novartis to move to the commercial development side."

Michail Shipitsin is another interviewee who made the transition from academe to business, and, while continuing to be involved in the science side of the firm, has risen to a managerial position as Senior Associate Director of Biomarker Development at XTuit Pharmaceuticals. After completing bachelor's and master's degrees in molecular biology at Novosibirsk State University, he came to Tufts University School of Medicine, where he completed his PhD in biochemistry and cell biology in 2004. He was a postdoctoral fellow at Dana Farber Cancer Institute for nearly six years and then made his first foray into industry with Metamark. In 2015, he joined the startup, XTuit Pharmaceuticals. Shipitsin explained his journey from basic science to significant involvement in management: "I didn't see myself too much as an academic person. So after twelve years in pure research, I actually had wanted to join a company for a while. I got the position at Metamark when a professor at Dana Farber was looking for people when founding his company. I was the sixth employee, and I got a bit of equity. So Metamark is now at a stage where they're trying to expand the selling of their product. We brought it all the way from basic research to the commercial phase, and I actually helped to do some commercial work as well.

"Then the XTuit opportunity came. I really like it, and it's a great experience. I think I feel this way because we started the company from zero, and it was great." He concluded: "The main difference between science and business is that the goal in business is to produce the product rather than a paper. Basically, your whole goal is to develop the test, and even if you see something interesting, you're not going to work on it. You will work only on things that are relevant to the product and the test. That's the main difference between academia and industry. I also got some management experience in academia because I had to manage a group, initially a single person, and eventually about twelve or thirteen people when I was interim head of R&D." Shipitsin explained well the difference between science and business and how he made that journey to the business side after twelve years in science.

Alexander Aristarkhov is a biotech consultant who has had a varied career in science after having been a research scientist at Moscow State University, where he received his PhD in biochemistry and molecular biology and worked for several years. He came as a postdoc sponsored by Harvard Medical School, where he spent eight years, changing his field from plant biology to genetics. His wife, who accompanied him from Moscow, also joined Harvard Medical School, where she continues to work as a research scientist. Aristarkhov said: "I was talking to a professor at Harvard and was thinking of what to do next. Things were becoming worse and worse in Russia. So at that point, we extended our postdoc agreements and we felt that maybe we could stay longer and get our green cards through Harvard.

"I was thinking, 'Should I be in academia or in industry?' And I was sort of thinking about technology, and I just moved to industry, to Applied Bio-Systems that is a large company in Boston. I joined as a scientist and rose through the scientific ranks to manage research projects. I worked eight years there. It was a real adjustment from academe. It's definitely a different mindset from being a scientist, where I worked with things just because I'm interested, to the industry side, where you do a project and need to be on time and on budget. And I was not prepared for that. But people here, you know, offer help to their neighbors. Then I moved to another company, and I have now been in industry for eighteen years. I went into the consulting business, and it's very hard to break into that because you have to convince customers that you're capable of doing what they need. My consulting approach is to do it online, and I get requests from hedge funds and from pharmaceutical companies to discuss specific points. I'm hired by a hedge fund that wants to enter the startup field rather than invest in established companies. So it's scientific, technological, and business expertise – everything together."

Irina Fayngersh, Senior Product Brand Manager at Natera, came as a child with her family from Zaporizhia, Ukraine. She is one of a number of interviewees with deep science backgrounds who moved into a broader role within biotech and pharma businesses, usually into a more business-oriented activity. After completing a degree in molecular and cell biology at Berkeley, she earned an MBA at UCLA. She explained: "I started out in pre-med since all of my four grandparents were doctors, and my grandfather, who we immigrated with, was a well-known oncology surgeon in Ukraine specializing in breast cancer. He was a huge part of my life, and he was very inspirational. I wasn't really sure if I wanted to practice medicine, but I was so inspired by his stories. And I really enjoyed studying biology in high school, but later I didn't find myself that passionate about it. And then I did some volunteer work at a hospital, actually in ER [emergency room], and I realized this was not for me. I don't want to be around blood. I don't see myself being drawn to actually practicing medicine, but I was still interested in how medicine has evolved.

"I took a few business classes, and I really liked it. When I graduated, I was able to actually find work in strategy consulting, which allowed me to combine my interest in medicine and business. I did consulting for a firm that did market research for biotech and pharma companies. After a few years, I got my MBA, and I wanted to work in marketing for a biotech or pharma or a medical device company. I came across Medtronic, and I loved some of the products they developed and the innovation, and I liked the idea that I could work for a large company and actually train and learn marketing and the business side of things.

"So I was fortunate to find that exact role at Medtronic when I graduated. I went to work in pipeline marketing, where we focused on identifying new markets and new product opportunities, then doing the voice of the customer and working with R&D and other functions to develop a product that the customer really needs and that the business needs." After six successful years at Medtronic, Fayngersh moved to Natera, where she has been working for about a year leading brand strategy and creating marketing programs: "I came here because I was interested in going to a fast-paced organization with a really innovative product. I already had strategic marketing experience and clinical research experience, but I felt I still had a gap regarding direct to patient marketing and also digital marketing. So it was perfect for me, it was a startup that wasn't in the early stage, so it was growing very, very quickly with a truly innovative product, and now we're uniting all panorama marketing under me. There are a lot of growth opportunities with a company that's growing, so I'm kind of taking it as it comes, and I expect to continue growing."

Leah Isakov, Senior Director of Biostatistics at Pfizer who came from Moscow, is another interviewee who had thought about pursuing a degree in medicine back in Russia, but as she explained: "I wanted to be a doctor, but at the same time I didn't like to deal with blood, and I was also very good in math, so I wanted to become a biostatistician from a very early age. I had some programming and a very good idea of what I wanted to do, which was in health care. So it's what I wanted to do, and it's what I did. In 1993 or 1994, I started working in a clinical research organization in statistical programming just because I had a programming background. And that was pretty soon after I arrived in the Boston area. My educational background in the Soviet Union was good, but I didn't choose applied math and went for computer information systems. I got my PhD in 2011, from Boston University, in biostatistics. I was working for a small company, so I rose through the ranks very quickly." Isakov was already senior director at Agenus, where she remained until early 2014, almost a decade. She referred to that company as "a pioneer in immuno oncology that is now a hot topic, but twelve years ago was not at all that, and actually there was a lot of skepticism. We were the first to produce a therapeutic cancer vaccine, 'therapeutic' meaning that you treat people who are already sick."

In 2014, she then became senior director of biostatistics and biosimilar statistical lead at Pfizer, and, during that period, completed her MBA at MIT's Sloan School the following year. She noted: "In any company, large or small, my work is basically the same thing. I work mostly in late-stage clinical trials where we design studies and develop a statistical part of the protocol in the early development of a drug and produce a report based on this data and interact with regulatory authorities for submission, but statistical analysis and data reporting are the basis for this. It's critical for approval and it's critical for business development because if the company wants to buy another company, you really have to look at the data and have a clear understanding of what's the path for this drug, for example, and what kind of evidence they have by the point of entry. But you still look at the data, and you still try to make a connection with other trials that were run. So statisticians are everywhere. They are not the most important people, on the surface, but they're behind a lot of critical decision making."

Maria Eliseeva is another interviewee from Moscow with a deep scientific background who decided to pursue a career that utilized her science background without remaining a scientist. She chose to become a lawyer specializing in intellectual property. She is cofounder and patent and trademark attorney at Patentbar International LLC, an intellectual property law firm specializing in patent and trademark protection for US and foreign companies in the United States and internationally. Back in Russia, Eliseeva earned a master's degree from the National University of Science and Technology in materials science, inorganic chemistry, and metallurgy, where she studied under Nobel Prize laureate, Alexey Abrikosov. After earning a master's in physics at Georgia State University, she decided to leave pure science and obtained a JD at the State University of New York at Buffalo; she was then hired by Hodgson Russ LLP, a large Buffalo law firm. She noted: "I think I was and still am a good patent lawyer, and the main competitive advantage I have is a very good scientific background, which my husband and I both brought with us from the Soviet Union. In 2003, I founded a firm with another lawyer, and that partnership endured for nearly a decade until 2012, when I cofounded another firm." Much like the experience of her husband Alexey Eliseev, described next, in her latest firm Eliseeva succeeded in utilizing three important elements of her background – science, law, and her Russian heritage and network.

Alexey Eliseev, Managing Director of Maxwell Biotech Venture Fund, chose a career in biotech different from most of the scientists covered in this section. He started as a bench scientist and research manager, later earning an MBA at MIT, and he transformed himself into a scientifically oriented businessman and eventually a venture capitalist. Eliseev received a PhD in biochemistry from Moscow State University and came to the United States as a postdoctoral fellow at Emory University, later becoming an assistant professor at the State University of New York at Buffalo. Eliseev's first experience in entrepreneurship was in the biotech startup, Therascope, in Germany that was cofounded by French Nobel laureate Jean-Marie Lehn and was focused on the drug discovery technology that Eliseev had developed in Buffalo.

He and his wife, Maria introduced in the preceding paragraphs, then settled in Boston, and he decided to learn more about business: "So I decided to take a break and actually went to MIT for one year to the Sloan Fellows program. About half the people in my cohort had small-company entrepreneurial backgrounds. It was a great experience, and sort of changed my outlook. I learned there is something in life besides science. Among the eighty people, there were about five from the biotech-pharma area. This was a challenging time for me, you know, moving from science to some form of entrepreneurship business. I had decided that I would still very much like to be in science in some form, but I realized that my base was like two legs, a scientific background plus newly acquired business education. I thought that adding a third leg, my ethnic background, would differentiate me more positively, so I got in touch with a group at Partners Healthcare, that was funded by the US Department of State, that distributed grants to Russian scientists. So I started working with this group, traveling to Russia, and then we founded a company called Boston BioCom that needed funding, so we raised $10 million from Pfizer. We ran into a problem with the regulatory system and the way that Russian scientists approached biotech and drug development there, but they actually commercialized some of those drugs in Russia.

"While I was at BioCom, in 2009, I got to know Dmitry Popov from Maxwell, which is one of the first Russian venture funds fully dedicated to life science investments and was initially funded partially through the Russian government, but 51 percent was from private capital. Some of the problems were similar to what we ran into at BioCom, but the difference was that we had more money. So we invested in two Russian projects, forming new companies and starting the development from very, very early stages of drug candidates. We also decided to develop the fund by benefiting from Western innovation, through taking advantage of Russia's established network of clinical trials. This is one area where the Russians are as good as anybody – in conducting well-defined, well-controlled clinical studies. Also, it's easier to bring a drug to the market in Russia than in the United States and Europe, so we licensed drug candidates at different stages of development from Western companies and continued their development in Russia according to the design of the Western company. It's too early to say if it's a viable model because we've been developing it for five years, which is not a long time on the drug development scale. We've invested in eleven projects – ten drugs and one device – and maybe four or five are developing reasonably well. I saw time and again that the US environment is easier to work in than the European, so six or seven of our projects have come from the United States."

Margarita Hunter-Panzica (born Kiseleva), Global Product Marketing Leader at GE Healthcare, is another interviewee whose earlier education and career were in science, but who later moved to the business side of biotech. She came from Kaliningrad, Russia, as she noted: "I had a master's of science in manufacturing engineering from Kaliningrad State Technical University, so before I went for my MBA at Northeastern University, I was working in technical roles in the biopharmaceutical industry in the US. I was in assay development/ quality control and I found it extremely lonely because you're in a lab by yourself most of the time. I needed to work with people. I needed to do something more engaging and strategic. That's when I decided to take a risk and quit my job and become a full-time student, not certain if that was the right move. I really enjoyed it, even though it was challenging because I had never studied finance or marketing. In retrospect, it turned out to be a great decision since Northeastern has a co-op program which led to the start of a successful career in biotech marketing. I got a job at Wolfe Laboratories, a small biotech company in Boston, to run the marketing department. It was a startup, and they didn't have a marketing department, so they just hired me and said, 'Build the department from scratch.' There was not a large marketing budget so it required great creativity. I reported directly to the CEO and enjoyed success in this part of my career for five years.

"My next position was with Thermo Fisher Scientific, but, a year later, my business was acquired by GE, so I was selected to be part of the transition to GE Healthcare. They are both big companies, and my desire was to be part of a big, established leader in the market. I had joined Thermo as the lead of the Asia-Pacific market. At GE, I had the same job in an even bigger organization. I then got a promotion to lead the product marketing team for the $300 million cell culture business where I manage a small global team. Speaking Russian has given me opportunities in global marketing and has facilitated my involvement in Russia and Eastern Europe and opened many doors. I'm happy where I am in my life right now. I always wanted to be successful and have been motivated and inspired by people who are even more successful than me. I will continue to shoot for the stars."

Wave Two 1987–1999: Research Scientist in Software, Internet, Communications, IT

Anna Lysyanskaya is Professor of Computer Science at Brown University, specializing in cybersecurity. Raised in Kiev, Ukraine, she came on a full scholarship to attend Smith College. She then earned a PhD in computer science and electrical engineering from MIT in 2002. Lysyanskaya described her primary research area: "Cryptography is the study of protecting communication and computation against malicious users. Of particular importance in my research are privacy-enhancing technologies that allow individuals to go about their daily online lives without disclosing unnecessary personal data. I am most proud of my work on anonymous credentials and electronic cash."

She also expressed interest in potentially translating her research findings into commercial applications: "Right now, I'm more on the academia, theoretical side, but I'm actually very interested in getting more involved in the business world. I'd like to see if my expertise, which is more foundational, algorithmic, and methodological, could be helpful to entrepreneurs. I think that there's a bit of a disconnect between what is possible and actually quite doable, from the point of view of all the cryptographic algorithms, and what's actually being used. So computer systems can be a whole lot more secure without requiring a huge theoretical breakthrough, just by using things that already exist off the shelf. I kind of want to get into the world a bit more." Among her honors are an IBM faculty award and a Google faculty award, and the MIT *Technology Review* magazine included her on their annual list of 35 Innovators Under 35. Her work has been incorporated into the Trusted Computing Group's industrial standard, has served as the theoretical foundation for IBM Zurich's Idemix project, and informed the National Strategy for Trusted Identities in Cyberspace. With her much-needed expertise in cryptography, she has the potential to contribute even more substantially to important areas of the US innovation economy.

Wave Two 1987–1999: Industry Professionals in Software, Internet, Communications, IT

A Wave Two interviewee who has been very successful in software in industry as opposed to academe is **Alexander Pashintsev**, Vice President of R&D at Evernote. Pashintsev completed a master's degree in radio technology and computing at Moscow State Institute of Radio Engineering, Electronics, and Automation and came to Silicon Valley to work on the Moscow-based ParaGraph's contract with Apple for the Newton project. He had been working as a project leader at ParaGraph that created the first usable handwriting recognition software and which Motorola sought unsuccessfully to acquire for $40 million. Pashintsev was one of three holders of the patent, and he continued to excel in that area in increasingly more complex technologies. After a short stint at Silicon Graphics, he then became a program manager of the Calligrapher Technology Group at Vadem. There, he worked on technologies related to calligraphy until the company was sold two years later to Microsoft. He described his activities with Stepan Pachikov, ParaGraph's leader: "Together with Stepan we moved fifteen people to a company that was designing a new kind of computer, a reversible tablet laptop that was very much ahead of its time. In two years, we ran out of money and sold our technology to Microsoft. We had gotten the award, The Best of Comdex, that was the biggest software show at the time, in Las Vegas. In four years, Microsoft just bought everything, all our technology. Windows still is probably using our code and running our recognition software as one of their engines for their tablets, phones, and desktops."

Pashintsev then joined Pen&Internet, Parascript, LLC, a company in Boulder, Colorado. He and his team, however, continued to work in Silicon Valley rather than move to Colorado. Meanwhile, as he noted: "Stepan didn't go with the team to Parascript, but moved to New York and started working on a little idea that was left behind from our earlier company. We were still thinking that we needed to revive the Newton legacy," referring to the handwriting recognition software that ParaGraph developed for that Apple product, which was dropped from Apple's product line when Steve Jobs returned to the company. "We wanted our contribution to be to keep this Newton legacy alive. So we had this application called Evernote – notes forever, saving all the notes. We had been working on this before Stepan left for New York around 2000, and he was already working on that project. He had been holding discussions with us, and, by the beginning of 2004, we actually merged our division and his company. He had two people and our division had about fifteen people here and twenty in Moscow. So we started with the name, Evernote, and Stepan was leading the team because he has something that makes people listen to him and follow him. So, for the next three years, we were looking for money, and we had more people, at least half the company here in Silicon Valley, with the other half in Moscow.

"Phil Libin was brought to Evernote as CEO with a mandate to look for new financing. He had come from Russia at the age of five and liked the idea of Evernote and moved here from Boston. By that time, we had a mobile version for phones, we had Windows applications, we had a server, we had all the concepts of what Evernote does. In nine months, we started a new service with new clients. I'm sure if Phil and his team hadn't joined, none of this would exist. It wouldn't have happened. There's a famous saying in Russian that a Russian would not take, 'yes,' for an answer. So that really slowed us down. It's important to trust people in a way that says, 'You can do this,' and let him explain. Stepan learned this, too."

Pashintsev described his role as Vice President of R&D: "Our team is called R&D. It's a good word combination because we start with some ideas, technologies, some science, and we are trying to solve some practical problem, and that problem may not be requested from us yet. But if I can suggest it, it may be useful. It may be interesting, or maybe not. So part of our work is spent on this 'R' when we are looking, when we are trying, when we are prototyping, when we are trying to sell internally in the company whatever we can do. If we do, the 'D' phase kicks in and we need to produce a product-grade code, we need to integrate it with other technologies, other parts of the products we have here." Pashintsev's story shows how he has contributed his technical expertise, along with many others, to the software industry with his company's technology and product line.

Another interviewee who is contributing to the translation of technology for users and clients is **Anya Kogan,** in her role as User Experience Designer Lead and Manager for Ad Words Display at Google, a core enterprise product area for that company. Previously, she had been Senior User Experience Designer for mobile payments. Kogan came from St. Petersburg and obtained a bachelor's degree in mechanical engineering from Tufts University and then moved on to the Georgia Institute of Technology, where she earned an MS in human–computer interaction. Before joining Google, she had a number of shorter term positions in similar fields and then spent three years as an interaction designer at HUGE, where she worked with a twenty-person design team on the next generation of online radio for ClearChannel, as well as assisting customers in choosing data plans with CellularSelf.

The experience Kogan gained during those years led her to Google in 2011, where she continues to utilize her extensive background in technology and human–computer interactions. She explained: "I have four direct reports and two more researchers who I kind of manage. I feel like I create a very familiar feeling amongst the team. I don't know that I'll be doing design ten years from now. I'm more focused on the soft skills now, I think, and I don't know if I'm meant to be a designer for another ten years. But doing a startup on my own, probably not. I guess maybe I haven't seen enough examples in my family, no one in my family is an entrepreneur. I'm probably not actually going to do that. I always want to have a goal ahead of me so that I'm doing things for a reason." Kogan is another interviewee with a strong technical background who has moved more toward management and a deeper interest in developing soft skills, the importance of which is emphasized in Chapter 7.

An interviewee from Minsk, Belarus, is **Eugene Boguslavsky**, who works at Facebook as a Release Engineer after having worked at Philips Healthcare and Salesforce, among other firms. He has a bachelor's degree in computer and information systems from SUNY Buffalo, as well as bachelor's and master's of management information systems degrees from the University of Wisconsin, Milwaukee. He described his career: "I have over twenty years of experience in working with technology. My primary focus is release engineering for large, enterprise-level companies like Facebook that have millions and even billions of users, and deliver changes to their users at massive scale with ludicrous speeds. I enjoy developing and managing tools and processes that ensure the safe and speedy journey of the code from the developers to the front-end users. I do not specialize in any specific platform, development environment, or scripting languages, and thrive in learning and using new technologies to automate and improve processes. Nothing gives me more pleasure than seeing engineering fruits flow through checks and balances and arrive safely to the users.

"I basically got hired at Facebook as a release engineer doing a lot of the front end, managing the front-end release process. During that time, the company has transitioned fairly fast to mobile platforms. With mobile, you have to deal with Apple and Google platforms, and they do things differently, especially Apple, that reviews every single release we have had. You can do as much as you can and throw it over the fence to Apple, and then it's up to them whether they're going to release it and how fast. So it's out of our control, and that can be a little frustrating, especially since our motto at Facebook is, 'Move fast and try out different things.'" Boguslavsky's work is another example of the utilization of technology in business and its importance in company products and customer applications.

Sergei Sokolov is Director of Product Management at the real estate software firm, Altisource Labs. Sokolov described himself as a "flexible product leader with a passion for excellence who can orchestrate collaborative team efforts towards strategic and immediate goals, usually in concert with developing partner alliances to achieve business impact and customer satisfaction." Sokolov came to the United States after receiving a master's degree in general and applied physics from the Moscow Institute of Physics and Technology to attend the University of Massachusetts, Amherst, where he earned a master's in electrical and computer engineering. He then spent his entire career in leadership positions in software development, focusing heavily on customer service and satisfaction with company applications. He first joined a startup, Sente, as an engineering manager and then moved on: "So my next transition was from the software for electronic design automation into Parasoft, a private company with a Polish owner in California, where I became Manager of Professional Services. I decided to switch from software engineering to something more software related, and I grew into product management.

"When I decided to leave Parasoft, I went to SmartBear Software, where I stayed for four years, becoming Vice President of Product Management. That was the only employment that didn't end up on my terms because there was a RIF, a reduction in force, following a change in the CEO. The product was doing reasonably well, but apparently something didn't quite fit the model of the new leadership. So that was the first time I actually had to look for a job in the US. I have thought about starting my own company. I guess I haven't because of either lack of guts or lack of a great idea. I'm quite independent, and it would suit me very well to build a company. Plus, I think I know a fair bit about quite a bit of stuff because I've worked in software engineering, I've worked with products, I've worked with sales, and I've done some business development and partner development that was reasonably successful. So I sort of know how things run. But that's not enough. I'm not an idea person, I'm more of an execution person with better attention to detail. So I think I'm a better facilitator than an idea generator, even though I do have a patent for technology that is an algorithm for power estimation for semiconductor chips. Even though the patent wasn't really for an idea from the ground up, it was an improvement over something that existed, yet a very meaningful improvement, and I'm proud of it. It's still incorporated in the products of the company I worked for at the time. So I'm not ready to do a startup yet, but I wouldn't exclude it, in fact, I'd love it."

Wave Three 2000–2015: Research Scientists in Biotech, Pharma, and Medical Products

Shamil Sunyaev, a scientist from Moscow, is a Professor of Biomedical Informatics and Medicine at Harvard Medical School, where he holds a Distinguished Chair of Computational Genomics. He is also a research geneticist at Brigham and Women's Hospital and is affiliated with the Broad Institute of MIT and Harvard. He received his PhD in molecular biophysics from the Moscow Institute of Physics and Technology. The Sunyaev Lab serves both Harvard Medical School and Brigham and Women's Hospital, focusing on genetic variation and other aspects of genomics. It develops computational and statistical methods for sequencing studies in these areas to predict the functional effect of mutations, and the lab also deals with projects in cancer genomics and applied human genetics.

In addition to working with basic science in an academic and hospital setting, Sunyaev connects with industry, primarily through consulting, but also through membership on scientific advisory boards: "I have a bunch of contacts with industry like sitting on science advisory boards, sometimes permanent and sometimes on call. They want your opinion on a direction they want to go in or some possibility they are considering. With startups, I think they want you for a different function. They come to you and present the project, and they want your frank opinion or idea. And sometimes there is follow-on consulting with them. And then there are occasional ideas from colleagues and firms who say, 'Why don't we do a startup?' Sometimes when I wonder if startups are a better way to fund research, I realize there are certain things you can do in academia that will not be found in industry because they will not make money, at least in the short run. And a lot of my interests are very abstract. But there are some that are more applied, and I think that's good, especially if with something more applied you can work with your colleagues at the hospital, where you may turn it into a company. But I don't like ideas like, 'Oh, this is how you can make a good startup and make a lot of money. So why wouldn't we do that?' I don't think I should do something just to make that happen. I should do something that I want to do." Sunyaev clearly stated his desire to remain in basic science in order to test out his ideas, which might take a long time before seeing a commercial application.

A medical researcher whose team's medical device has led to a startup is **Nikolay Vasilyev**, an Assistant Professor of Surgery at Harvard Medical School and a Staff Scientist at the Boston Children's Hospital. Previously, he had been at the Cleveland Clinic in Ohio and at the Bakulev Center for Cardiovascular Surgery in Moscow. Originally from Gorki, Belarus, he was educated at the Sechenov First Moscow State Medical University and was a researcher and practicing medical doctor. His appointment at Children's came from visiting there while attending a conference in Boston: "I visited the cardiac surgery department, spent a day, and we quickly found common ground in terms of the scientific project, and, to my surprise, at the end of the day my host, the department Chairman Professor del Nido, said, 'Why don't you join our group?' I knew it was a great department and the number one congenital cardiac surgery department in the country and in the world. In 2005, I joined the department first as a fellow and then as a research associate, and immediately got into a very interesting community. We had been developing very interesting devices for image-guided cardiac surgery, and the philosophy was that the clinical community would collaborate with engineers and also with industry. And that's the project I have been working on for the last ten years. We've developed the devices and computer algorithms for image processing for cardiac surgical procedures. But my goal always has been not only to develop the devices for the sake of research and publications, but to bring these devices to clinical practice, to help real patients, because I am a medical doctor by training and I have always been interested in helping each individual patient with new technology.

"Over the years, using NIH funding, we developed the new concepts and new devices that led to scientific papers and scientific contributions to the community, and we're now at the stage where we have strong laboratory results. The next step is to bring these results and these devices to clinical practice, and, in order to do that, you have to have a new and very different level of funding. You have to create a startup company, you have to license the technology from the hospital, and you have to secure the funding. And that's what we've been doing for the past few months. We've created a company, and we're in the process of licensing the technology. We cofounded the company with my chief, Pedro del Nido. Both of us will stay working at the hospital, and the CEO of the company will be responsible for the day-to-day business, while we will stay as scientific advisors. Our several devices are for minimally invasive heart surgery, and the idea is to perform surgeries inside the beating heart under image guidance, without stopping the heart."

Wave Three 2000–2015: Industry Professionals in Biotech, Pharma, and Medical Products

Marat Alimzhanov, originally from Russia and raised in Tselinograd, Kazakhstan, is a Lead Staff Scientist specializing in oncology at Acceleron Pharma. He has a master's degree in biology from Moscow State University and a PhD from the Engelhardt Institute of Molecular Biology in Moscow, followed by postdoctoral appointments in Germany and the United States. Prior to Acceleron, he performed the same role at AstraZeneca for about seven years. He explained: "At Acceleron, I am doing some early discovery projects making fusion proteins that involve extracting a part of particular molecules that makes it antibody-like. With its recognition capabilities, its function is to block the signalling of a receptor on the surface of cells. What I like about Acceleron is that they're doing this biologic thing that, from my experience, is a much more successful strategy than others, and I also like the fact that the environment here is open-minded. Acceleron's approach allows you to start with basic research and then look for applications.

"The environment is similar to what I found at AstraZeneca, and it fits my own mentality. For instance, I think what drives a lot of discoveries, a lot of new things, is the mentality of 'why not?' The mentality I enjoy is the environment where it is alright to say, 'Let's try it, why not, and see what we get from it.' It's a bit of a risk-taking approach, but I always had the feeling that if something excites me, something looks interesting, I really want to explore it. It's just like simple curiosity and pushing things as far as you can. With drug discovery, it's always a balance, and when you talk to managers, you always get this perception, 'OK, we have a limited budget and we can only support so much,' so you always have to make a decision as to which project to support or not to support. But, at the end of the day, you have to be honest with yourself. You have to push it, but if you see that it doesn't work, you have to say, 'We should probably stop it.' But I think one of the problems of big pharma is that once people get vested into their projects, they cannot let them go." He reflected on his change of career direction: "I was skeptical about moving from academia to a large pharma, but what I experienced was that it was really open, it felt like biotech. It was really open to explore new things, very risky things, things that might fail. The companies I've been with were ready to take chances."

Sasha Proshina, Financial Manager at Sanofi Genzyme, was originally from Tomsk, Siberia, and came after completing her bachelor's degree in management and economics at St. Petersburg State Institute of Technology. She then obtained a master's degree in regional planning at the University of Massachusetts, Amherst, followed by an MBA in finance at Boston University. After five years as an economist at UMass, Amherst, she moved into the biotech space, working in marketing research, market analytics, and global financial management first at Vertex Pharmaceuticals for five years, and then as a manager in those fields at Sanofi Genzyme. Proshina explained: "So now I've been at Genzyme for three years and I like it here. I think it's a good organization. It's a big organization, but somehow they manage to do things well in the Genzyme way. There are two divisions, one for rare diseases, and the other, for multiple sclerosis, is very young so it's almost like a startup. Since I started, we've launched in about forty countries. And I'm the one who's checking global sales, patients, and volume. So I get to connect all of it, and that's pretty cool because as we realize, we are US-based and that has a lot of complexities in the health care system, and forecasting is very hard. But the global aspect is fun. And the finance team is pretty good, too, and I think they have good leadership right now, and I think that Sanofi is doing a great job since they acquired Genzyme. They put French expats, very smart guys who are willing to travel, into key positions, and they kept everybody else. That's why I was so excited about working here. It's like working in Europe, but I'm actually living in America. I got a new boss who is a super-nice French guy, so I'm very satisfied." In contrast to most Wave Three interviewees, Proshina was not a scientist and had no scientific background. Her story illustrates that it is not always necessary to be a scientist or a technically oriented person in order to make important contributions to the technology sector.

Jane (Evgeniia) Seagal is another interviewee who has made a highly successful transition from being deeply involved in science in the academic realm to a similar role in industry. Born in Novosibirsk, Siberia, she is a Senior Scientist III at AbbVie, where she has been leading the hybridoma generation group since 2012, after joining the company in 2010. Like numerous others, in addition to her scientific research work, she is also a manager. Upon taking a leadership role, she made dramatic improvements by changing how throughput would be done in the lab, which has been equipped with state-of-the art technology, including robots, to conduct highly promising research. Seagal left Novosibirsk to study in Israel, where she received bachelor's and master's degrees in microbiology and immunology at Tel Aviv University, and a PhD in immunology at Technion-Israel Institute of Technology, often referred to as Israel's MIT. Following her PhD, she moved to Harvard Medical School. She recalled: "At Harvard I worked with one of the star scientists in the field of immunology, Klaus Rajewsky, and if you're in the field, you know who the top scientists are. Right after my PhD, I sent him an email saying who I was and what I was interested in, and he invited me for an interview, and that's how I decided to join his lab. I stayed there for almost seven years and then I joined Abbott Laboratories, now AbbVie. I think this is a great place for me because there are so many things to learn, so many things to understand. I'm lucky to have a job where I can help so many people in the future. I think it's important to know that what you're doing with your life has an impact and can help others. As to my aspirations, I would like to develop and learn new things, and I think my current job is great because I have these options. But I think, too, that I would like to learn more about other aspects of the company and the management of the business."

Wave Three 2000–2015: Industry Professionals in Software, Internet, Communications, IT

The vast majority of Wave Three interviewees in software were not involved in basic research. **Davit Baghdasaryan**, who came from Yerevan, Armenia, is involved in Internet security as Senior Security Engineer at Twilio, which provides a cloud communications platform of building blocks to add messaging, voice, and video in web and mobile applications. He came to the United States after graduating with a master's degree in computer science from Yerevan State University and working as a software security architect at Hitegrity, a company that provides software and hardware research, development, and consulting services to companies in the United States. Baghdasaryan came in 2008 to become a senior software security engineer at one of Hitegrity's partners, Validity Sensors, a company specializing in security authentication. During his time there, he received a certificate in advanced computer security from Stanford University in 2010.

After about a year and a half, he moved to Nok Nok Labs, a Palo Alto startup that again specialized in Internet security, where he started as a software systems engineer and rose to become Director of Advanced Technology. In recounting that experience, he noted: "I was part of the founding team. I wasn't really a cofounder, but technically my ideas made me a very important contributor to the company. I wasn't a CTO, I was like a systems architect. I was the main architect, and then I was Director of Advanced Technology. I had actually been working on an idea with a former boss, and the thing that I designed, actually, has become a very big thing. Very big companies are employing that technology now. Samsung, Google, and others use it, and there is a consortium of around 200 big-name companies that are also endorsing it. I have my own patents on it, but they belong to the company that I worked for. One is on secure transaction systems and methods, and the other is on secure user authentication using biometric information, like fingerprints. But, yeah, I'm the original designer of that product, so I'm very proud of that success.

"The second company I worked for was based on that idea. My boss had decided to start a company based on it. We got money and exceptionally good people, and we started Nok Nok Labs. So I spent three years there, and it was a great time. The company is still there, and

I have a very good relationship with them. The time to move came, and I had a good opportunity in San Francisco. It's a very different technology, very different areas, and I wanted to change industries. I did receive some equity at Nok Nok Labs, and I'm holding onto it because it's still a private company." After three years at that firm, he joined Twilio, in a different industry with different technology. He explained: "I thought that Twilio would be a good challenge in terms of the products they are building, and the ambition that the company has. I decided to change industries and it was a big, tough decision for me. But it turned out well, and I don't have any regrets. I love this company. It's great, so I'm looking forward to the future. In engineering, I'm the only architect for this company, and it's a pretty big company. My boss's boss came from Skype, and he is VP of Engineering here. We like former Skype engineers since what they have created in the past is very relevant to what Twilio does. They are great professionals, and so it makes sense to hire them. When I was at Nok Nok, I managed five engineers, whereas here I'm more of an individual contributor. So that's one good thing about our profession – it's portable and you can work from anywhere. This is because of how the technology is developing as well as communication and other things. You can be any place in the world and continue working together." Baghdasaryan is another example of interviewees making important contributions, including patents, as well as adding substantial value to numerous companies and, in some cases, starting their own, in the critical Internet security industry.

Ksenia Samokhvalova, who was born in Gorky, Russia, has an impressive background in science that she has utilized in transitioning to a supporting role as a Senior User Experience Specialist at MathWorks. That company is a developer of mathematical software used by engineers and scientists. Its products include MATLAB, a language of technical computing, and Simulink, a graphical environment for simulation, both products being industry leaders. Samokhvalova earned BS and MS degrees in physics at the State University of Nizhny Novgorod (formerly Gorky) before coming to the United States in 2003 to pursue a PhD at MIT, which she earned in applied plasma physics in 2008. She stayed at MIT for a while as a research assistant and visiting scientist before joining MathWorks as an Applications Support Engineer. She went on to become a usability specialist and then a Senior User Experience Specialist in 2013, a position she still holds. She explained her rather circuitous journey:

"I was trying to figure out what to do, and considering whether to be a scientist like my mom was in Russia. When I was in school in Russia, I liked everything, and I was good in math and OK in physics, and I was a straight A student. So when I was at MIT, I got interested in software, and I was looking for a software job that wouldn't require me to know a lot about coding.

"So that's how I ended up at MathWorks. I did their tech support and project type of job in a group called the engineering development group, and I did that for two years. I would be talking to people on the phone about the problems they had with the software package called MATLAB. I was thinking that I had spent eleven years in hard-core physics, and, as you can imagine, just doing user experience seemed really weird, but that's kind of what happened. I was learning a lot of new things and learning a lot about communications and how to talk to people. I was an applications support engineer for two years, and then, in the middle of that, I figured out that the user experience thing sounded really awesome. So I tried that. I felt that my background in physics was helpful because a lot of people I work with have science as their background, and then they became developers, or they were engineers first and then they became developers. My coworkers are like family. And, as a Russian woman working in a field primarily of men, I have always felt a need to prove myself. Still, it's hard for me to imagine working anywhere else."

Dmitry Kovalev, a Consulting Systems Engineer at Cisco, is also deeply involved with his company's customers, as was Ksenia Samokhvalova. Kovalev had ten years of experience in enterprise networks, information security, and data center technologies, and he was involved in designing and implementing new technologies to cut costs and improve performance. He has bachelor's and master's degrees in computer and information systems security and information assurance from the Moscow Engineering Physics Institute, and he is pursuing an MBA part-time at the University of California, Berkeley. He described his background: "After I finished my master's program, I was already working in Moscow, and I really liked what I was doing. I was with a small consulting company doing integration projects for major Russian banks, and the projects were really interesting, both in terms of the technology and in terms of the expected outcomes because they essentially helped the companies to protect their assets. But I was looking more for international experience around the areas that were adjacent to what I was doing at the time. And since Cisco was the major vendor of the equipment, it was definitely on my radar of the companies I would like to work with. They offered me a position in Amsterdam, kind of an extended internship where they hire top new graduates and take them for a year of very extensive studies as well as on-the-job training. So, after that program, I guess I was quite successful and they offered me to join Cisco here at headquarters in San Jose, so I relocated from Amsterdam. Working in Amsterdam was an eye-opening experience because, before that, I had always worked in the Russian culture and got used to it. And then suddenly I had to deal with a mix of cultures and also to deal with a company culture that was a completely different thing. A major difference that really struck me was that when I was working in Russia, the main goal was just to do the projects, and pretty much the projects were your whole life, your whole goal of existence.

"When I joined Cisco, it was a different organizational culture because it's not just the work that you're doing. You have to be good at that, but you also have to look around and see what else you can do to make the company better, maybe not even related to your particular area. But, basically, having fun is part of the culture, recognizing that you can't work all the time or you'd be burnt out. Another difference is this corporate approach to doing things, like when you need somebody's help, you go to the person and ask for their help, and they might tell you, 'Yes, I will help, but I have to consult my manager, or you should talk to my manager first.' And another thing that was probably the best thing of all is that Cisco is a technology company, and so we have lots of technology tools that allow us to better work together. It's really amazing how the technology that the company has helps you in your work. I'm in my sixth year with Cisco, and it's interesting. I could never have imagined that I would be working at Cisco headquarters here in San Jose when I was starting my career in IT in Russia. My work really involves a lot of working in sales, so, for me, a failure means that I will not get my commissions. But, overall, at Cisco, if something doesn't work out, it's not the end of your life. You have to learn and make sure you're not making the same mistakes as you're moving forward. It's a good environment, much like sports. I was playing serious basketball in Russia, and you can't always win the game that you are playing. But if you lose the game, it doesn't mean you have to give yourself a shot in the head. It means you have to go and train harder and have to get ready for the next game. So it's the same approach here."

Anton Voskresenskiy chose the research and development side of industry, serving as Chief Research Officer at Northern Light, a leading search and analytics company serving large firms in the pharmaceutical, IT, banking, and insurance industries. Voskresenskiy is responsible for maintaining the company's cutting-edge technology in text analytics, machine learning, automated analysis of research documents, and new delivery platforms such as mobile. He joined Northern Light in St. Petersburg in 2003, after ten years of software engineering experience with a leading offshore development company building applications for US companies. He later moved to Northern Light's Boston headquarters as Vice President of Engineering and CTO. He earned a BS and MS in mathematics from the Maritime Technical Institute of St. Petersburg.

Voskresenskiy described his career in software: "Toward the end of my studying at the university, I was looking to find some work as a software engineer. In university, I was already working part time for a Russian company, but I don't consider that a software development job, it was more of a computer-associated handyman, and I was doing everything. But after I graduated, I needed to leave those guys and find a proper software development job, which I did with the American company, Animation Magic [founded by interviewee, Igor Razboff]. Animation Magic in Russia collapsed around 2001, because it was not healthy financially. The Russian offices had to be closed, and I was briefly unemployed. But then I had some friends who offered me a job as an individual contributor with an American company. So I was basically working at home, delivering things over the Internet. I was working with some people that I knew from Animation Magic who were already living in the United States, and that went on for a couple of years. One friend was working on this Northern Light project, and they were looking for someone to make sense of the software they had bought, and I was hired as one of those guys. It was around that time that I decided I wanted to move to the United States, and had a deal with Northern Light to do so, and that was in 2006. The team here was about the same size as the one I was working for at Northern Light in Russia. It was really the same team, and it fluctuated at around ten people. A lot of the time I ended up talking directly with the CEO, but officially I reported to Igor. I immediately fell in love with US culture, and I loved living in Boston. So now, at Northern Light, I'm really a contractor and they are my one client. We have a very good relationship. I don't have any employees at the moment, but I do have a bunch of contractors, some in Pakistan, Tunisia, and Ukraine, and work with them mostly on Skype. My last big project for Northern Light lasted for around five months, and, at the maximum, I had six people. Basically, Northern Light is my client, and I am delivering completed projects to them, and then they deliver them to their customers. So far it's working great."

A similar story was recounted by **Sergey Kononov**, who is Senior Delivery Manager in Cambridge at EPAM Systems, a global product development and digital platform engineering services company. Originally from Kharkov, Ukraine, Kononov came from Moscow, where he had worked for Luxoft, the major competitor to EPAM. At Luxoft, Kononov rose from risk management project lead to project manager. He had joined the company shortly after graduating from Kharkov National University in 2004 with a master's of mathematics and informatics, specializing in applied mathematics. Kononov manages and monitors the quality of work and client deliverables to ensure they are within forecast and budget, and he also works with EPAM's development team in migration of systems from development to production.

Kononov explained his move to EPAM: "There are two big companies in Eastern Europe, Luxoft and EPAM. I started working at Luxoft in 2005, first as a Java developer as part of a portfolio management team developing web-based applications. I then moved to a project with Deutsche Bank as a risk management technical lead, driving a team of seven to ten of the craziest talents you could imagine." He explained that Luxoft is a global IT service provider with more than 11,000 software engineers. He became employed there working on projects in English, although he said he could not really speak English very well at the time: "One of my friends told me I should apply for a position as a support engineer, but I didn't know English. He insisted, 'You have to apply,' and they made me interview in English. There was a client who came from New York to interview me who asked me first about technology, and I was OK on the technology side. They kept talking to me, and they kept being patient, and they accepted me. I don't know what miracle happened, but they accepted me for a junior developer position. So there was this guy who was the project manager, and he was the best project manager I've ever worked with. And he said, 'Take the lead in a small project of three people with more responsibility.' I had some problems because I sometimes didn't understand coworkers in London, Frankfurt, New York, and Singapore when I was in Moscow. I committed to delivering the projects too soon and couldn't deliver, and that was a big problem. Later though, I wouldn't say, 'OK,' or give a commitment, until I understood what someone was saying. Then people became very patient with me, although the first month was not easy. People at Deutsche Bank began saying that I was the best project manager. I'm very proud of that."

Kononov turned philosophical: "Sometimes you get depressed in Russia, like when you move from being an engineer to project manager, and part of the problem is that you stop seeing some of the valuable work that you yourself produced. If you're an engineer, you see some of your work result in functionality, or your program works, or you can make the demo. As project manager, what have you achieved? Nothing. Project plans, talks, talks, talks, is all you do. All the results are achieved by your team. You do nothing. And this is so depressing, and I had to keep doing that stuff. I kept hiring and building the team, but at some point I realized that these are my achievements. When you build a good team, I realized that's what I can do well. I understand technology and I can give directions, but typically I hire or grow engineers who can do things better than I can, and that's one of the criteria. That's when I realized that this is one of my achievements." Kononov was insightful in presenting the sometimes difficult psychological adjustment involved in making the change from individual contributor to manager.

Maxim Matuzov, who came from Murmansk, Russia, is a Search Program Manager at Apple. Although his educational background is a master's degree in international business and foreign languages from Murmansk State University, most of his career has been in software development, which he learned primarily on his own through Internet courses, as well as in various related positions. His first job in the United States was as a technical project manager for ZoolaTech in Silicon Valley, where he stayed for three years before moving on to become a program manager at Google. He noted: "I liked it at first, but it was not an interesting job for me. So once I saw the opportunity at Apple in Search, I took it after about a year at Google. I like Search so far, I always liked it. I had always read about it, and I had some experience in it working with Yandex back in Russia," referring to Russia's leading Internet search firm.

In contrasting the cultures at Google and Apple, Matusov noted: "At Apple we talk about nothing here; absolutely no information goes outside. Google talks about everything. That's the biggest difference I've seen so far." He continued: "I'm a fan of Apple, but I'm kind of biased because I work for the company. But just compare laptops on the market, and what's the difference in Apple products? It's perfection, great attention to detail that calls for pushing yourself to be as close to perfect as you can be. I've thought about being an entrepreneur, but I don't think I'm there yet. I mean, people are telling me, 'You should go out and start your own company, with the resume you have.' I don't think I'm ready yet because I want to still learn things from, you know, very inspirational people. I sometimes talk to them and I think, 'Wow, what an idiot I am because they are so smart and so interesting to listen to. I've only been here at Apple for a year. If there's an opportunity, for example, to work for Apple in, let's say, Australia or Canada or Dubai, then I'd try something new. The thing is, I can't be stuck in the same company in the same business forever, because I like to develop myself. My goal right now is to become a Director before I'm thirty, and then maybe go to a smaller company. Who knows, I'm still young."

An interviewee who appreciates the startup culture in Silicon Valley is **Shalva Kashmadze**, a Product Manager at Pocket Gems, a mobile gaming post-startup that he joined after graduating from Stanford Business School in 2013. He came from Tbilisi, Georgia, "in the USSR, not the state," he clarified with a laugh. He had worked for the Bank of Georgia after earning a degree in banking as well as a law degree at Tbilisi State University. With his educational background and his current career in a tech startup, he refers to himself as a "recovering banker," a field he chose in Georgia to provide financial support to his parents and other family members. He described his job search in the United States: "The Stanford network helped me get a good job when I graduated. The job search is a scary, scary process that just freaks out everybody. It was definitely long and tough, and more so for people not being from the US. It was hard in my case since I was trying to start something that I had never done in my life. I'd never in my life worked in tech, never been a product manager, and never worked in a startup.

So I was new to lots of things for the first time, and I could never have done it without my Stanford network. When I contacted Pocket Gems, they said, 'Hey, we're this mobile gaming company up in San Francisco. We were founded by Stanford GSB [Graduate School of Business] people a few years ago.' And they were happy to talk with me because there was this commonality, and they felt that I was part of the Stanford GSB family versus someone from outside.

"Tech companies are usually good at sponsoring your visa because they have to bring engineers from abroad all the time. I was the first Georgian at the company, and also the first Georgian at Stanford Business School. I think less and less of going back to Georgia. In San Francisco and Silicon Valley, all the news I'm hearing is so exciting, and I'm hearing about these companies that are doing amazing things. Companies that have been founded by friends, or that my friends work for. Every day you learn about some new, crazy, awesome companies and ideas. And you feel that you are a part of it. You can affect it. And you can build the next Facebook, or just build the next something. I'm really fulfilled being here, and I think that Stanford gave me the sense that I can go wherever I want. I also think that, at some point, I might get involved in a startup that I think is an amazing learning opportunity, but you also need to be at a certain stage of life and have certain talents to help build a startup. So, in my mind, I need to know a certain number of things before I get there. For now, I need to be a small part of something bigger, probably several times before I learn all the rules and learn how to do it on my own. And then, once I have these things in my back pocket, I can be a member of a founding team and bring all these things to the table. So I think it's going to happen. It's not my explicit goal to be a member of a founding team, but I will be really happy if it happens at some point."

Nick Bilogorskiy is Director of Security Research and a member of the founding team of Cyphort, a Silicon Valley cybersecurity startup. He emigrated at a young age from Kharkov, Ukraine, to Canada. There, he completed his bachelor of science in computer science and philosophy at Simon Fraser University, followed by a stint at Fortinet, a Vancouver security company, where, as he noted: "I began to learn about malware. I went through a transition where I joined them as a quality assurance tester, did well, and was promoted to QA [quality assurance] manager. I started learning more about the technology. And their antivirus researcher left the firm, so they had an opportunity to promote someone, and I was that person. I had to transition from QA to actual research, learning about malware and reverse engineering. I really learned how to be an analyst." He left there for his first job in the United States at SonicWALL, another cybersecurity company: "I was promoted a couple of times. I joined as an engineer, became a manager, and they promoted me to director. So it all went well, but as the Russian saying goes, 'The better is the enemy of the good.'"

Bilogorskiy is not satisfied with the status quo, also obvious from his three patents aimed at improving various aspects of cybersecurity. He continued: "So when an opportunity to join Facebook presented itself, I had to take that chance, and I left SonicWALL after four years. I went to work at Facebook as Chief Malware Expert, which was another big break for me. It was my second time at an IPO company, and I joined them in 2010. At the time I joined Facebook, I had already made a name for myself as the computer virus guy. There are not too many of us around, and we all know each other. It's kind of a close circle. So I went to a conference and that's where I met up with the Facebook crew. We went to dinner and some parties and got to know each other socially. A couple of months later, they decided to become one of the first social network companies to have their own antivirus guy on staff. So they recruited me, and I didn't even have to interview. I went in and I talked to the boss, and it wasn't a technical interview. It was more like, 'Sit down, Nick, and tell us what you think about Facebook. How would you work to make it more successful? If you're dealing with a hacker, how would you figure out who it was?'

"It was a very different kind of job than I was used to. I was used to a job where a security company made a product and I wrote signatures to detect malware. It was very technical engineering work. Whereas at Facebook, there was this big company of around a thousand people, and it already had hundreds of millions of users, and you are working to secure the whole thing for everyone. So it was more working with law enforcement and suing bad people, figuring out who the bad guys were and working with the FBI and Secret Service, working on deconstructing cyber attacks, and attributing them to real people and fighting back, not just playing defense but playing offense against them. It was very, very interesting. At Facebook, I had to learn a lot and work at a very high speed. It was a young company, and at the time, I was twenty-nine or thirty. In my previous companies, I used to be one of the younger guys, but at Facebook, I was one of the older guys because everyone was, like, twenty-two, twenty-four, twenty-five. I think we were very successful in that Facebook continued to proliferate, and users never quit Facebook after those attacks because we cleaned them up pretty quickly. So I spent two years at Facebook, which was a very exciting time.

"And then I left them to become part of the founding team at Cyphort. When a recruiter reached out to me to ask me to work at Cyphort, I realized that this startup company had the same idea of behavior-based malware detection that I saw as the future, and I would get to contribute on the ground floor. Facebook had really taught me that it's exciting to be at a startup, and financially it was a good experience, and it was a great education. I felt like I worked for the smartest people in the world, but I also found out that many of them had left, and that's when I started thinking of leaving. Facebook does burn you out with how hard you work there.

"I came to this startup called Cyphort that, in a month, will be four years old. We built it up from two people to about seventy now, and we started selling our product and getting millions of dollars in revenue. It's been very exciting to see the whole thing grow from the beginning. Our success is nowhere near the trajectory that Facebook once had, but it's still so much better than thousands of other companies that just go flat and die, or go bankrupt." Turning to his own career, Bilogorskiy commented: "I have started teaching myself a lot more about management, about entrepreneurship, about how the business is run. I work with customers now, I work with sales. I'm really looking forward to running my own business eventually and target the CEO role. I want to move from the technical field that I have been in for the last fifteen years to the business field. That would be something I would definitely see as a potential future, to run my own company. I have been running my own nonprofit organization for the last year, Nova Ukraine, that provides needed goods to people there. That has also been a great experience and has a mission I'm very passionate about, and that's one reason you want to start your own company."

This chapter described the career aspects of interviewees whom we selected as interesting and important illustrations of their contributions to the US innovation economy, primarily in roles other than as entrepreneurs, although some had been involved in startups during their careers or planned to do so. Chapter 5 dealt with the entrepreneurial experiences of interviewees as they founded companies in the United

States, sometimes after having also done so in the former USSR. Chapter 4 covered the migration experience, providing a segue to Chapters 5 and 6 dealing with experiences in the United States. The next two chapters illustrate the cultural and workforce adaptations needed as foundations for such accomplishments in the United States. These include, as will be described in Chapter 7, overcoming challenges and drawing upon sources of support such as mentors and role models as well as networks. Chapter 8 focuses on essential soft skills to succeed in the US innovation economy, including teamwork, managerial and leadership styles, and communication styles, as well as building trust.

7 | Cultural Adaptation
Challenges and Sources of Support

Our interviewees understandably experienced challenges in adapting to their new environment – the culture of American society, its innovation economy, and the organizations with which they became involved. These included educational institutions as well as workplaces, including companies and nonprofit organizations. The process of adaptation involved facing challenges that arose, as well as seeking and securing sources of support to ease that process. We begin with challenges in adapting to a new language and culture specifically in the US educational system and the workplace, followed by challenges with new living conditions and circumstances in daily life. Interviewees then recount their experiences with the major sources of support they drew upon – mentors and role models as well as networks. These individuals were entering a society that required acquisition of a new language, adaptation to a very different culture, and, for most of them, a reconsideration of their own sense of identity, a process we introduce in this chapter and explore in depth in Chapter 9.

Adapting to the US Educational System and the Workplace

As discussed in Chapter 3, the educational system in the former USSR contrasts sharply in content and process with the US educational system at all levels. Even for the few who had not been educated in the former USSR, interviewees were generally aware of that system through stories from family members. The contrast in systems was bound to create challenges, particularly at the primary and secondary levels, often due to the deep imprinting of institutions that occurs during those early formative years. **Yulia Witaschek**, who came from Moscow in 1994 at age seventeen, shared her memory of the initial adjustment at Connecticut College: "It was probably the hardest thing ever. For the first two days, it was utter misery. Because obviously there was a cultural clash, because nobody would talk to me. I wasn't dressed the right way, I didn't speak English very well, I was too slow, I wasn't really cool by the American standard. I spent the first year being the nerdy kid because I went from being top of the class, most popular, and knowing how to do everything back home, to I didn't know how to use the phone, and I had no idea what this whole picking your own classes thing was."

Another interviewee who came in 1994 shared intense recollections of adapting as a teenager: "When we moved to the United States, I hated it so much that I told my parents that I was just waiting until I turned eighteen and then I would leave and go back. The first challenge was language, even though I had gone to a specialized English school in Moscow. I was very shy in terms of speaking the language, and a combination of that with the teenage years and other frustrations made it hard. Also, because my parents were going through their own difficulties with finding jobs and being stressed out, and also going through a cultural shock, they couldn't help me. The first few years, I felt like I couldn't make any connections with Americans because it was so much about this individualist culture. And in Russia you had such tight friendships, and you're always helping people and sharing everything, that this came before respecting the law. It was just secondary because there was no respect for the law in Russia. Whereas here it's the other way around: it's respecting the law, being an individual, and only then connecting to people. It's not a criticism; I think it's just how it evolved."

Polina Raygorodskaya, who came with her family from St. Petersburg at age four in 1990, recalled her childhood in the Boston suburbs: "It was definitely very challenging in Holliston because that's where I spent most of the age of six to twelve, and I stood out like a sore thumb being an immigrant and half Jewish. The majority of the town was white and Catholic. I sang in a choir because that's what my friends were doing, so it was definitely tough being there as a Russian, half-Jewish immigrant where there was only one other Jewish person, and not a single other immigrant. So that was definitely challenging as a child because I felt like I was always kind of on the outside." **Anna Winestein**, who immigrated with her family at age nine from Leningrad in 1991, also expressed the sense of being an outsider: "We arrived in March, and, after a couple of weeks, I entered school. And it was not a pretty experience, because I came in towards the end of the year. So here's this kid who turns up dressing funny, talking funny, and not understanding how to behave because it was a shock to me. Obviously, I wasn't in the boonies of the Soviet Union, but even in Leningrad where I came from, the modes are different. It took me a long time to understand how much kids really ran the schools and how little in control the adults were. I was used to my school in Leningrad, where the teachers and other adults were in control. And there was a different kind of maturity expected. And the social dynamics were to me sufficiently different, even though I learned certain things relatively quickly."

Nataly Kogan, who emigrated with her family from Leningrad in 1989 at age thirteen, spoke candidly about her emotional memory of that time as a young teenager: "My short explanation is that my immigrant experience was extremely difficult. It was just very heavy and very dark, and I forced myself to learn how to speak English without an accent. I truly believed that my classmates were just going to get the better of me if I didn't. So I went crazy learning English. I went to the University of Michigan language labs, watched and imitated tons of TV shows, and I stopped speaking Russian. Not knowing English made me feel like an idiot in school, and I'd lost my identity of being a top student. I felt lost and overwhelmed. And, as Russian Jews, we didn't really have coping mechanisms. You know, you don't sit around your family table and talk about your pain. You deal with it inside. You swallow it up, and you keep pushing, and what I did after that, my kind of achievement overload, as I call it, was a complete and clear desire to escape from the inside. So, it's not that I didn't want to create all the companies that I did, but I had so much pain that I just decided to put a 'happiness Band-Aid' on it. I literally decided at seventeen that 'I know how to deal with this. I'll get to the American dream, I'll be happy if I achieve a lot of things.'"

Isaac Fram, who had come two decades earlier from Riga, Latvia, at age twenty-one to study at the University of Illinois, Chicago, arrived during a time of great social change in America: "My adaptation was very good. I really enjoyed it, but for me it was kind of a shock because 1969 was the time of the Vietnam War. So the first thing I saw on campus was the 'flower children' on the lawn playing their guitars and telling me how great communism was." Given his experience living under communism, Fram's shock was clearly understandable.

A more common issue for interviewees revolved around language. **Anna Scherer**, who came from Ukraine in 1994 at age eleven, explained: "In Sevastopol, I started taking English classes, but they were mostly useless when I came here. They taught us British English, but then I was immersed in the Brooklyn accent, so for the first half year, I had no idea what people were saying. And I'd translate the homework assignment into Russian, and translate it back. It was a long process." **Michael Ostrovsky**, who came from Moscow to study math at Stanford in 1997 at age seventeen, described how language was his biggest challenge while academics were relatively easy: "What was harder the first year was social interactions with others. And it was purely for language reasons. There maybe were some cultural issues, but not really. Even when you grow up somewhere else, you still watch American movies, so a lot of things are not surprising, not shocking. But the new language is hard. The worst feeling is when you go for lunch with a bunch of other people and someone starts telling a joke and everyone laughs and you feel like a complete moron, because you didn't understand what they said. You don't know what to do. If you laugh, you're an idiot. If you don't laugh, you're an idiot. I mean, that's terrible. Because when you talk one on one you can ask, but not in those group interactions. But I realized that I had to go through it. After about a year, it went away and became much better. So the social side was much harder than the professional side.

"In terms of the professional side, I would say there was almost zero transition. I was actually surprised how easy it was. I didn't speak good English when I came here, but in terms of math and other hard classes, those were actually not hard because if I didn't understand something while the professor was explaining it, I could just go read the book. If I didn't understand some word, I could go and look it up in the dictionary. It's really easy, and, in terms of written assignments, again, if it's math, even if your language is not the most beautiful, if the formulas are correct, the professor can see that you understand what you are doing. So that was easy. I had to take some writing classes, which I hated at the time. I just absolutely hated it. Not because they were hard, but because I thought they were useless. I was here to study math and hard stuff, and they made me study writing. Ex-post, I realized that was the most useful class by far."

Financial executive **Aziz Mamatov** initially came to study in 1994 at age twenty-one and returned as an immigrant two decades later, at age forty. He discussed the feeling of initially being marginalized in his social connections after having been a member of the professional elite in Tashkent, Uzbekistan: "I was in a state of culture shock. It was 1994 or 1995. I had so many differences with American students, so I had a chance to speak only with socially awkward, marginalized people who would talk to people like us, foreigners. I'm not saying that we were marginalized, but we didn't fit in, and so we were speaking with misfits who we didn't appreciate. And there were a couple of really great guys who were on the top of the social ladder, but we didn't have a chance to speak with them because they were either too busy or they were snobbish, or whatever. In our own culture, we were on the top of the social ladder, we were the leaders. Everyone liked us, but here we were just marginalized. I think the first couple of months, I was a bit shocked and then I was in a depression, so I think that my first four months were not entirely functional."

The obstacle of language was a theme not only for those who entered school, but also for those entering the world of work. **Andrey Doronichev**, who came to Silicon Valley from Moscow in 2012 at age thirty, also experienced language and culture shock particularly in the business environment: "For me, the first shocking exposure to the US culture and the general business community happened the first week I joined Google. That was my first exposure into this world. Trying to learn a new language was hard, understanding simple phrases like 'catch up' or 'touch base,' or whatever, that show up in countless emails a day. I'd never used a calendar in my life before, and the US is a very calendar-driven culture. In Moscow, you plan one business activity a day and that is the best you can do, what with the traffic and all."

Adapting to New Living Conditions and Circumstances

Interviewees encountered profound differences in conditions of life, both positive and negative. **Alex Bushoy**, who came from Chisinau, Moldova, in 1991 at age thirty, expressed the contradictory feelings that arose from the new social conditions he encountered: "It was a very happy time. Really. Most people can't understand that, but for me it was a very happy time because I came from that chaos in the Soviet Union. And here it was so quiet and so peaceful, and I didn't have to think about anything. And you get your welfare check, and your refrigerator is full, and you're not afraid of anything. It was like a haven." Yet he reflected on the difficulties of finding employment in those years: "The year 1991 was not a good year for the United States economy. It was extremely difficult to get any job. We couldn't get anything, not even a department store job. Especially if you didn't speak English. So we started with just doing cleaning jobs at houses. And we went to the synagogue for English classes, but I got bored very fast because we were sitting with a seventy-five-year-old in the same class, and it was kind of a waste of time for the first few weeks. I decided I couldn't take it any more, so I left and created my own program for learning English myself." **Alexander Aristarkhov**, who was originally from Novosibirsk, Siberia, and who came in 1990 at age thirty-four, had a similar experience: "For us it was really a big shock to actually enter into that different world. At that time everything was different – stores, renting, the banking system and credit cards. We got a lot of help from people who invited us to the US, but they just helped with advice."

Anastasia Khvorova, who came from Moscow in 1995 at age twenty-six, spoke about other unexpected aspects of life in the United States: "Another completely shocking experience for me, after coming from Europe, was understanding the taxation system. So, even if my fellowship was $23,000 a year, it sounded reasonable to me, because when you are entering the country as a nonresident, you're not allowed to claim any deductions, you have to pay taxes at the maximum. So, after paying the taxes, my actually monthly stipend was only $1,100. And the second thing I didn't realize was that the public kindergarten was from 11:30 to 2:30. You have to provide daycare before and after, and there is no infrastructure in place. So with a salary of $1,100 and a daycare bill of $800, you were supposed to survive on $300 a month." **Leah Isakov**, who came from Moscow in 1992 at age twenty-six, also spoke of the difficulties of being both an immigrant and mother of small children while pursuing her profession: "My English was less than mediocre, and I couldn't drive, and we ended up in rural Massachusetts. It was difficult with two little kids, six months and three years old. I had my hands full."

Natalie Hill, who came from Minsk, Belarus, in 1995 at age twenty-one, reflected on her early experience with low-level employment and finding herself alone in a new environment: "My first job was not so much fun. My boss was not the nicest person. I've hated Taco Bell ever since then. Boston Market was a much nicer place from the management perspective and the people. And then one day I just talked to the person at the bank next door and she offered me a job there. I was so happy to leave that fast food business, and working in the bank was pretty good. I was there for a year before I got my internship with SIAC."

Simon Selitsky, who came from Moscow in 1991 at age twenty-one, spoke about the initial family support he had on arrival in the Russian enclave of Brighton Beach in New York and the difficulty in moving away from that base: "Everything was different. It was like going to space. I had initially a soft landing in Brighton Beach. I was very lucky that my aunt and uncle and my grandparents were alive at that time; they'd already started there. So I spent a few months there before I moved to Washington Heights. And it felt like being alone in space for me. I was alone by myself, no brothers or sisters. I was all by myself, and it was all about survival. I really felt like I was climbing a high mountain, Mount Everest maybe."

Another interviewee learned a hard lesson in high school in the United States about cultural differences toward helping others: "A couple of Russian friends were not prepared, and I would pass on the answers to them. But I was caught doing that and almost thrown out of high school. I knew that here in the US it was the wrong thing to do, but not where I was from, because there you're supposed to help your friends. So it wasn't a moral conflict for me – helping my friends came first." The individual later adapted to US norms against helping others in such situations.

Alexander Vybornov came in 1993 at age eighteen, from St. Petersburg and experienced immersion in American life as a young college student: "You know, it was obviously a shock in terms of interests, being born in one country with a very different political structure, just the worldview of things. On the other hand, I would say it was equally a shock at my first summer job at the university, because I was on a student visa and was restricted in the type of work I could do. My first job during freshman year was actually installing water meters, so essentially plumbing if you will, in Brazil, Indiana, population 5,000. I worked in a city hall garage, and the type of folks there were of very blue-collar backgrounds and from very red states. So I obviously had some preconceptions and was worried starting to work there, because here I am, speaking broken English, a young guy from a former communist country. But I was very amazed by the openness of the American people I worked with. I think all of it hinged on work ethic and integrity. As long as you exhibited these qualities, you'd get accepted across the board. And I found this actually paralleling very nicely into the academic and high-tech environments throughout my time here. So that's probably the biggest pleasant surprise."

Adapting to Language and Culture in Daily Life

As the immigrants settled into life in the United States, they continued to encounter adaptation challenges. **Anton Rusanov**, who came from Karaganda, Kazakhstan, in 2012 at age twenty-eight, recalled adjustments he had to make to feel culturally comfortable: "The craziest adjustment was to be really careful about what I am saying. Because the culture is different and what is considered totally normal in Russia can be treated as maybe sexism or sexual harassment or whatever here, even if I didn't have any of those intentions or any actual bias. So I have to express myself really cautiously." **Maxim Matuzov**, who came from Murmansk, Russia, in 2009 at age twenty-one, took advantage of opportunities to create a healthier lifestyle for himself and experienced little in the way of culture shock at work: "I adjusted my lifestyle to a healthier way, which is a good thing. I mean, my body right now is much better than it was even in school. The food we eat here, the lifestyle we have here, is much better for you than the lifestyle we had in Russia. I think that's really an important thing. As for cultural adjustments, so far my work proves different. Since I've been promoted a lot of times, I'm thinking it's really working, and I can't think of really a big, big adjustment of myself. I haven't actually had any shock because I probably was expecting something like that." **Shalva Kashmadze**, who came from Tbilisi, Georgia, in 2011 at age twenty-five, also commented on embracing the California lifestyle: "If California is the first state you see, you are really lucky because it's so different from many places in the US. I think people are more open. They're from different places, they've travelled, they know foreigners, and they don't ask you too many questions even though they see that you have an accent. So, for me, it was a super, super, smooth transition being here and interacting with these people. I didn't feel any sort of barrier between us." Fellow Georgian, **David Boinagrov**, who came from Tbilisi to study at Stanford in 2008 at age twenty-two, found the transition somewhat more challenging than Kashmadze: "I guess my thoughts were: 'OK, it's my second time moving to another country, so I thought it was going to be as easy as when I went to Russia, but actually it was much harder. I found out that I didn't know English well, especially the informal, everyday language. I also had to get myself more familiar with the culture, social norms, and values of the US and Silicon Valley in particular. Maybe I'm not particularly the best communicator, probably like many physicists or technology people, so in that sense it was harder than in Russia, but it wasn't all that bad. But at Stanford I had a lot of work to do and didn't have much time to think about that. And being successful in my PhD work gave me confidence and strength to move on."

David Gukasian, originally from Yerevan, Armenia, came in 1995 at age seventeen. He spoke about the process of finding his way in America: "Cultural integration into the society was a challenge because everything is brand new and you have to learn on the spot. Nowadays, I think it's a lot easier because there are a lot of people who went to Harvard, MIT, Stanford, who were Russian speaking, and they created a website for people listing what to do, where you need to go, here is the paperwork you need to get. So everything is pretty much laid out. Back then in the '90s it was, basically, you just show up, everybody's in the same situation, everybody is looking for something to do. So you needed to really understand which direction you wanted to go, what you wanted to do with your life, with your career. I kind of just got lucky, and I found the right friends who were able to push me towards something so that now I have a career and a future." **Oleg Rogynskyy,** who came from Dnepropetrovsk, Ukraine, in 2003 at age seventeen to study at Boston University, also encountered cultural challenges due to a lack of information: "I didn't know what to expect. The scale of BU was very surprising. The whole kind of bonding together orientation part was very interesting because nothing like that exists in universities in the Eastern Bloc. Also what was surprising, as I later found, was that even though BU was trying to guide international students well, they didn't assess what we knew and what we didn't know about the way that Western schools operate. There was an assumption where, 'Hey, you should know at least the basics,' kind of thing. So I ended up taking graduate-level classes my first semester because I had no idea how it worked. Nobody told me you shouldn't do that."

Sources of Support

To cope with the challenges of adapting to American life, many interviewees drew upon various sources of support both in the former USSR and later in the United States. Perhaps foremost among these was their own spirit of entrepreneurship, creativity, and innovativeness, characteristics that had typically been discouraged, if not punished, in the former USSR. Their strengths in those areas could be an indication that they had avoided deep imprinting from that environment, and/or they had experienced new layers of imprinting in the United States, including from educational institutions and work organizations. We first explore support that came from mentors and role models. And because networking was also an especially important source of support, we devote the remainder of the chapter to that crucial topic. We emphasize these sources of support since they helped provide interviewees with a sense of confidence, creativity, and entrepreneurial spirit that could ease their transition.

Mentors and Role Models

Many interviewees credited mentors and role models as sources of inspiration for their own entrepreneurial spirit, creativity, and innovativeness. We categorize them into three groups: parents and grandparents, teachers and professors, and workplace colleagues. In many cases, interviewees did not distinguish between mentors or role models. Thus, we treat the terms interchangeably, recognizing that the identification of the person or persons was more important than categorizing them according to strict definitions. In fact, scholars have noted that there is no consensus around the definition of mentors,[1] and the same applies to role models.[2] We view mentors as individuals who take an active role in another's personal and/or professional development by engaging in active listening and providing guidance and alternative courses of action. Role models provide aspirational examples for individuals, whether the role model is personally known to the individual or not. We had both of these contexts in mind when asking our interviewees about who might have been their mentors or role models, and we did not provide a

---

[1] David E. Gray, Bob Garvey, and David A. Lane, *A Critical Introduction to Coaching and Mentoring: Debates, Dialogues and Discourses* (Los Angeles: Sage Publications, 2016).

[2] Maximilian Anton Maier, *Key Issues in Innovation Management: Supply Chain, Creativity and Organisational Role Models* (Erlangen, Nürnberg: Friedrich-Alexander University, 2015).

distinction to the interviewees. Thus, we treat mentors and role models similarly.

Parents and Grandparents

A large number of interviewees credited their parents and grandparents as being mentors or role models to them. This should not be surprising because parents in the USSR typically took exceptional interest in their children, often because there was only one child in the family, and tried to provide them with a happy childhood and shelter them from the harsh realities of the Soviet system. Also, interviewees who emigrated as children were able to reflect on the difficulties their parents faced during the immigration and resettlement process. The strong influence of grandparents is also understandable since they often lived in the same household and raised their grandchildren while the parents were at work since the Soviet system required most women as well as men to work until retirement. In other cases, children were sent during the summer to live with their grandparents, sometimes a long distance away. Interviewees' parents and grandparents were typically highly educated and often worked in prestigious technical and academic professions. Interviewees expressed admiration and affection for their family members, and credited them with being role models and mentors. Parents and grandparents would likely have been important sources for imprinting the values and behaviors that appeared overwhelmingly positive in interviewees' recollections and were fundamental to the development of their identities.

Social media entrepreneur **Nataly Kogan**, who was born in Baku, Azerbaijan, and came from St. Petersburg in 1989 at age thirteen, spoke glowingly of her parents, including her entrepreneurial father, Semyon, also an interviewee: "I've always been really creative, and you know, that came from my parents. My father is trained as a polymer physicist. That's what his PhD is in, but he's a scientist and he's equally trained in chemistry, mathematics, and physics. So he was an engineer and scientist, and my mother was a pianist and a piano teacher. As a testament to them and to how hard they work, my mom is still a piano teacher and performs sometimes, and my dad is still a scientist who runs his own companies. As my mom says, 'He might have a "disease" of starting companies, and why can't we just have calm jobs?' Kogan summed up her parents' influence on her life: "It's a really interesting question, and I don't know if until recently I had what I would call active mentors.

I would probably say my parents are the two stars in my sky. They are my heroes by every scenario. I worship them both, look up to them both, consider them godly creatures. I really do."

Polina Raygorodskaya, who came from St. Petersburg in 1990 at age four, also credited her parents as role models, her mother having been a documentary film director and her father a marine biologist. She explained: "Maybe both my parents influenced me with their emigration experience. Building a business is very similar to moving to a new country and starting from scratch, and building a family and a career. So it's having exposure to that, being able to see that people can live the American Dream and have anything they want if they work hard and figure it out." Her statement reflects how her parents were able to immigrate to the United States and succeed, and she compared that experience to starting her own online travel booking company, Wanderu. Another interviewee explained how her parents supported her choices: "I think my parents gave me a good base because they believed in my siblings and me and trusted our decisions, giving us the opportunity to fail and not to hold back. I think it gave me a way to think about myself: I'm mature, and I can do anything, whether it's entrepreneurial or something else."

Alexandra Johnson vividly recalled how her parents encouraged her in the many activities she was interested in and instilled in her a love of travel and diverse cultures that influenced her educational and career choices. Born in Vladivostok, an important port in the Russian Far East, she came in 1990 at age twenty. She recounted: "My mom and dad really got me started, but throughout my life there was always somebody who would show me the way. The story goes that at some point my parents had a really tiny apartment and I was sleeping in a suitcase when I was a baby. So that's how I got on the road the whole time. And I know I was conscious of who I was when I was three or four years old. I didn't consider myself – I wouldn't say an ordinary girl – but I always felt like my world was much bigger than other people's because I had the luxury of seeing the enormous country of the Soviet Union first-hand. It was quite unusual. I think I was the only kid in our apartment complex who could have a conversation about all kinds of things because I'd seen it. So I remember all those connections I was making as a child, and I still remember that I connected with people regardless of where they lived, regardless of what their nationality was. I came and had fun with them. That is something which I guess was an imprint since the day I was born, and that's what I continue to care about throughout my life."

Johnson talked about her father as being a huge influence: "He has been instrumental really in shaping a lot of my outlook on life. He put himself through school, became a ship's captain. So he would be gone like nine months out of the year. He had a remarkable presence in my life, and he would write me letters and those letters were so deep. And families would get together at the radio stations, and you could do these radio letters to your father. But then when he was home, it was always a holiday, because we would get on planes and travel around the country. I think I had the best childhood. He obviously was my hero, and I could share anything with him." Johnson also credited her mother as an important influence: "But my mom was the one who made sure that I stayed steady, that I performed. And I think I was really blessed that nobody would tell me 'no' to any of my interests. They were always welcoming to whatever I wanted to do. I really believe that, as a child, the way they raise you, the family, is much more important than anything else. But you know, one of the most important things that my dad taught me was that I can do anything I want in this world, as long as I work hard. And that's what he said, 'do your best and it will happen.' And still, to this day, that helps me so much in my business world because I'm surrounded by men."

Tatiana Saribekian (born Gavrilova), originally from Moscow, who came in 1999 at age thirty-four, shared an amusing anecdote about her parents not taking credit for her ambition, hard work, and achievements that included founding a lumber company and later an investment firm: "My parents were always laughing and saying, 'You're not from us, you're from a neighbor or someone else,' because I have a different personality. My father was very artistic and talented, a very sensitive person. He was not business-oriented at all. And my mom, she had a good career as an economist in the Ministry of Defense, but she was never on the front lines and she always wants to be in a more supportive role. In school, one of my professors said, 'If I could say only a few words about each person, I would say that Tatiana never takes 'no' for an answer.' And I think this is true."

In describing mentors, **Mikita Mikado Teploukho**, originally from Minsk, Belarus, and who came in 2006 at age twenty, began by speaking of his mother: "I think my mom was a mentor when it comes to the very basic stuff that humans have to be taught, like being kind or being a human, basically being a good man. That came from my mom." He clearly had thought about the difference between a parent's guidance and other active mentors: "But I have a lot of people here in the US who are also my mentors. I wouldn't say I had somebody who helped me throughout my entire life and has been a guide for me. But yes, I had a lot of people who helped me. A lot of them."

Many interviewees mentioned their fathers as being primary influences when asked about sources of their entrepreneurial spirit and creativity. One interviewee credited her father with helping build her endurance, recalling that he used to say: "If you fall, you have to get up and do more, and if bad things happen tomorrow, there will also be good things. So you have to let it go." **Eugene Shablygin**, who came from Moscow in 1992 at age thirty-two, credited his father with significantly influencing his career: "I would say that my father influenced my life very significantly because when I was a kid we lived not far from Moscow State University. I remember we were walking around the then relatively new buildings and my father was telling me that this was the place, Moscow State University, which is the best university in the world, and about its department of physics. He said, 'Physics is the best science you can imagine. It's the most important thing.' When you listen to this from a very early age, for me there was obviously no other choice but to become a physicist." His father had been the Chair of the Physics Department at a textile institute, and Shablygin followed in his footsteps, graduating with a degree in physics from Moscow State University.

Dmitry Kerov, who came from St. Petersburg in 2009 at age thirty-seven, echoed that view: "The best mentor was my dad. He was a mathematician who worked for the Mathematical Society of St. Petersburg. He was invited multiple times to Europe, the United States, and Canada to teach in universities and to work on mathematical problems. He was very wise, and he helped me a lot in finding my way and teaching me how to behave." One interviewee whose entrepreneurial father had a direct influence on his own future was entrepreneur **Mark Kofman**, originally from Tallinn, Estonia, who came in 2011 at age thirty: "What kind of business didn't he do? He's done manufacturing of clothes, he's done the furniture business, and he likes retail a lot. You never know where your entrepreneurial spirit comes from, you probably learn from some small things from each of your parents. So, yes, definitely, what your parents do influences you a lot."

Tatiana Novobrantseva, who came from Moscow in 2001 at age twenty-eight, credited both her parents, especially her father: "I would certainly say that my parents had a huge influence on me, particularly in the way they were never aligning with the official version of things. I can give you a very good example. When we were growing up, we were all taught that Lenin was our 'grandfather,' and that's exactly the word that was used, and we would hear lots and lots of stories about our 'grandfather.' But when I would come home, my father would tell me, 'You know, you have two grandfathers, and here are their names. They have nothing to do with Lenin, so don't be fooled. He has no relationship to you.' That was at the age of six or seven."

Some who mentioned their parents also referred to their grandparents. **Eugene Khazan,** who came from Riga, Latvia, in 1991 at age thirteen, recalled: "Obviously, my parents and grandparents. A lot of things in the USSR were a little backward, but actually, one of the things that was not was women's role in society, which is very topical here in the US today. I'm not sure if this is just my family or a broader trend, but both my grandmothers were medical doctors. And this goes back to the generation of the 1920s and 1930s. So I grew up with three of my four grandparents being medical doctors, and for me it was just par for the course that my grandmothers were both doctors." Similarly, **Stas Gayshan,** who came from Moscow in 1992 at age ten, spoke of the importance of both his parents and grandparents, as well as friends: "I would have to say that my parents are my most consistent mentors in life, as important as mentors in business. And I think also my grandparents. I go back to family for the most consistent things. I think with the family, they've always been there, and they've been enormous shapers of who I was as a kid and who I am as an adult. And then for individual subjects, I have a lot of friends whom I have really leaned on over the years to give me advice, good and bad."

Some interviewees specifically mentioned their grandmothers. **Greg Rublev,** who came from St. Petersburg in 1993 at age fourteen, recalled the trauma that his grandmother endured during World War II: "I think my grandmother was kind of a role model. She lived through the blockade of St. Petersburg. She was a very determined person." It was not lost on Rublev that surviving the 900-day siege of Leningrad by the Nazis during World War II was a display not only of courage but also the will to survive, a lesson that apparently influenced Rublev's career as a serial entrepreneur. **Anton Manuilov,** who came from

St. Petersburg at about the same time as Rublev, in 1992 at age fourteen, also mentioned his grandmother: "My grandmother was always
there for me as I was growing up. She was a medical nurse. I also had
my mom's friends who were medical doctors, and I was just around
them, and I was really thinking about pursuing a medical degree
back then. So they were my mentors." There was little doubt that
these role models greatly influenced Manuilov's choice of a career in
life sciences.

Iryna Everson (born Yurchak), who came from Kiev, Ukraine, in
2008 at age twenty-six, also mentioned her parents' friends as being
role models: "I would look at some of my parents' friends and the
ones who were successful and the ones that I liked. I realized that the
ones I really liked were successful and confident. So I'd ask them what
school they went to and why. And I thought, 'I want to go to that
school, too.' So I would say this person who was a family friend kind
of inspired me. He probably doesn't know about it, but he's more like
a role model, not a mentor. I didn't have a mentor as a teenager,
unfortunately." Everson's comments showed her understanding of
the difference between a role model who probably was unaware of
his influence on her versus having a mentor, whom she seemed to
visualize as being a more active participant in her personal and professional development.

In this section, we have presented experiences and stories from
interviewees about how their parents and grandparents influenced
them personally and professionally. Their stories varied, as would be
expected, but contained the same thread of the importance of family in
their development. Also as expected, interviewees focused primarily on
the development of their values and character that formed the foundation of their creativity and, at times, entrepreneurial spirit. Some noted
the importance of family members in their professional development,
primarily as scientists but also as entrepreneurs. We continue exploring
important role models and mentors, now focusing on teachers and
professors.

Teachers and Professors

Teachers and professors were another group recognized by interviewees as having had important influences on their creativity and spirit of
entrepreneurship. These included teachers and professors in the former
Soviet Union and professors in the United States. All of these educators

were seen as having provided not only knowledge but also inspiration and support that became important during their time together as well as later in life. This recognition is a testimony not only to their positive educational experiences in the United States, but also to the excellent education they received in the former Soviet Union at both early stages and during their university experiences. As noted in Chapter 3, these experiences often provided them with academic capabilities to succeed in the US innovation economy. Many interviewees were also fortunate to receive superior education in leading American universities, including Stanford University and the University of California in the San Francisco Bay area, and MIT and Harvard in Boston-Cambridge. Recognition of these various influences and advantages will be seen in their stories. In a more or less chronological order, we will first present their comments on mentors in their earlier education in the former Soviet Union followed by their university education there, and finally, their educational experiences in US universities.

We start with **Vadim Gladyshev**, originally from Orenburg, Russia, who came in 1992 at age twenty-six. He recalled the influence of a dedicated high school chemistry teacher: "From early on, I participated in olympiads, science olympiads, and I would say the major influence on my becoming a scientist came from my high school chemistry teacher. I could never forget the way she was teaching us. Her teaching was just amazing. She would pick one person from each grade who was the best in chemistry and would form a club. After classes we would do some experiments. She would cook for us, bring us to her home, and it was like a small family. She would bring us popular science books and various articles. So we would go as a team to the regional olympiads. We would all be first place in every grade. It was just a regular school, nothing special, but her teaching was just amazing. So I finished high school with a gold medal. And then I went to Moscow State University. From Orenburg, it was not a typical kind of transition because Moscow State is the best university in Russia, and so very few people could go there."

Greg Rublev, who came from St. Petersburg in 1993 at age fourteen, recalled a valuable lesson he learned from his high school biology teacher: "I had a biology teacher who said something that just stuck with me. He said, 'An honors student knows what they don't know.' I thought that was really interesting, so if you want to be in an honors class, you have to understand the big picture and know what you don't know. If you need to know it, you can go back and look it up. That's how I understood it. And he gave me a C minus. There was no D in that class. It was either a C or an F, so that time he gave me a C minus. That was like a second chance, and I had really good grades after that. So he was a role model for me."

Professors back in the former Soviet Union were also remembered by a number of interviewees as being mentors or role models. **Mike Sandler**, who came from Moscow to Canada in 1991 at age twenty-six and later to Silicon Valley, recalled two professors at Phystech particularly: "They both were very well-known scientists related to the space industry. One of them went into medicine later, while the other was still in the space sector. He recently visited San Francisco where he had a project with Berkeley. I guess they're sending a space mission to the sun." Another person who mentioned a rocket scientist in the former USSR was serial entrepreneur and investor, **Max Polyakov**, who came from Zaporizhia, Ukraine, in 2012 at age thirty-five: "My major mentor was Dr. Khamin who was a famous rocket scientist I met in Ukraine. He was one of the top three Soviet rocket scientists. He invented ballistic missiles to launch small satellites, mostly military ones. But he became scared about all that. It doesn't change the world for the better. One lesson he taught me is that the world you leave is the one you create yourself. He taught me to look back and understand what was your rocket that you built. Not literally a rocket, but what you create, big steps. He became head of science in one of my organizations and is now seventy-seven years old."

Alexey Eliseev, who came from Moscow in 1992 at age twenty-seven, stated that his main mentor was his PhD advisor in whose chemistry lab he worked at Moscow State University. Eliseev continued his postdoctoral education in the United States at Emory University and later earned an MBA at MIT's Sloan School of Management. He continued to collaborate with his mentor through the early 2000s, even after his mentor became a professor at a university in Mexico. Their relationship was a long-term one: "We actually built a relationship that lasted for decades." In contrast, **Alexey Wolfson**, who came from Moscow in 1997 at age fourteen, spoke of the one who influenced him most as being a person we would describe as a role model: "He was not my professor, but I had gotten into the best department, the biology department at Moscow State University. The chair of molecular biology was Alexander Spirin, who is a famous Russian scientist. I believe he had a tremendous influence on me." Wolfson became a research scientist at the University of Colorado, Boulder, where he received his executive MBA degree that was likely a harbinger of his successful career as a biotech company founder.

Sergei Burkov, who came from Moscow in 1989 at age twenty-one, mentioned fellow students as mentors back in the USSR: "First I think of other students who were a couple of years older than me when I joined one of the research groups. It was at the Landau Institute of Theoretical Physics, and these were the smartest people I ever associated with." Coincidentally, **Dennis Bolgov**, originally from Astrakhan in southern Russia, who came in 1998 at age twenty-five, said that Landau himself was his role model: "Landau is pretty much considered the father of Soviet physics. Role models, it's tough to say. I would say, however, that in my childhood I read a biography about Landau's life, and it made a great impression on me. So I would say he was my role model."

A substantial number of interviewees mentioned professors in the United States as being important mentors. Many such interviewees came as children or young adults, the latter typically on F-1 student visas to attend university, often graduate school. **Timur Shtatland** came from Kiev, Ukraine, in 1988 at age twenty to attend MIT where he graduated in chemistry. He later attended graduate school at the University of Colorado, where he obtained a PhD in biology. He noted: "All three professors with whom I studied there were great. The one that I ended up working with, Larry Gold, is amazing. He is one of those rare people in science who combines enthusiasm for science, a great scientific mind, and business acumen. He's super smart. So he is all of that in one person. He is also, last but not least, a great person. I wish I could be as generous as he is." **Leonard Livschitz**, who came from Kharkov, Ukraine, in 1991 at age twenty-four, noted that a professor at Case Western University, Ken Loparo, was an influential mentor to him: "My English comprehension was pathetic, absolutely pathetic, and I failed the language competency test to become a teaching assistant. But, you know, Professor Loparo awarded me a research assistantship. He gave me all these interesting topics, and he put me in touch with his colleagues. I got probably one of the best research topics I could ever have dreamt about." And one of the Wave One interviewees, **Alexander Gorlov**, who came from Moscow in 1976 at the age of forty-five, noted the importance of his department chair when he was a faculty member at Northeastern University: "Periodically, I worked with members of our department. The chairman, John Cipolla, worked with me and coincidentally nominated me for the Edison Medal that I was fortunate to receive." The medal is the most prestigious award given for innovations in electronics and electrical engineering in North America and is awarded annually by the Institute of Electrical and Electronics Engineers (IEEE).

Another interviewee who credited a professor in the United States was **Anna Lysyanskaya** who came from Kiev, Ukraine, in 1993 at age seventeen. After graduating from Smith College, she had a positive mentoring experience in the PhD program at MIT: "I found mentors at MIT and particularly Ron Rivest, but not just Ron himself. It was the whole Theory of Computation Group there, and it was part of what was then called LCS. Actually, Ron who was my PhD advisor, had a lot to do with putting this group together, and so the reason it was such a wonderful, warm group is in part because of him. But it was also a group of people who are intellectually curious and just wanted to solve problems, and everybody's door was always open and everybody was ready to contribute. It was just sort of a community, and I was very lucky."

A very positive experience was recounted by **Alexis Sukharev**, originally from Grozny, Russia, who came in 1990 at age fifty-four after having been a professor of mathematics at Moscow State University: "Meeting with Joe Traub was a turning point for me. He really made a difference in my life and in starting my business. Joe Traub was chairman of the Department of Computer Science at Columbia University, and he became chairman of Computer Science and Technology at the National Research Council." **Ilya Strebulaev**, originally from Moscow and who came in 2004 at age twenty-nine after receiving his PhD from London Business School, described the mentor he found there: "I obviously worked hard, but I was also lucky because I met my advisor, Professor Stephen Schaefer, who is now my close friend. I was his assistant for the first year and then he became my advisor. And since then he's been very important to me."

In many cases, there was overlap across the three groups we noted at the beginning of this chapter, as well as within the groups, as when both parents and grandparents were mentioned by interviewees. In this section on teachers and professors, both in the former Soviet Union and the United States, we present several more such multiple influences.

Yury Lifshits, who came from St. Petersburg in 2007 at age twenty-three, recounted: "So my second 'father' would be the person who was leading my early childhood mathematical training during those six years in my math circle." As noted in Chapter 3, Lifshits went on to win two gold medals in the International Mathematics Olympiads. After receiving his PhD in theoretical computer science at the Steklov Institute of Mathematics in St. Petersburg, he studied at the California Institute of Technology as a postdoctoral student, where he had an influential mentor. Lifshits stated simply: "Professor Bruck is my third 'father.'" Undoubtedly, both his mathematics coaches in Russia and Professor Bruck of Caltech were instrumental in Lifshits's work as a research scientist at Yahoo and his subsequent career as a serial entrepreneur.

Similar circumstances awaited **David Gamarnik** who came from Tbilisi, Georgia, in 1990 at age twenty-one. He described the influences of two mentors in the context of his own philosophy of mathematics research, the first in the USSR and the second in the United States: "I've been surrounded by great people, and I think the common thing among all my mentors, and I have several, is their complete, uncorrupted love for research. That uncompromised love of research for research's sake, that difference was very engaging for me, very attractive to me. So it's really given me a creative feeling rather than just a career. You know, that's the type of people with whom I've been lucky to interact and who have mentored me starting with my advisor in Georgia, who is now in Israel. He just had an absolute love of mathematics for its own sake. I caught the fire, as we say, and then when I came to MIT, that love resurrected in me primarily because I started interacting with my advisors and felt the love of research for research's sake, and pure curiosity, and the drive and desire to achieve something. It was not just working toward tenure and making sure one would get tenure. So working, doing research for research's sake, ambitions, that's what I found in my mentors, and so I got the fire from them."

Workplace Mentors

In addition to family members and educators at various levels, interviewees cited managers, coworkers, and others in their workplaces who were instrumental as mentors and role models. Entrepreneur **Vlad Pavlov,** who came from Dnepropetrovsk, Ukraine, in 2010 at age thirty-six, helps explain this symbiotic relationship. He noted that the process is a two-way street requiring not only receptivity by the receiver, but also the capability to recognize, merit, and utilize that important gift. He recounted: "Of course, I do have mentors. I have good mentors. For example, at my current startup, my advisory board includes Bob Iannucci who was a senior vice president and chief technology officer of Nokia, and Bjarne Stroustup who is the author of C++ programming. Also, Vivek Wadhwa. I don't even know how to explain who he is to people who don't know him, but if you read his papers or just heard about him, there is no need to explain. Yes, I have good mentors. But it's a question of what comes first: the chicken or the egg? I am who I am because I have good mentors. And I have good mentors because I am who I am." Pavlov adds an important nuance to mentoring, noting that one must be open to mentoring and also worthy of having good mentors. His insights explain well the two-way street involved in the mentoring process.

Another interviewee who answered in a philosophical tone was **Andrey Klen**, who grew up in Smila, Ukraine, and came in 2014 at age twenty-six: "I think I got a lot from my friends, and I got a lot from people I work with. I just observe and watch how they do things and try to take the best from them and try to learn the best from them. Communication is crucial, meeting new people is crucial. Meeting the best is crucial because you raise the benchmark to the level those people achieved, and you try to follow their steps and you try to fill their steps." **Sten Tamkivi**, who came from Tartu, Estonia, initially in 1994 at age sixteen, stated: "I haven't had any formal mentors. I've worked for some really great people, and they've given me a lot of space. When you work for Niklas Zennstrom in the early days of Skype, you don't see much that you would call mentorship. You work with people, and you sort of learn from them along the way, but you don't always see it. I don't have clear role models as if it was like, 'OK, there's this guy I want to imitate and want to be more like.'"

Kate Torchilin came from Moscow in 1993 at age twenty-two with a degree in chemistry from Moscow State University. She then earned a PhD in biochemistry from Tufts University, followed by an MBA from Harvard Business School. Her primary influence was her father, Vladimir Torchilin, also an interviewee, who is an internationally renowned professor of pharmaceutical sciences at Northeastern University. She said of him: "He's a role model in any sense of the word, for pretty much anything, a universal influence on my life." She then mentioned another professor as also being a role model: "I think in terms of professional role models, Dr. Bachovchin, my professor in the doctoral program at Tufts University, was really good. I remember that I had actually read some of his work and that was the main reason I applied to be in his lab. There was an absolutely groundbreaking paper he wrote that was published when I was still in Moscow that was really rethinking an important area of enzymology. I reached out to him, and he helped me arrange an interview, and he was very open minded and creative and entrepreneurial himself."

Ilya Yaroslavsky credited workplace colleagues as well as a professor. Originally from Saratov, Russia, where he earned a PhD in laser physics at Saratov State University, he came in 1998 at age thirty: "A professor at Louisiana State University named Harold who I worked for introduced us to the American culture and way of life, and he opened our eyes in many ways to how we look at the country and the society." Yaroslavsky also credited colleagues in two companies where he worked: "The founder of Palomar Medical Technologies and the former CTO, Mike, helped me to become acclimated to New England, which was obviously a successful experience since he rose to become vice president of advanced research before leaving the company after it was acquired." Again showing his appreciation and willingness to learn, Yaroslavsky spoke of a coworker at his new company, IPG Photonics: "Gregory was very important for me as a teacher because he had also been in a situation where he was formed as a person and a professional in the Soviet Union and then came here and made a successful career. He was an example of this tradition for me, and he taught me a lot in many different ways." These comments reflect Yaroslavsky's recognition of the forces of imprinting that many like himself experienced back in the former USSR that often had to be recognized and dealt with in the new personal and professional environment of the United States.

Many other interviewees identified people in various roles in the workplace as being mentors and role models. **Alex Petetsky**, originally from Berdychiv, Ukraine, who came in 1991 at age twenty-four, clearly differentiated between the two types of influence: "My father was my mentor. He was well respected, very handy, and perhaps more loved by my mother's family than by his actual mother. He was an amazing man." He then referred to a colleague at work as being a role model: "I worked with one guy, he is an Indian, one of the smartest people.

There are people that have certain smarts, like they're very good at math, they're very good at physics, they're very good at something. He was different, he was very good at everything, a very well-rounded executive-type person who I learned a lot from. We worked together at a leasing company, and we still keep in touch. It's not what he taught me, it's what I got from him. I learned how to present myself at meetings. I learned from him how to get your point across. I would just look at his demeanor and how he would walk into the room and what he would say and how he would react to certain things. I'd think: 'You know what, this is interesting, this makes sense, I've got to try this.' He was more of a role model than a mentor. I don't think I ever had a mentor at work."

San Francisco venture capitalist **Alex Gurevich**, who emigrated from Moscow in 1990 with his family at age seven, noted: "I consider my partners as my mentors. First and foremost, they have been really good about taking me under their wing and mentoring me and giving me feedback. So they're probably my first line of mentors, which is good when you work with people you consider to be your mentors."

A different twist regarding workplace mentors was provided by software entrepreneur and investor **Grigore Raileanu**, who came from Moldova in 2012 at age twenty-nine. He recounted the importance of a customer in his professional growth: "One of the customers of my first company, Remsys, runs a successful business in the United States. He came from Russia or Ukraine, I don't know exactly which, but he moved to the US around 1990, like twenty-five years ago. Yet he managed to grow his business, and he had sold some businesses in the past, so he has been very successful. We often meet at different conferences, and I ask him for his opinion on products, business ideas, or how I can attract certain customers. So, yes, I consider him an advisor, and he still helps me."

Two other interviewees had numerous mentors, finding them exceedingly helpful. Entrepreneur **Alexei Dunayev**, originally from Kiev, Ukraine, who came from New Zealand in 2007 at age twenty-six, said: "I've had countless people mentoring me. I would do a disservice by even mentioning some, just because there are so many. People in the startup community, in the technology space, are very open. They're very interested in helping, and they're also very knowledgeable. It's really a case of drawing upon people's expertise and reaching out to people for advice and for mentorship. So lots and lots of people have helped me in countless ways." Similarly, **Umida Stelovska** (born Gaimova), originally from near Samarkand, Uzbekistan, who came in 2009 at age thirty, recounted lightheartedly: "For me, just about everybody who speaks English was a mentor." More seriously, she added: "That's really true, because that's how I learned. But I was also really lucky. I am able to discuss things with people who have really made it, who are behind multibillion-dollar companies. What I do is go to lots of conferences, meeting people, and approaching people."

The experience of being mentored was not only valuable in itself, but was also a cultural adjustment for many interviewees. **Iryna Everson** (born Yurchak) learned much from her mentor and, in fact, from the Upwardly Global organization to which she was so grateful for helping develop her skills to secure employment and function well in the work environment. Such adaptation almost always included the necessity to reinvent oneself. In a very real sense, this required, in many cases, erasing or modifying imprinting from their home countries and adapting their identities to a new cultural situation, a process undoubtedly enhanced by the mentoring they received. The lessons we take from our interviewees are that mentors and role models can be found in many circumstances and that individuals are receptive to and appreciative of such people for their guidance and inspiration.

Networking

We highlight networking as potentially the most important type of support for adapting to a new environment. Networks can be valuable sources of social support and convey norms of social identity.[3] Networks involve not only a person's relationship with others, but also the multifaceted relationships among people having mutual interests, backgrounds, or other professional or social bonds. Networks are one of the most powerful assets for individuals as well as their firms by offering the potential to gain access to valued resources including power, information, knowledge, advice on tasks and careers, and access to capital.[4] Networking involves interacting with individuals

---

[3] Joel M. Podolny and James N. Baron, "Resources and Relationships: Social Networks and Mobility in the Workplace," *American Sociological Review* Volume 62, Number 5 (October 1997), 673–693.

[4] Daniel J. Brass, "Being in the Right Place: A Structural Analysis of Individual Influence in an Organization," *Administrative Science Quarterly* Volume 29

and firms with whom one has established a relationship or strong ties. Those relationships can provide entrée to broader networks of weak ties that could give access to even greater levels and different types of resources.[5]

Networking was an essential skill in the former USSR, and virtually all of our interviewees, except those who emigrated at a very early age, would likely be experienced in utilizing that practice there. People drew on their tightly knit and highly trusted networks of strong ties with family and friends to gain valued resources. Those strong ties then introduced individuals to their family and friends, who then became connected with weak ties to those individuals, the whole process extending an individual's overall network to gain information and access to a broader range of influentials. A likely challenge, however, would be to adapt that skill productively in their new American environment. This could well require adaptation to new norms of networking, again requiring some modifications of the imprinting underlying that potentially useful skill. Individuals might accomplish this through their own initiative or through the activities of organizations like American Business Association of Russian-speaking Professionals (AmBAR), the Global Technology Symposium, and Silicon Valley Open Doors (SVOD) in Silicon Valley, and the US–Russia Chamber of Commerce of New England (USRCCNE), New England Russian-speaking Entrepreneurs (NERSE), and other such organizations in the Boston-Cambridge area that provide introductions and connections to a more diverse network.

We sought to determine how these professionals utilized networks in work and social environments. We also wanted to uncover the extent to which these networks generally consisted of people from the former USSR, the United States, or other countries. Our major conclusion is that their personal or social networks included primarily Russian-speaking individuals from the former USSR, while professional networks were far more varied. We see this as not surprising and, in fact, quite positive. Networking with others speaking a common language and sharing similar cultural experiences provides a measure of

(1984): 518–539; Tom Elfring and Willem Hulsink, "Networks in Entrepreneurship: The Case of High-Technology Firms," *Small Business Economics* Volume 21, Number 4 (2003), 409–422.

[5] Mark S. Granovetter, "The Strength of Weak Ties," *American Journal of Sociology* Volume 78, Number 6 (May 1973), 1360–1380.

familiarity and comfort in a social environment that eases the demands and challenges of a new culture for immigrants. Also, many of our interviewees had families, and this type of network could certainly ease the adjustment process for their families.

On the other hand, associating primarily with one's own cultural group could limit workplace adjustment and career progress. Thus, broadening one's network and associations in the workplace and broader business environment could be beneficial in developing language and communication skills as well as teamwork, general trust-building, showing responsibility, and gaining insights into the local culture, including norms of ethical behavior. Many of these skills that would be acquired through a broader, more culturally diverse network could often lead to finding mentors who could facilitate the adjustment needed for career development, which could include advancement not only in the requisite soft skills, but also in vital hard skills. In presenting the stories in this chapter, we focus on interviewees' professional networks, first discussing those that predominantly involved people from the former USSR. We then examine professional networking, with networks providing connections with Americans and those of other nationalities. The final section discusses the founding of networking organizations in Silicon Valley and Boston-Cambridge that focus primarily on Russian-speaking professionals.

We begin with the words of **Dmitry Kuzmenko**, originally from Ukraine, who recognized the importance of networking while working in Latvia as well as in the United States: "Latvia is totally reliant on networks for doing anything. I was looking for a software developer job back in Latvia, and I got that job. By the way, it was very easy. I had just graduated, and I sent an email to my friends. The same evening, I had a beer in a bar with one of them and told him what I wanted to do. The next day, I had an interview, and the following day I started. Looking for a job is much simpler and much less formal in Eastern Europe and the FSU [former Soviet Union] countries." Kuzmenko also noted the importance of networks in the United States in applying for jobs, which initially surprised him: "Here a network is very useful, too, as I learned when I left Microsoft. I had three months of just floating around and wanted to take time to think. When you apply through friends and acquaintances, even if you don't know them very well, they kind of push your case and often personally hand your resume to the hiring manager. That usually gives you at least a 50 percent chance of getting a phone interview. So the level of reliance on networks is substantially different. In Ukraine, you just can't survive without networks. Otherwise, perhaps you might get a very basic job with a limited career. The United States is a country where you can climb the ladder with your own hard work, even if a network isn't there. However, a network still can help a lot as it significantly speeds up the job search."

Professional Networking Primarily with People from the Former USSR

A number of interviewees noted that their professional networks were comprised primarily of Russian speakers. **Dmitry Skavish**, who came from Yelets, Russia, in 1999 at age twenty-nine, said of his professional networks: "Well, it's like 60 percent Russian speaking and 40 non-Russian speaking. The people I communicate with on Skype or in person in my business, most of them are Russian speaking." Like many others, Skavish's involvement with a Russian-speaking network was based on the people he dealt with in his startup, Animatron, that included programmers back in Russia. Another entrepreneur, **Igor Gonebnyy**, originally from Krasnoyarsk, Siberia, and who came in 2014, stated: "My network is definitely Russian speaking. I need the people in the Russian-speaking community because if you want to play in the local market, you have to talk to local people because there are too many options that you don't know about if you're not in the system." His statement lies at the heart of networking, which is very much an information system that can be critical to one's professional career and business. And because he was new to Silicon Valley, it was natural for him to turn first to the Russian-speaking community for guidance. **Eugene Baron**, who came much earlier from St. Petersburg in 1990 at age twenty-one, recalled: "In my last job, I think one of the reasons they took me is because I speak Russian and they have a development team in Ukraine. The person who was interviewing me was from India. This was a big plus so they were really excited." Baron also recalled a more deliberate instance of getting a job through his Russian-speaking network: "I self-contracted with an outsourcing company owned by a friend of mine, and the whole development team was Russian speaking. I was flying to St. Petersburg on business trips. It was much more fun to work with a Russian team than with a non-Russian one."

Entrepreneur and investor, **Evgeny Medvednikov,** who came from St. Petersburg in 2014 at age thirty-four, also had many Russian-speakers in his network: "I don't select them particularly. I think it's just because of my connections. If I had more connections with Americans, I would probably invest in companies founded by Americans. So I don't especially choose Russian natives, but it happens." **Davit Baghdasaryan,** who came from Armenia in 2008 at age twenty-four, also demonstrated an understanding of the cultural foundations of networking: "Most of my professional network is comprised of mainly Armenians, just like my social network. Most of my friends here are Armenians and are professionals. They came here after the Soviet collapse, so mostly within the last twenty years, and even the last ten. I guess coming to the US has become easier, and all of them are very talented software engineers. So, yeah, we are very connected."

An interesting perspective on networking with professionals from the former USSR was provided by **Yelena Kadeykina,** who came from Moscow in 1997 at age eighteen and became a citizen eleven years later. She is the founder and CEO of two companies involved in education and professional training. Startup Access is a Boston-Cambridge firm that provides consulting services to entrepreneurs and investors from the former USSR who are seeking to bring their companies to the United States or to join an American company. Kadeykina explained why such a service is so important: "About 50 percent of the US-based people who help with this initiative are Russian-speakers. One of the reasons is that I like to help develop our community. I believe it is the right thing to do. The Russian-speaking community in the US is often more challenged than others because we are not sufficiently united. Look, for example, at immigrants from India – it's a very different story. Most of the people who have gone through my programs are very hardworking and bright, and sometimes all they need is experienced advice at the right moment. My other project, Hermiona Education, has a similar goal: to help striving individuals from the former USSR realize their own American Dream by pursuing educational pathways in the United States. We have developed a network of successful new Americans from the former USSR who have graduated from top American schools and universities and are excited to leverage their experience for the benefit of their countrymen through long-term mentoring partnerships. I believe that building a community of people with common roots works well for many types of businesses."

Professional Networking Including Russian-Speakers but Not Predominantly
Some interviewees conveyed stories that people from the former USSR were included among their professional network members, but were not the primary members. **Eugene Boguslavsky,** who came from Minsk, Belarus, in 1989 at age fifteen, commented: "It really depends on the workplace, and I work with Russians all the time. Still, it depends on the situation. If I'm in touch with somebody who I think is Russian, and if we have time to chitchat, I try to find out about their backgrounds and where they came from, just to kind of build a sense of camaraderie, if there is time for it. In most of my daily dealings, though, lots of stuff is happening so fast, so mostly there is little time for personal chitchat because we're trying to get through an operational issue, or we're trying to get a new software release out." His comment on time being an obstacle was noted by a number of others. Entrepreneur **Katya Stesin,** who came from Moscow in 1991 at age twenty-three, commented: "I don't know. I never focused on Russian-speakers, but I do have a lot of Russian-speaking connections, usually from Russia, because they wanted to connect through LinkedIn. So I have a lot of them who I know, but I also know a lot of VCs [venture capitalists] who have nothing to do with Russia." Like Boguslavsky, Stesin's network connections go well beyond those from the former USSR.

As a representative of interviewees involved in large US corporations, **Sergey Markov,** who came from Moscow in 2007 at age twenty-eight, stated: "At Facebook, I do not have any Russians on the ad team, but we have a few Russians on the Instagram team, and we work very closely with other Facebook ad teams that also have quite a few Russians." Putting the potential networks in context, he added: "I think Facebook has about 550 Russian-speaking people, so that's quite big, and most of them are engineers. There was a similar group at Microsoft when I was there of about 700 or 800 people." Like the next two respondents, it was inevitable in a large US tech company that Markov included Russian-speakers among his professional network members, but his network extended well beyond that group. Similarly, **Anton Rusanov,** who came from Karaganda, Kazakhstan, in 2012 at age twenty-eight, commented on Russian-speakers at Google, noting their use of email to communicate among that group: "I do subscribe to their mailing list and read it daily. I respond when I can give valuable input or advice to someone asking questions. I've been to a couple of events organized by people in that group that they call Russian-speaking social hours. But I don't do it that often because I have a family and I prefer spending time with them." In a similar vein, **Dmitry Kovalev**, who came from Moscow in 2008 at age twenty-four, responded: "There is a big Russian-speaking Cisco community. There is a special mailer for that group, and they do it regularly, just to mingle and socialize. I don't often attend those events, but at the same time, I do have my own network of Russian-speaking peers who I see quite regularly." Kovalev's comments add to those of others who note workplace professional networks, both Russian-speaking and also non–Russian-speaking networks. Two themes emerge from these interviewees' comments: the existence of large Russian-speaking groups in major US tech companies and the time pressure limiting their use as networks. **Valentin Komarovsky**, who came in 1995 at age fourteen, recounted another interesting workplace perspective: "I'm from Russia. Another man here is from Azerbaijan, and we also have people here from Ukraine. But our network engineer is from the US, and he was born here and doesn't speak Russian or understand it. But he likes the culture, and he's a brilliant man, but doesn't really understand the culture to deal with many of the Russian-speakers here." He ended with an insightful comment: "But I really think it's more about the people. You get good people, and they help grow your business."

Professional Networking Primarily with Non–Russian-Speakers

The preceding sections illustrated that many interviewees chose other Russian-speakers predominantly or as important members of their professional networks. Others networked primarily with non–Russian-speakers for various reasons. One group had come to the United States at an early age and received much of their education there. A second group basically said they never had the desire to seek out Russian-speaking professional networks or that they even deliberately avoided them because of their deep dislike for the former USSR, typically in a political sense, while one interviewee felt that the Russian-speaking community in the United States was not welcoming to him. A third group noted either explicitly or implicitly that the nature of their jobs led them not to seek out Russian-speakers.

An example of the first group is **Anya Kogan,** who came from St. Petersburg in 1990 at age eight. She recounted: "I want to participate in some Russian-speaking events at Google, and I almost went to one last week. I didn't and often don't because I get kind of embarrassed since my Russian is good enough to fool some people, meaning that I have some fluency for sure, and my accent is pretty good for someone who grew up mostly in the US. But, at a certain point, it's very hard for me to keep up. My Russian's just not good enough. And, more importantly, I don't understand Russian jokes anymore." Another interviewee who came at a very young age expressed somewhat similar views. **Alexander Chekholko,** originally from Krasnoyarsk, Siberia, who came at age eleven in 1994, expressed little interest in networking with other Russian-speakers: "It's not really a thing I'm looking for. If I want to hang out with someone, I want to have some shared interests. And other than our similar backgrounds, I don't see that now with other Russians. Still, at meetings at work, when I see people with Russian names, I might nod and smile and go up and chat with them afterward. But overall, it's not really a big deal for me."

The second group who indicated that they either did not have the desire to seek out Russian-speaking professional networks or deliberately avoided them had distinct reasons for doing so. **Michail Pankratov,** who came from Moscow in 1974 at age twenty-six, stated: "I have very limited contact with Russian-speakers. As I said, I came at a time when I kept telling everyone that I left Russia or the USSR because I didn't like anything, including people who made it happen. You see them living under that system and praising their leaders, so I've really stayed away from most of them. We do have some Russian friends, but very few. And even professionally, it happens that I am involved in and engage with some Russians, but it's not because they're Russian but because they need my services and I'm interested in what they're doing. But otherwise, no, I'm not really involved with any Russian network." His perspective is all the more understandable if we hark back to his story in Chapter 2 about his harsh life growing up in the former USSR and the discrimination that he faced. A somewhat similar story was recounted by **Vladimir Torchilin,** who came from Moscow in 1991 at age forty-five: "I know there are some Russian-speaking networks, but no, I don't join them, although I get invited to be involved. I don't want to. I just don't want to. One is called the Russian-American Scientists Association. I run away from them because I don't like to work for the people who might be behind them, who I don't like."

The largest group of interviewees who said they did not deal extensively with Russian-speaking professional networks almost always tied it to being the result of their business circumstances and activities. Most noted that they were open to being involved with Russian-speaking professionals but that it was not a priority. An interviewee who came in the early 2000s recounted: "I would say my professional networks are not Russian-speakers, but personally and socially they are mostly Russian-speaking. It's probably a cultural thing, where you relate better to someone who speaks your language, and I can open up and express myself equally in either language. So it makes no difference to me anymore. But there is something comforting about being able to speak in Russian. And sometimes I think I'm better understood. But professionally, with the experience I've accumulated as I've moved along, I don't distinguish much because it's all about the other person and who they are." Similarly, **Andrey Kunov**, who was originally from Zhekazgan, Kazakhstan, and came initially in 1994 at age twenty-two, noted: "My business associates are not primarily Russian, while my social network is. They're very different, like having friends and running a business are very different things. You'd better not mix friendship with business. That's another painful lesson I learned from my life. If you want to keep your friends, don't make them your business partners. It's a dangerous way of both running the business and keeping friends." **Mike Sandler**, who was originally from Moscow and came to Silicon Valley after emigrating to Canada in 1991 at age twenty-six, noted that the way his business was growing, he was becoming much more involved with non-Russian speakers: "We're now hiring more non-Russians because we were previously involved mostly in R&D where Russians were excellent. Now our company needs marketing, sales, and technical writing, and that's not where Russians' strengths are." **Natalie Hill**, who came from Minsk, Belarus, in 1995 at age twenty-one, responded: "Well, professionally, I don't seek out Russian speakers so much. My profession is really, I feel like it's all English. I'm in computer science programming, and I feel that when I'm at work, I live in a non-Russian world. It's even hard for me to switch to Russian when I speak about my job because I've learned all of it only in

English, so it's a hard translation for me to make on the spot. So, professionally, I don't really have a Russian-speaking network."

The previous interviewees indicated that their professions and job orientations were not conducive to, and certainly not dependent on, establishing or taking advantage of Russian-speaking networks. Rather, they worked primarily with collaborators, colleagues, or clients who could add value to their businesses and careers. Our last such comment is from **Mikita Mikado Teploukho**, also originally from Minsk, Belarus, who came in 2000 at age twenty-six. He described his professional network: "It's mostly international people, Silicon Valley people. There really isn't one nationality. American is probably the dominating one, but there is really not one in particular. There is a bunch."

In summary, interviewees who were not reliant on or heavily involved with Russian-speaking professional networks had varying explanations for that situation. For some, it was a matter of having come to the United States at a very early age and being educated primarily in the American system. For a few, it was a conscious decision not to engage with such networks. Most, however, were those whose businesses and careers led them to be involved with a broader network, sometimes heavily American, but for some, even more international, with people from many nations. The nature of professional networks itself is essentially pragmatic in that it is usually aimed at advancing one's career, business success, or even job prospects.

Many found that these goals were best accomplished through a predominantly Russian-speaking network. Others relied on a more diverse professional network, noting that Russian-speakers were not prominent, although they often added that they were important in their social networks. This distinction should not be surprising since members of immigrant communities tend to seek support from others of that community. But, as our findings show, this tendency can begin to dissipate, professionally at least, as an individual becomes more established in his or her career and business undertakings. This often leads to becoming less dependent on one's ethnic network and evolves toward a more diversified and eclectic network of individuals who become more important to career and business success. The same tendency was seen in Silicon Valley and Boston-Cambridge organizations. We cover those organizations next.

Founding of Professional Networking Organizations

By the early 2000s, the number of Russian-speaking professionals in Silicon Valley had reached a critical mass. That development led to the creation of networking organizations for Russian-speakers in the business community, life sciences, and academe. Over time, some organizations evolved to include participants regardless of ethnicity. These networking organizations include the Russian Technology Symposium, AmBAR, SVOD, the TEC Club, and an Armenian entrepreneurs' group, all based in Silicon Valley. In the Boston-Cambridge business and scientific communities, formal networking organizations include USRCCNE and NERSE. Regarding academic organizations, leading ones are the Russian-American Medical Association (RAMA) and the Russian-American Scientists Association (RASA). We turn to interviewees who had been involved in the creation of such organizations to relate stories of their founding and evolution.

Silicon Valley Networking Organizations

One of the earliest organizations directed toward the Russian-speaking business community was the Russian Technology Symposium, cofounded in Silicon Valley in 2003, by **Alexandra Johnson.** The annual investment conference on venture capital, technology, and entrepreneurship initially focused on doing business with the Soviet successor states. Over time, the organization was renamed the Global Technology Symposium, having evolved to serve a more broad-based clientele of many nationalities, particularly those from emerging markets. Johnson explained: "Well, first of all, I think it's really important for Russian speakers to have centers. After I'd been in this country for about fifteen years, I think I didn't need to prove anything more. That's why I wanted to look into something that would show there are opportunities business-wise for Russian-speakers because a part of me always wanted to connect the two worlds of my homeland and my current home, because of my special relationship with the country that produced me and was my home. Then, a couple of years later, it was clear that what we created was important for more than just one region because we discovered how much interest there was within Silicon Valley to see what's happening outside, so I just expanded my horizons a little. I like everything about the events of the Russian Center, but I'm just answering my calling. But

I also like global people, and I love talking to those from emerging markets." The Global Technology Symposium continues to thrive.

Around the same time, Russian-speaking professionals working at firms including Google, Sun Microsystems, Intel Capital, and Silicon Valley venture funds came together to found AmBAR, the American Business Association of Russian-speaking Professionals. **Anna Dvornikova** recalled: "Back then there were groups of Indians, Israelis, Irish, Chinese, and so on. But Russians had no network. I think that a lot of Russian-speaking professionals started moving to The Valley at the end of the 1990s, and, when they arrived, of course, they were preoccupied with their own stuff. They needed to find jobs, to secure their income. They were not thinking about networking. But then, after five or seven years, things were going right, they became more or less stable, and the economy was booming. And suddenly it all crashed. In the depressed economy, people were losing jobs and were looking for networks to secure their lives. So AmBAR was organically created. We would have networking events, professionally focused, with interesting speakers, once or twice a month."

Another early member of AmBAR recalled the challenges of starting the organization, attributing some problems to the founders not having any experience in doing so. The organization seemed to suffer from its own success in that multiple professional opportunities eventually led to conflicts among the founders: "I was still in the management of the organization when the problems started. We just didn't manage to scale it professionally. It grew too much, too fast, and we had several very successful projects. But we didn't take the corporate governance with enough seriousness and professionalism. We really didn't have any experience with such a huge success. A lot of people from Russia were coming and needed our help and services to do business here. And there was a lot of unemployment at the same time here, so some people said, 'Why are only you people doing this? We want to do it too.' So it was complex due to competition for the organization and that's what created the problems, I think. The AmBAR name became a big platform." The conflicts were eventually resolved and AmBAR continues to be a vibrant organization for networking and learning, with a membership base of more than 50,000 business professionals of all national origins.

Anna Dvornikova went on to found another networking organization, Silicon Valley Open Doors (SVOD). The technology investment conference initially brought together entrepreneurs and investors focused on startups in Russia and Ukraine. The conference later expanded to include entrepreneurs and investors from all over the world, eventually with 2,000 people attending the annual conference. Dvornikova and her business partner, Stas Khirman, also lead The Entrepreneurs' Club, a networking group that again began in the Russian-speaking community and expanded to include other groups that held meetings in Silicon Valley, New York, and Moscow. The precursor organization was TEC Club, founded by Khirman in 2003, that was an online network of several thousand high-tech professionals, investors, and angels from around the world.

Another organization catered to the Armenian community in Silicon Valley and was founded by **Nerses Ohanyan** who came from Yerevan in 1998 at age fourteen. He recounted: "I had started a networking group in The Valley that was focused on Armenian entrepreneurs and technologists in the Bay Area. By that time, we had some connections to entrepreneurs from Armenia and they had come to visit us several times. So at that point, we started thinking about going to Armenia for entrepreneurial events there."

Boston-Cambridge Networking Organizations

The USRCCNE was founded in the late 1980s by Americans. It is an active organization for Russian-speakers as well as others with an interest in Russia, but it was focused on doing business in Russia rather than developing a network of Russian-speaking business professionals in the United States. Also on the east coast, **Daniel Barenboym**, who came from Moscow in 1990 at age fifteen, started an organization in 2015 called NERSE. He explained: "We get together for monthly meetings, and we talk about our startups in Russian. It's all about technology. There are no real estate agents or life insurance salespeople involved. So it is literally all about technological startups when we get together and talk about our ideas, and we often pick on somebody to defend their ideas."

Some interviewees in the medical and life sciences communities also were involved in organizations focusing on Russian-speakers in their fields. **Nikolay Vasilyev**, who was originally from Gorki, Belarus, came from Moscow in 2003 at age twenty-nine. He recounted his involvement with the Russian-American Medical Association (RAMA) and its journal, as well as RASA: "When I arrived in the US, it was a very interesting time and I met a lot of Russian-speaking scientists and physicians in Cleveland. When I was there, I became involved in RAMA, the Russian-American Medical Association, and I became a member of the Board of Directors. I also founded the English-language RAMA Journal and became its first Editor-in-Chief. A lot of Russian scientists still were afraid to send their results to English-speaking, foreign journals. There is some psychological barrier. They think that their English is not good enough, or that their paper will be rejected because they are Russians, and things like that. We helped a lot of people have their first papers published. I also cofounded the Russian-American Science Association because there was no such organization that united the Russian-speaking scientific community in the US. There were five of us, and we founded RASA here in the Boston area in 2009. We wanted to preserve and develop the Russian-speaking scientific community as part of the global Russian community. And we think great science is possible where teams from different scientific backgrounds and from different countries are involved. I think when people bring diverse scientific expertise to address difficult scientific problems, those problems will be successfully solved."

It has not escaped our attention that Russian-speaking professional organizations have flourished more readily in the Silicon Valley-San Francisco area than in Boston-Cambridge. AmBAR attempted to expand to the Boston-Cambridge area some years ago but could not gain enough traction to establish itself there. On the other hand, Indian, Chinese, and other ethnic groups of technical professionals seem to have been successful in both locations, such as The Indian Entrepreneurs (TiE), originally created for the Indian diaspora but more recently opened to all nationalities. This situation was analyzed in a 1996 work asserting that the institutions and cultural orientation of the Massachusetts technology scene, as exemplified in the Route 128 economy, was less well-established and more conservative than that of Silicon Valley.[6] Whether this premise is still valid could be questioned, but the comparison between Silicon Valley and the Boston-Cambridge hubs regarding Russian-speaking networking organizations is still valid. Another explanation based on reviewing the innovation-entrepreneurship scene in the Boston-Cambridge area could be that there are so many non–ethnic-

---

[6] AnnaLee Saxenian, *Regional Advantage: Culture and Competition in Silicon Valley and Route 128* (Cambridge: Harvard University Press, 1996).

based organizations in this space that the need for ethnic-based ones has now been obviated. We note that even those organizations in Silicon Valley that began by catering to that clientele have now broadened their perspective to include members irrespective of national origin. This evolution is reflective of the maturing not only of ethnic immigrant groups, but also of the organizations that served them, as well as the increasing competition from numerous organizations that offer the same services regardless of ethnicity.

We have positioned this chapter as the first of two that depict the challenges and sources of support that interviewees experienced in adapting to their new environment in the United States. Their stories began back in the former USSR since that was the source of some impediments to adaptation, as well as some of the support that helped them overcome such issues. Challenges of adaptation were manifested in the United States as interviewees sought to assimilate into a new culture, with problems varying in type and intensity, as their stories illustrated. Adapting to the unique culture of the United States could be a rough road, but, overall, our interviewees learned to reach back to sources of support from the former USSR, as well as to seek out such sources in their new country, and we noted sources of support from mentors and role models in both locations. We also described areas of difficulty in adapting to the US educational system, living conditions, and social circumstances, as well as a new language. We also emphasized networks of various types as major support vehicles. In addition to the challenges of adapting to American culture, our interviewees also had to adapt to their new workplace environments and strengthen their soft skills that would be necessary to flourish in the US innovation economy. That essential adaptation is the subject of the next chapter.

8 | Workplace Adaptation
Developing Soft Skills

This chapter emphasizes the need for immigrant technological professionals to take actions to acquire new skill sets appropriate for cultural adaptation to the US work environment. We introduce the subject of soft skills and explain why they are so important, particularly in the twenty-first-century innovation workplace. Future 500, a nonprofit organization focused on innovation and sustainability, found in their 2015 member survey that developing soft skills for tech professionals was the number two concern of their member organizations.[1] A similar conclusion was reached by the *Future Work Skills 2020* report.[2] Hard skills are "teachable abilities or skill sets that are easy to quantify," such as knowledge acquired from a university degree, computer programming skills, and the ability to speak another language. In contrast, soft skills "are subjective skills that are much harder to quantify. Also known as 'people skills' or 'interpersonal skills,' soft skills refer to the way you relate to and interact with other people."[3] Soft skills include teamwork, communication, flexibility, patience, persuasion, and motivation. The *Future Work Skills 2020* report's list of top ten skills that will be essential in the 2020 workplace include three that fall into the soft skills category: social intelligence, cross-cultural competency, and virtual collaboration. As noted in that report, social intelligence is "the ability to connect to others in a deep and direct way, to sense and stimulate reactions and desired interactions." Cross-cultural competency is "the ability to operate in different cultural settings. In a truly globally connected world, a worker's skill set could see them posted in

---

[1] www.future500.org
[2] Anna Davies, Devin Fidler, and Marina Gorbis, *Future Work Skills 2020* (Palo Alto, CA: Institute for the Future for the University of Phoenix Research Institute, 2011) http://www.iftf.org/uploads/media/SR-1382A_UPRI_future_work_skills_sm.pdf.
[3] Alison Doyle, "Hard Skills vs. Soft Skills," *The Balance* (2016) https://www.thebalance.com/hard-skills-vs-soft-skills-2063780.

any number of locations – they need to be able to operate in whatever environment they find themselves." Virtual collaboration is "the ability to work productively, drive engagement, and demonstrate presence as a member of a virtual team." Employers especially value employees equipped with effective soft skills in addition to hard skills: "While certain hard skills are necessary for any position, employers are looking increasingly for job applicants with particular soft skills. This is because, while it is easy for an employer to train a new employee in a particular hard skill (such as how to use a certain computer program), it is much more difficult to train an employee in a soft skill (such as patience)."[4]

Many technological and scientific professionals globally, although typically possessing sophisticated hard skills in a technological area, tend to exhibit less-developed soft skills. This is true in the United States as well as elsewhere, and thus such skills must be developed for these professionals through further education and workplace experiences. This is true whether they manage overseeing interdisciplinary teams, are startup founders requiring such teams to attract capital, or are venture capitalists evaluating such teams. To be effective often necessitates retooling or acquiring new soft skills, or at least appreciating them. We recognize also, as did Future 500, that engineers and scientists may not have had opportunities to hone such skills through education and experience in their earlier scientific endeavors. As such, these skills are generally gained later in the careers of such talented professionals. In summary, technological skills can bring success in technological roles, but may well be insufficient in team, managerial, and leadership roles. Given the potential weaknesses in soft skills that such otherwise talented professionals possess, it is important to emphasize potential mechanisms for dealing with those weaknesses. Among them are the organizations in which they work, including the culture, and training that is often provided, as well as the mentoring and networking activities described in Chapter 7. Our interviewees validate these sources for improving the range and depth of soft skills. In order to benefit, individuals need to be open and willing to change their attitudes and behaviors and thus receive a new layer of imprinting in their adopted environment on top of layers from prior experiences in the former USSR.

---

[4] Anna Davies, Devin Fidler, and Marina Gorbis, *Future Work Skills 2020.*

Soft skills that are effective in the US innovation economy were typically not well developed in the former USSR, even in the managerial sphere. Soviet managerial and leadership styles were fundamentally transactional rather than transformational and mechanistic rather than humanistic.[5] They focused primarily on job completion rather than on incentives and motivations that might better lead employees to higher quality results – in essence, getting the job done rather than getting it done well. Thus, effective leadership is a crucial skill set that immigrants must acquire in order to be successful in that role in the United States. With this background, this chapter's topics will include teamwork, managerial and leadership styles, communication, and trust. We will present relatively briefly the types of experiences interviewees likely had in the former USSR, as well as challenges they likely faced in these skill areas necessary to succeed in the United States. We also will emphasize how their multiculturalism could enhance their ability to succeed in their new workplaces. As noted in Chapter 1, research has shown that multicultural work teams and multicultural organizations can affect a person's cross-cultural competence, which refers to the ability to work effectively with others from different cultural backgrounds.[6]

Yet immigrants still must erase the possible imprinting that could have occurred during their time in the former USSR, particularly among those who had worked in organizations employing the Soviet managerial and leadership styles just mentioned. These experiences often muted creativity and taking responsibility and thus stifled any

[5] Daniel J. McCarthy, Sheila M. Puffer, Oleg S. Vikhanski, and Alexander I. Naumov, "Russian Managers in the New Europe: Need for a New Management Style," *Organizational Dynamics* Volume 34, Number 3 (2005): 231–246; Daniel J. McCarthy, Sheila M. Puffer, Ruth C. May, Donna E. Ledgerwood, and Wayne H. Stewart, Jr., "Overcoming Resistance to Change in Russian Organizations: The Legacy of Transactional Leadership," *Organizational Dynamics* Volume 37, Number 3 (2008): 221–235.

[6] Günter K. Stahl, Martha L. Maznevski, Andreas Voigt, and Karsten Jonsen, "Unraveling the Effects of Cultural Diversity in Teams: A Meta-Analysis of Research on Multicultural Work Groups," *Journal of International Business Studies* Volume 41, Number 4 (2010): 690–709; J. P. Johnson, T. Lenartowicz, and S. Apud, "Cross-Cultural Competence in International Business: Toward a Definition and a Model," *Journal of International Business Studies* Volume 37 (2006): 525–543; A. Joshi and H. Roh, "The Role of Context in Work Team Diversity Research: A Meta-Analytic Review," *Academy of Management Journal* Volume 52, Number 3 (2009): 599–627.

inclination toward entrepreneurship. Additionally, there was little emphasis on genuine teamwork, and individuals typically preferred keeping a low profile and staying out of trouble. Inherent in such circumstances was not just a basic lack of generalized trust, but also a reliance on particularized trust among members of an individual's network. Relatedly, and of essential importance, is the topic of ethics which, as in many transition economies including those of the former USSR, can be very different from what would be expected in the United States.[7] All of these topics are germane to the overall area of soft skills that implicitly deals with interpersonal relationships that, to be legitimate, must embody mutual trust, effective communication, and a shared sense of ethical behavior. We begin with the topic of teamwork.

Teamwork

In some respects, the views expressed by our interviewees regarding teamwork reflect their comments on networking in Chapter 7. Characteristics of openness, communication, and the ability to forge commonalities with others are inherent in both domains. As we reviewed comments regarding teams and teamwork from the interviews, we noted several common themes. These included the various roles and types of teams, the leadership and managerial styles of team leaders, team membership, and the often necessary transition from being an individual contributor to becoming an effective team leader. Some interviewees commented on multicultural teams as well as on how leaders or members from the former USSR reacted to teamwork. Managerial and leadership style is an important element of teams, and thus many comments will address those topics as well.

In the former USSR, team members in software and other fields were typically unaccustomed to focusing on customer needs and instead worked to meet specific criteria and standards. Yet developing software for the United States and other developed countries often requires

---

[7] Sheila M. Puffer and Daniel J. McCarthy, "Finding the Common Ground in Russian and American Business Ethics," *California Management Review,* Volume 37, Number 2 (1995): 29–46; Daniel J. McCarthy and Sheila M. Puffer, "Interpreting the Ethicality of Corporate Governance Decisions in Russia: Utilizing Integrative Social Contracts Theory to Evaluate the Relevance of Agency Theory Norms," *Academy of Management Review* Volume 33, Number 1 (2008): 11–31.

including customer insights and developer creativity and flexibility to achieve successful results. Serial entrepreneur and investor **Max Skibinsky**, who came from Moscow in his twenties in 1996, provided deep insights into the experience that many technical developers from the former USSR had that inhibited or slowed their transition to becoming successful offshore developers for US companies: "I knew back then how to manage a remote team like that, literally an out-sourcing team composed of all Russians. Lots of them were coming from the same scientific community typical for Moscow State University, kind of that whole scientific class of Soviet Russia, and they all tell this story like, 'Hey, we're from Russia and we have this strange magic that makes our work special.' The Russian team abso-lutely loved top-down, Stalin-style dominated management. So, for them, giving them creative freedom absolutely freaked them out. They had no clue what to do with creative freedom. They were like, 'Hey, no, give me the specs and I will judge my work on how well I comply with your specs.' Unfortunately, it was a culture of total obedience and domination.

"So the second part was that it was literally that Russians had no clue back then what services were. The USSR wasn't really focused on services for citizens. So that part of the culture was completely missing. In contrast, when people were starting computer software in the US, they took the habits of the real-world service industry and put it into the software. That's sort of the initial ideology of Apple, like, 'Hey, how can we be nice to you, how can we be convenient to you? Yes, you're using a computer, but how can we make this a pleasant experi-ence for you?' And in Russia, this whole part of the DNA was com-pletely missing. We immediately understood that these guys needed specs, they needed to work on something scientific. You give them an extremely difficult technical challenge, and they can solve it and they will be delighted. Managing them has to be done from a point of view like, 'Hey, we're in the army, we have a major. Tell us what to attack and we will attack that, and anything outside is not our responsibility.' They didn't have the feel of people working on a startup team, but it wasn't by choice. That was the only thing they knew in life, that was their work DNA. The company that we built the software for launched something like three or four products, primarily built by this Russian team. I think one of my managers of the Russian team was relocated here by that company, and it was bought by Hewlett-Packard a few years after. It was a success story for all involved." Skibinsky provides insights into the cultural differences between Americans and those from the former USSR that resulted in very different approaches to software development. These differences had to be recognized and dealt with by team leaders and managers like himself. His insights suggested how imprinting in the USSR had shaped the Soviet team approach to software development.

An important milestone for technical professionals is making the transition from being a technology-oriented individual contributor to a team leader. Cofounder and CTO of CoachUp, **Gene Shkolnik,** who came from Moscow in 1992 at age seventeen, recounted: "I was in a number of different startups where I was leading smaller teams, so I had some opportunities to learn the necessary skills. But as far as leading larger teams, Paul was a big help, and he was mentoring me a lot, and so was Giorgos. At some point, I made a conscious decision that management was what I wanted to do. I remember sitting down and having this internal dialog with myself. At that time, while leading a pretty large team, I still was trying to code. I understood that if I continued doing that, I'd just be doing a mediocre job at both coding and management. So realizing that I had to make a choice, I've decided to just focus on being a good manager and a leader." Shkolnik helped clarify the difficult transition from being an individual contributor to becoming a team leader, an especially difficult role in a startup.

Eugene Baron, Product Manager at ROkITT, who came from St. Petersburg in 1990 at age twenty-one, described his successful transition from individual contributor to team leader: "I'm very social and I like working with people. I like managing teams, and I very much prefer managing people rather than managing computers. And I also enjoy working with customers, and I enjoy working with the people who worked for me or who didn't necessarily work for me. Basically, I like working with people, and that's the best part of my job." When asked whether there was something in his formative years that helped him develop his interpersonal skills, he replied: "Yes. When I was in eighth grade, I read Dale Carnegie's *How to Win Friends and Influence People*.[8] It was a translated, tenth carbon copy in manuscript form that

---

[8] Dale Carnegie, *How to Win Friends and Influence People* (New York: Simon & Schuster, 1936). Although banned in the Soviet Union, this book was widely and secretly read by those interested in American popular culture.

was passed around underground among friends. You know how that worked in those days. It made a big impression on me at that age. I think if I had read it when I was older, it might not have had such a big effect on me."

AbbVie Senior Scientist III **Jane (Evgeniia) Seagal,** originally from Novosibirsk, Siberia, came in 2004 at age thirty-two. She had learned to adapt her managerial and communication style to work effectively with different personalities, and she clearly articulated sentiments shared by a number of interviewees: "In this company, everyone works in teams, and I really like to work in teams. I have a very great team of people that I'm working with now. My function is supporting every project team in the company that needs the molecules we are making. I think, during the year, we are probably working with over a dozen different project teams. It's not always easy, but I think it's fun because you have some shared responsibilities and some shared goals. And you have to make it happen, right? Well, I'm very direct. It's just very much my style. And I know that not everyone likes it. But for some people I can be direct in different ways, and I'm learning how to soften my approach. If something goes wrong, and if there's an issue, people should know about that. I also get feedback that many people like my style because they see I'm not hiding something or trying to turn in a little different direction by saying something softer."

Other insights into the workings of teams were added by MIT Professor **David Gamarnik,** who came from Tbilisi, Georgia, in 1990 at age twenty-one to study at New York University before going on to earn a PhD at MIT. Now an associate professor of operations research there in the Sloan School of Management, he had worked for eight years at IBM as a researcher at the T. J. Watson Research Center Department of Mathematical Sciences. Gamarnik described his team experiences at IBM: "The nice thing about IBM, about being in a research lab there, is that relationships are built by teams and not by individuals. So you have a team and team leaders, and you have a group of people, and you all work together as a team." Gamarnik compared teams he works with at MIT with those he had worked with at IBM: "The team experience at IBM and here at Sloan has been somewhat different, but I think the common thing is that, in a team, it's not just about me. It's about everybody, in that each member of the team brings unique skills and knowledge. In the case of the kidney exchange research project, because it required knowledge in fields of certain algorithms and optimization techniques, that's sort of my edge. That's something I knew a lot about and not something that other members of the team were too familiar with. There's a mixture of economists and medical doctors on the team. I guess I had knowledge that my colleagues didn't have, and it was useful and relevant."

When asked about interpersonal relationships in teams, Gamarnik explained: "By and large, my experience on teams has been quite harmonious. But sometimes you have to become more assertive and argue for your point of view to convince people that your view is the right way. Once you realize that the direction being followed is not promising, the sooner you realize it the better. You save people's time, including your own. So that's when I became more assertive and said things like, 'Look, we should just drop this direction.'" Gamarnik illustrates the importance to teams of domain expertise, and also indicates that the soft skills of persuasive communication and even argumentation can add value to a team and help bring it to a successful outcome.

Sergei Burkov, serial entrepreneur and Founder and CEO of Alterra who came in 1989 from Moscow at age twenty-one, provided a view on teamwork at Google and noted that coming from the former USSR was not a hindrance in that environment. When asked whether the educational system in Russia was not particularly oriented to working in teams, he agreed: "Yes, but it worked OK. I think there were many problems there, but Google is actually not very strong on teamwork either. So Google is comprised of very strong professionals, and because many of them rely on their own resources and strengths, they do not ask for help. In some sense, asking for help at Google is not cool. That's one of the weaknesses of Google. If you're asking for help, it means that you're not professional enough, that you don't know your stuff. So maybe this is an exaggeration to some degree, but it is true that a background of Russian individuality wasn't outrageous."

An instructive contrast between the potential effects of Soviet and American cultures on teams came from **Yulia Witaschek**, Global (TAM) Strategic Customer Engineering Manager at Google Cloud, who came from Moscow in 1995 at age seventeen. Although she was employed at Google, her story is set in a very personal context: "Culturally, there are a couple of leftovers from Soviet culture that I have to battle with every day. I call it my positive side. The way in Russia that you approach things is, you say, 'No,' three times. And if somebody really thinks it's worth doing, they'll come round a fourth time and keep asking, and then you can say, 'yes.' Whereas in America, you're supposed to be positive and think about how we can make things happen, versus how this is not going to work. The way Russians think is, we start from a glass half empty and work our way to a glass half full. Americans start at a glass full and work themselves down to a glass half empty. The end points are the same, but how we get there is a little bit different. I'm very aware of the fact because people have been commenting that, even though I might have the right focus and did the right things, it can sometimes be viewed as very negative. I don't see it as negative. But now I'm aware of the fact that some people think it's negative, so I try to catch myself." Witaschek provided insights into the potential conflicts that Russians and others from the former USSR might have working in teams in the United States because of their different cultural backgrounds and approaches to interpersonal relationships that lie at the heart of soft skills. She made it clear that understanding another's point of view is fundamental to success in that realm.

Dmitry Fonarev, Senior Vice President of Development at SmartBear Software who came from Kharkov, Ukraine, in 1990 at age nineteen, noted: "So there certainly are cultural differences between many countries and cultures, and that's why they're called that. I've noticed that interactions with folks in Russia are different than interactions with folks in Sweden or Romania or other countries, and I've done business with folks in China, India, and Vietnam as well. I guess I can appreciate a bit more the causes or roots of some of the Russian specialties. I can relate to them a bit better. Our team communicates well, but I've seen many Russian teams who won't speak until spoken to, especially at first. As a business person, I have to do what needs to be done. Russian people are more reserved. For example, when I came aboard at SmartBear, I had to go to meet my new teams in Florida, Stockholm, and Tula. The reactions were rather different. I could immediately connect and interact very freely with the folks in Florida and Stockholm. But it took many months to establish the same level of trust and communication with our Russian counterparts. With Russian teams, the people are more reserved, and the boundaries between boss and employee are more pronounced than in the Western workplace. I'm trying to help them to be more – I don't know – open and free with me and less official. Western teams are typically more vocal, proactive, and independent."

Fonarev's observations were similar to others, including **Dennis Bolgov**, Founder and CEO of Tocobox, who came from Moscow in 1998 at age twenty-five. Regarding friendships with an Indian and an Asian colleague, Bolgov noted: "We worked together, hung out together, and I've worked with one of them since 2005. When you work with someone for that long, he becomes a friend of yours and a trusted person." Bolgov provided an insight into how difficult it is sometimes for Russians to trust others, as he expressed the length of time it might take to develop a new friendship and trust, likely a reflection of the difficulty they have in trusting others that we emphasize in this chapter. Taking substantial time to develop trust may well be a reflection of the imprinting that occurred among some interviewees, but clearly among members of software development teams back in the former USSR, as just explained by Dmitry Fonarev. He faced this situation with his Russian teams, even though he himself had grown up there. Perhaps a reason that he seems to display no such imprinting could be that he emigrated at age nineteen and has been in the United States for more than twenty-five years.

Another insight into the Soviet mindset came from entrepreneur **Alexey Eliseev**, Managing Director at Maxwell Biotech Venture Fund, who came from Moscow in 1992 at age twenty-seven. He illustrates how imprinting can be an advantage or disadvantage depending on one's new environment: "Here in the Boston area, most scientists and entrepreneurs try to take advantage of the very broad network. It is interesting that our Russian, or Soviet, mindset sometimes works against that. Maybe it has to do with the way we were educated, but many Russian scientists are quite individualistic, and I'm not an exception. Many of us tend to work on our own in one research area. That works very well in academia. It's perfect there, because every faculty member is unique. Many of my friends, unlike me, continued their careers in academia, and they made great progress. This is where the Russian mindset is helpful. In contrast, in the high-tech entrepreneurial environment, it may or may not be helpful. For example, the way we founded our company was based more on established personal relationships than on engaging in a broader network. A broad network always has synergistic effects. You get together at an event, like a seminar or a cocktail party, where you're talking to several people and some ideas come up. If you're interested, you follow up with one, and then with another one, and all of you circulate around. This is what Boston is good for. In contrast, in Russia, you often see that serious business interactions emerge within very small groups of people that have a long history of personal friendship or family relationships. It is surprising how many 'father-and-son' type high-tech startups founded by Russian expats you can see in the United States. Maybe this is because we grew up in an environment that lacked institutions for an entrepreneurial community and the inspiring environment such as you see in Boston."

Having seen various perspectives on managing teams and working in teams, we now proceed to a discussion of multicultural teams based on insights from two interviewees. One shared how he came to appreciate the importance of working in teams, especially multicultural teams, whether in leadership or member roles. **Anton Manuilov**, Senior Scientist II at AbbVie, who came in 1992 at age fourteen from St. Petersburg, said: "I'm very open to suggestions, and every day I'm still learning new things. So if I have someone on a team who happens to know something that I don't or has a better suggestion for doing things, I usually take those suggestions seriously and try them out. People seem to be very nice, at least where I work. In my industry, most people are not from the United States. It just happened that way. Actually, my friend and close colleague is an American. He was born here, got his education here, and was trained here. Once he looked around and said, 'You know, out of these ten people, I'm the only one who is an American.' I answered: 'Yeah, but that is not stopping us from getting where we need to go.' Myself, not being born here, although I did get my education here, I enjoy working with people from other countries, other cultures. Not because I'm forced to, but I just enjoy listening to stories. People go home to India, China, Romania, and other countries, and they have something very exciting to tell when they get back. I really enjoy that." Manuilov adds extra insight into the potential value of multicultural teams, an important lesson since many teams in the Boston-Cambridge and Silicon Valley innovation hubs are made up primarily of people who were not born in the United States.

Another interviewee who discussed multicultural teams and managing individuals from different cultural backgrounds is **Sten Tamkivi**, Cofounder and CEO of Teleport, who came from Estonia initially in

1994 at age sixteen and who had been the first General Manager of the Estonian startup, Skype: "When I first left the office in Tallinn, there were about 400 people. I think there were about 30 percent ethnic Russians who had grown up speaking Estonian as well, or who had learned Estonian. Basically, the rule at Skype was that in all our communications we always speak the language that works for everyone involved, and what we write down is in English. On a day-to-day level, I see no difference in managing a team with individuals of different cultural backgrounds. We had twenty-seven nationalities. But I don't think that managerial differences would necessarily be run on the basis of nationality."

Tamkivi provides background on the multicultural teams at Skype, which could reflect what happens in other companies with multicultural teams. The prevalence of many non–American-born professionals in the companies and teams in Silicon Valley and Boston-Cambridge requires an effective method of communication among those diverse nationalities. English will often be required for general communications, but allowing other languages among smaller groups could ensure effective communication in such circumstances. We noted in Chapter 1 that research has shown that multiculturalism can be an asset within teams since it can encompass diverse backgrounds and points of view. We believe that immigrants like our interviewees can enrich discussions and decisions given the different perspectives they might bring to the companies and teams in which they work. In this sense, the results of earlier imprinting can well have positive outcomes for tech professionals like our interviewees, as well as for the companies where they work.

Managerial and Leadership Styles

We begin this section with an interviewee who discusses managerial and leadership styles and also considers geographically dispersed teams in different countries in the context of his startup headquartered in Berkeley, California. **Alexei Dunayev**, Cofounder and CEO of TranscribeMe, came in 2007 at age twenty-six from New Zealand, where he had moved with his family at age fourteen from Kiev, Ukraine. After specializing in entrepreneurship and marketing in his MBA program at Stanford, he got together with his cofounder, also born in Ukraine, who had come to San Francisco at age four.

TranscribeMe specializes in voice recognition software in multiple languages. Dunayev recounted: "We now have close to forty-five people at TranscribeMe, and the immediate team I work with every day has about ten people. We have two more offices, one in New Zealand and one in Belarus. We do our software development in Belarus and our R&D operations in New Zealand. Having employees in different countries and on different continents is great since it provides for almost continuous worksites for the company. We won a few competitions and got some major support from companies like Microsoft through our New Zealand connection. Also, New Zealand is a great place to form research and development teams since there is lots of great science going on. Probably less in terms of programmers and developers, but that's why we have Belarus because the guys there implement the R&D that the New Zealand team comes up with."

When asked about managing and motivating people in remote locations with different cultural backgrounds, Dunayev replied: "In my mind, that's a function and responsibility of the management team. The job of the management team is to make sure everybody in the company has the resources they need, as well as clear direction, and is given the ability to execute. So the management team actively supports everybody else in making sure that they can deliver. And this is very critical in startups because you typically don't have a lot of room for error. And I'm quite independent and driven, and I think that attracts like-minded people. Whether that's a leadership style, I'm not sure, but I think that's also quite important for getting the team to jell, because then everyone has to share this approach. They really have to want to make things happen. And I believe it's a leadership and culture style that works quite well." Dunayev's insights are based on his personal experiences in different countries, and he has developed an effective approach to working with dispersed teams.

Michail Shipitsin had a somewhat similar experience, although his multicultural teams were not geographically dispersed. Shipitsin came from Novosibirsk, Siberia, in 1997 at age twenty-two. He is Senior Associate Director of R&D and Project Leader at Metamark Genetics. In describing his style, he noted: "I got some management experience fairly early, initially managing a single person, and then eventually a group of people, and ultimately a large group of people. So I went through some pretty interesting development during those transitions. I had no formal management training but my style just sort of evolved.

As is usual in science here in the US, there was very diverse ethnicity, with only two or three who were actually native-born Americans. The rest were from Israel, Russia, China, and Bangladesh, a very mixed group culturally. And although it can be challenging if they're coming straight from their home countries, most of my group had already gone through some sort of filters like graduate schools or masters programs. So since they've been here for a while, it wasn't so difficult."

Explaining his managerial style, Shipitsin continued: "Most people respond to the carrot and stick, right? So you need to look for an individual's characteristics and try to anticipate their wishes and demands, and what they would like and what they wouldn't like. I've found that it's not so different between people of different ethnic backgrounds, at least after they pass through the filters of graduate school like I mentioned. I guess they had become a bit more, well, standardized. I tried to always be very professional in that regard, I think because of my family background. My mother had managed a large number of people in a large legal department in the Soviet Union, and her message was always that you have to be professional." Reflecting back, he added: "Well, I would never really use 'the stick,' but it was more that I would show my displeasure if I thought they were working too slowly or not doing enough or not doing the right things. Also, I do praise people at times, but not too much." We might add that his restrained use of praise might be seen as stemming from his own cultural imprinting of having been raised in the former USSR, where praise was typically not used as prevalently in educational and work environments as in the United States.

Ilya Yaroslavsky, Manager of Advanced Product Development at IPG Photonics, had a style similar to Michail Shipitsin in his approach to treating all employees the same, whether Russian-speaking or not. IPG Photonics, founded by the highly successful Russian-immigrant entrepreneur, Igor Gapontsev, manufactures sophisticated laser equipment for medical and other applications. Yaroslavsky, who came in 1998 at age thirty, from Elista, Russia, noted: "As a manager, it's important to match people's work with their talent. I try to be very open with all my employees and not withhold any information that does not need to remain confidential. I treat Russian-speaking employees different from others only in a cultural reference I might make with them, but never in the professional aspects of our work." He added: "I think it's important to understand the context. I had been with a company that had a very open culture, but we were acquired by a company with a very closed and top-down management culture. So when I came here to IPG, it was really a breath of fresh air. The whole atmosphere here is very open and very conducive to innovation and development of new things. That's what this company is about."

Another interviewee who exhibited a flexible managerial style is **Iryna Everson** (born Yurchak), Procurement Manager at Pattern Energy Group. She came from Kiev, Ukraine, in 2008 at age twenty-six: "I try to find a personalized approach to each person and kind of change my style accordingly. Some people just need reasoning, like why it's so important for them to deliver on time. With those, I just tell them, 'Hey, here are the implications. If you don't deliver, this other person cannot start their work and then we delay the whole project.' If I just explain this to them, more or less the consequences, they understand. Most people actually are like that. It's like if you give them the reasoning why it's important for them to deliver, they will try to do the right thing. So it's not as easy to motivate a person like this, one who tells you they have conflicts with their deliverables. Still, I have to work these things out, and it can sometimes be complicated by the fact that individuals come from different countries with different cultures."

Shamil Sunyaev, Professor of Bioinformatics and Medicine at Harvard Medical School, came from Moscow in 2002 at age thirty-one. He explained well the difficulties some technical professionals have in clarifying their managerial style, even for themselves: "I don't think I have any leadership or managerial style. I don't know. I attended a three-day management course that was supposed to help if you start your own lab, but a lot sounded trivial. I did think there was a human element, so you treat people like they're humans, and you talk to them and think about what they want. It reminded me of nineteenth-century psychology that you should classify people and understand that people are different. For example, some are more of a talker and others are more of a listener. But I never really thought about my style or anything like that. Maybe I'm wrong because some of my people complain that my time management is not very good, but many things are not well organized in science. So maybe I should think more about that, but I'm basically trying not to have a style, which may not be right. Maybe I should think more about it." When asked if he felt he was inclined to be indirect, he responded: "Yes, like, you know, California style, where people might say, 'Or maybe it would still work if you would do something differently.' It may be my personality, but I'm not sure. It's not something I'm normally questioning." Sunyaev seemed comfortable not having clarified a leadership or managerial style for himself, noting that in science many things are not well-organized from a managerial perspective. He was also clear in his understanding that being indirect was often used by managers in the United States, which would be different from the more direct style typical of the USSR.

One interviewee explained how his managerial style had developed through reflection in different organizational environments, including his current role managing people in varied countries and cultures. **Ilya Kabanov**, Global Director of Application Security and Compliance at Schneider Electric, a multinational with more than 140,000 employees, came from Moscow in 2014 at age thirty-seven. He received a PhD in technology and operations research at the Moscow State Institute of Electronics and Mathematics and subsequently earned an MBA at the MIT Sloan School of Management. Kabanov has vast experience in other large companies in Russia, as well as in a successful startup. He made the transition not only from being an IT individual contributor, but also from succeeding in more domestically oriented organizations in Russia to his executive role in a large multinational corporation. Much of his success was achieved by thoughtfully adapting his managerial style to the environment of whatever organization in which he was a manager.

Kabanov noted: "I was constantly learning how to build a career, what knowledge I should get, what skills to acquire, and what new habits I would have to get to succeed. It all involved understanding each environment first before making any decisions. Learn, ask, learn, ask. I did this at a Russian telecommunications company and then in transportation and logistics companies. Here at Schneider, I spend more time on learning because of the scale of the company, its complexity and globality. I've been lucky. I've worked in different companies, and I've been exposed to a variety of organizational cultures and managerial styles. Reflecting on what I saw, I was able to develop my style, how I like to read and interpret people. And it's not hierarchical. I believe in knowledge sharing and collaboration. You somehow have to perform effectively in a particular managerial structure. And you should help the company and your team to succeed. In the company where I work now, in a global environment, there are different time zones and languages and different places that create a lot of complexity, because

I have to work with Australia, India, China, the US, well, the world." Kabanov consciously analyzed his evolving managerial style and noted the importance of the organizational environment and culture in being effective in that environment. That view bears a resemblance to what Sunyaev also seemed to be saying about the less formal style that could be effective in his scientific environment.

A senior executive with experience at various prestigious financial institutions, including Citibank and Goldman Sachs, came from Russia in the late 1990s while in her twenties; she spoke of the learning curve in becoming a manager: "Managing was very difficult for me, the same way it is for every new manager. You know, handling the responsibilities while you're trying to manage other people. But I was fortunate and had a very supportive environment and excellent people. I guess my style is that, in general, I get along well with most people, and I usually don't have conflicts. I kind of find ways to communicate honestly with everybody and set expectations, and work hard in general. I don't know if it has anything to do with my Russian background, but I don't like meetings involving many people. I like one-on-one, so you know, my style is that I need to get the information to them, and I would prefer to just send emails so they're clear on what they need to know and share information that way. I would do this regularly because I managed a large portfolio of relationships. I found meetings wasteful. I did get good results because my portfolio was large and diverse, and I had to get my people to cooperate.

"So, in addition to managing my people directly, I needed to cooperate with many teams in the whole region. So I would go to Austin, Dallas, Denver, and Seattle. I needed to establish relationships and get a loan and position a product, like a line of credit that my group was managing with those teams. They really thought I did an absolutely outstanding job. I think I've had a very successful experience." Regarding decision-making, she added: "I found that it's very hard for me to build consensus. Here, you know every time whose decision it is, and you give feedback or advice, but you don't make somebody else's decisions. You know which decisions are yours." Like many interviewees, this executive has shown the ability to evolve her leadership and managerial styles to ones that were appropriate for her environment.

A number of interviewees described their style as being results-driven in that they focused on results, although they may have varied somewhat in their approaches to achieving goals. For instance, one executive who came from Moscow in 1994 as a teenager is a director at a leading global asset management firm after having served as a vice president there. Her work involves leading a global team of investment professionals in developing new products as well as investment metrics for risk management. She explained: "For example, if we need to implement a certain system, there would be many people working with me who do not report directly to me, but I would need to give direction to make sure that things come together. In fact, in my job as a risk manager, the only way I can be impactful is not if I come up with something smart, but something smart that I am able to convince people that they should be using it. Otherwise, there is no result. In convincing and persuading, I think it's just being really well-prepared for what you're trying to change. You have to be sure you have all the facts, and you've done the analysis, and you've thought about various positions that others might take and counterarguments they might present. So I guess I'm always trying new approaches and not giving up, but hopefully also not being pushy but being friendly and helping people arrive at certain conclusions that you want them to arrive at, by asking questions and kind of guiding them toward that." In doing so, she displayed the characteristics of leadership needed to reach group goals while achieving the results for the overall organization.

Maxim Matuzov, originally from Murmansk, Russia, came in 2009 at age twenty-one and managed a team at Apple focused on Internet search development. He explained the need for deadlines and adjustments along the way: "Well, if there's a project, there's also a deadline and things that need to be done. I guess that if it's not being done, we know we need to make some adjustments to our schedule, maybe. My team can always come to me saying, 'We have this problem, and we need to fix it. Or I have too much on my plate right now, and I might need some help.' My door is always open for them. But also I'm always meeting with other departments like UX [user experience] design, marketing, analytics, reporting, and so on. I haven't gotten any negative feedback so far, so I guess it's working. But I can't really say that there's been something special about my style. I'm only probably just copying lots of good managers that I've considered good managers from reading books, and about good management practices. Plus, we have here in this environment Steve Jobs being almost like a semi-god for us.

"I know I need development, and my biggest negative thing is perfectionism, even though it can be positive, too. What I mean is attention to detail, like with the iPhone. And that takes pushing yourself. I do a lot of weightlifting, and I'll go to a competition in August and I'll probably be at the semi-professional level. For my coach, I give him my all. I guess it's just about pushing yourself because there are things to do, and you need to prioritize them. If you work well with people, if you communicate well, and if you're willing to learn, those are probably the most important skills nowadays." In describing the difficult job of a team manager, Matusov showed a strong inclination to meet results, as well as a need to be open to his team members, accompanied by his willingness to learn. He showed openness to acquiring and utilizing soft skills that are so important for technical professionals as they move into leadership and managerial roles.

We conclude this section on a results-driven managerial and leadership style with observations by **Michael Barenboym**, President of Baren-Boym Company. He came from Moscow in 1990 at age twenty-four and, for two decades, has been president of his own medical and industrial design firm after having spent several years in engineering and managerial positions in other biomedical device companies. His company employs several physicists in the Boston-Cambridge area as well as nearly thirty specialists located in Russia, Ukraine, and Belarus who are mechanical engineers, many of whom have PhDs and use their knowledge to analyze mathematical problems. There is also a group of industrial designers there, and the employees in those locations work from their homes. Barenboym explained: "You don't need to have an office. You need to have smart people around you and organized projects. And we do a good job at this, working almost twenty-four hours a day because of the differences in time zones," sounding a theme similar to Sten Tamkivi, noted earlier in this chapter.

Barenboym continued: "Thank God for the Internet. It's a matter of participating in different groups, different activities, learning about what other people are doing, seeing who is talented in what areas. That's how it came about, and it's been happening over a long time. It doesn't just happen overnight. I'm not really just a manager, I'm a mechanical engineer, so they can't do something I don't know about or lie to me. I know exactly what I want, exactly what they need to do, and I coordinate their efforts. A lot of times, people want to do things that I don't want them to do, but I allow them to do it to keep them happy. Sometimes I explain things well to them, but I still feel they want to explore this because engineers and designers are artistic people. Sometimes you have to let them do it the way they feel and the way they see things. At the same time that I tell them, 'You can do it your own way,' I also say, 'You're going to do it my way as well.' So they do it two or three different ways, and we select the one that I like most.' I consider them not as my employees but rather my colleagues. And I just respect them. And so my style includes a system that I developed. I measure success by being able to finish what we start. We have lots of repeat business, so that's a real indication that we're doing a great job." In addition to his explanation of his leadership and managerial style, Barenboym also provides a segue to the next section on managing teams remotely, since the vast majority of his employees were located in Russia, Ukraine, and Belarus.

Several other interviewees also discussed managing globally dispersed teams, as well as allowing team members some latitude in their approaches to software development. For instance, Facebook Team Lead, **Sergey Markov**, who came from St. Petersburg in 2007 at age twenty-eight, commented: "When managing, in many cases, I want to do something one way and people want to do it a different way. Actually, it was a learning experience that sometimes it's difficult to get people to agree with you, even if you're right. So sometimes you need to let people go and do it their way, maybe fail, but still do it their way, and not force something on people that they don't want to do. It's just that not everyone thinks the same way as you do. That's one of the things that I learned back at Microsoft."

A somewhat different perspective on managing employees remotely was provided by **Dmitry Skavish** who came in 1999 at age twenty-nine. A serial entrepreneur, Skavish is CEO of Animatron, a Cambridge animation startup he founded in 2011. He described how managing virtual locations has sometimes not been especially productive: "On managing virtual work, it's actually been challenging at different points in time over the last three years. We have two major offices, one in Cambridge and one in Ukraine. So we actually have people there, and the rest of our team works in this location. And we found that most of the people here are not really cut out to work from home. At least we have seen problems, psychological problems and others, that I usually have with people who are working alone.

They always think that they're missing something, or they're probably not in the loop, or something like that. And it requires much more effort to keep them in the team. Basically, I would like to keep people in the offices here, but at least they can choose. I try to manage by keeping them close by in the office. If that's not possible, then we talk every day. Here, we do Google Hangouts with people and use lots of tools to communicate and keep track of things like project tasks. In addition to individual Hangouts with people, once a week we have Hangouts with the whole team. We believe this group Hangout gives access to the leader and everyone on the team has a voice. We talk about what's going on." Skavish's views reflect a managerial style based in personal, face-to-face relationships at least domestically, but includes other techniques he and his company developed to accommodate exceptions to his own preferences and that demonstrate flexibility in his leadership and managerial style.

We conclude this section on managing remotely with an account of the management style of a highly successful entrepreneur whose cofounded company, Octane Software, was sold to E.piphany for more than $3 billion. **Kira Makagon,** who came with her family at age thirteen in 1977, from Odessa, Ukraine, is Executive Vice President of Innovation at RingCentral, a Silicon Valley–based company that provides cloud business communications and collaboration solutions. The company was founded by two friends from the former USSR whom Makagon decided to join in growing their already successful company. Makagon leads product strategy, product management, R&D, engineering, operations, and marketing/business systems serving more than 300,000 customers in 2015. Makagon oversees large teams in Russia, Ukraine, and China, in addition to the Silicon Valley office. In discussing the remote organizations, she noted: "They're all engineering organizations, developers, managers, or technology leaders in one way or another, and they're all our employees. In managing them, I have really good people who work directly for me, and that really is the answer. I don't directly manage people in Russia, for instance. I go there and spend a week at a time, getting into a lot of details with local management. I'm not on the phone with these remote organizations. I'm on the phone with the heads or the general managers of the office. But the guys who work for me, and they're all guys, are really good at managing remotely. The projects are all joint projects, so there are weekly calls and weekly meetings so that it's really functioning like one organization. The idea is for people remotely not to feel like they're part of a different company. It's important that they know it's all one company, and that takes a lot of communication.

"This is a very much results-driven organization, so the method has flexibility in that managers have control over different cultures and different styles. Everyone is synched up to their goals and marching to the same set of priorities, goals and objectives, and time schedules. You can establish a culture to prevent things from falling through the cracks. I think that people, when they're proud, are proud of the results. But you still have to keep people on their toes. It's just human nature for people to slow down after a good result. You've got to have a fair balance about acknowledging people's work and results and yet continue to be very demanding. You've just got to keep people challenged, and I think that's true at any level." Makagon's style has allowed her to leverage her executive and strategic expertise throughout a large, dispersed organization, and she has clearly thought through the entire process of doing so.

Communication Style

As was true of teamwork, communication, both verbal and nonverbal, is another critical soft skill that could be another area for development by our immigrant interviewees in their new environments. Language would likely be a barrier for many in their early days, but perhaps even more difficult to address would be the style of communication that is fundamentally different from that in the former USSR. Many interviewees found communication to be confusing and challenging, and it often required cultural adaptation. Regarding verbal communication, many noted that the direct style communication that they grew up with in the former USSR seemed at odds with the more indirect style more commonly used in the United States. So, too, was nonverbal communication, particularly the American readiness to smile, although some came to appreciate smiling as a display of confidence and comfort.

Verbal Communication

The most dramatic difference in verbal communication is the directness of style in the former Soviet Union in contrast to a more oblique manner on the part of Americans. Insights into this difference are offered by a number of interviewees, beginning with **Anya Kogan**, User Experience Designer Lead and Manager for Ad Words Display at Google, who came in 1990 at age eight from St. Petersburg. She spoke of her experiences at Google: "I've worked with a number of Russian engineers, and generally I feel like the Russians that I've seen have assimilated pretty well. I could envision issues like the way I've seen Russian people argue, 'This is the way it is.' And then another Russian would say, 'No. This is the way it is.' Whereas in America, or the way I myself would argue is, 'OK, you know, I really think that maybe that is how it works. But how have you seen it work? Let's figure out why we have different perspectives and let's find a compromise.' I think it's a bit more open. I would advise somebody in that situation not to take it personally. I don't think that Russians are trying to attack character. It's just the way they communicate things. So don't be afraid to argue. You might need to do it in a slightly more aggressive way than you're used to." Kogan, in many respects, was providing advice that would likely be very similar to that of David Gamarnik, quoted earlier in this chapter, who found that he had to become more assertive in his arguments, even when other team members were not from the former USSR.

Kogan added more insight: "So, I mean discussion and communication are an integral part of teamwork. They're not really separable. I haven't really had issues with the people I've worked with. I've worked with a couple of Russian engineers recently here at Google, and we get along pretty well, and they're usually open to ideas and they have ideas." **Eugene Baron**, Product Manager at ROkITT, who came from St. Petersburg in 1990 at age twenty-one, elaborated: "There are more similarities than differences in people from different cultures. So I'm just going to talk about the differences. The differences are that Russians are much more personal and much more emotional, and you know, they are very, very honest. They will tell you right away if there is a problem, and they're usually not going to cover it up. If you need people to tell you everything is good, don't worry, the Russian team will tell you instead, 'This is wrong and this is bad, and we need to change that.' And you know, they appreciate frankness. That also helps formal communication. Having drinks together also helps a lot. But you know what? I'm more American in this way. I behave like an American. I'm not as direct as Russians are." Baron had already learned much about American-style communication during his time in the US technology world.

Other interviewees expressed a preference for the directness found in the former USSR and did not see the value in changing their personal communication style. For instance, **Ksenia Samokhvalova,** Senior User Experience Specialist at MathWorks who came to the Boston area in 2003 at age twenty-three, was well aware that her direct communication style stood out in her US workplace: "As a Russian, I don't have a good tolerance for 'bull.' I also feel like I'm too blunt sometimes." However, she noted that her first manager in the United States, an American man, encouraged her to use her bluntness to point out problems, and she has found it useful for that purpose. **Andrey Doronichev,** Senior Product Manager at Google who came from Moscow in 2012 at age thirty, also recognized the impact of his direct communication style and learned to adapt it to be more accepted by his coworkers: "I am much more ruthless than many other people. I am more direct and pushy in a way. People would say about me: 'His communication style is way too aggressive. I cannot work with this guy.' Now that they are used to me, they say: 'Oh, you know Andrey. He's pushy. He'll be pushing, but that's what we need here.' I've managed to calm my temper to the extent where it's manageable, where people can kind of work with it. But I didn't fundamentally change." **Eugene Boguslavsky,** Release Engineer at Facebook who came from Minsk, Belarus, in 1989 at age fifteen, explained why he preferred a direct and frank approach to communication: "I like it very well because how else do you know how you're doing? I think that the worst that could probably happen to you if you don't get any feedback is surprises. If you're not performing to your level or people have a hard time working with you, you don't want it to be a surprise. You want to know that right away so that you can adjust, or maybe realize that's not the right place for you. If people don't appreciate what you're doing, it's better to know it earlier than later. So I usually give 100 percent honest feedback to everybody."

In a similar vein, **Boris Berdnikov,** Staff Software Engineer at Google who came from St. Petersburg in 1995 at age nineteen, expressed his disdain for the "sandwich method" of feedback found in US workplaces, whereby a criticism or recommendation for improvement is inserted between two positive statements: "I don't like the 'sandwich method.' The direct Russian way is a much better way to establish understanding and communication because it's more authentic, because the message is much clearer. I don't mean the way where, like the Russian saying goes, 'I'm the boss and you're the fool.' That's the extreme of it. Yet I feel like it's much better overall. Not in the disrespectful extreme, but the directness. That's the Russian in me speaking. So this is hard for me, and it may be part of the cultural divide." Despite his frankness, Berdnikov clearly recognized that going to the extreme, as in the Russian saying, would be unproductive.

The direct and unvarnished style more typical of those from the former USSR at times led to difficulties with team members, peers, and subordinates in the United States, and, according to several interviewees, that direct and blunt style was similar to what they had also experienced in Israel. **Sergey Gribov**, Partner at Flint Capital who came from St. Petersburg in 1997 at age twenty-eight, didn't mince his words: "In the US, people are very polite and nice. In Russia we're not, and in Israel it's the same way. If an Israeli thinks you are full of 'bull,' he would just tell you." And **Stas Khirman**, investor, serial entrepreneur, and Managing Partner at TEC Ventures, who came from Kiev in 1997 at age thirty-one, noted other cultural differences in personal interaction that he had observed: "Americans substantially are more considerate of another person's space and another person's feelings than Russians and Israelis."

A significant exception to the use of direct and straightforward communication in the former USSR was with superiors. In that hierarchical, top-down management system, subordinates typically would avoid delivering bad news or pointing out problems for fear of negative reprisals, including being blamed or even punished. Thus, people from the former USSR often needed to learn to do so in their US workplaces, where such input was expected and encouraged. **Nerses Ohanyan**, Director of Growth Analytics at Viki who came from Yerevan, Armenia, in 1998 at age fourteen, explained the challenges of learning to speak directly and honestly to superiors and taking responsibility: "With somebody whose stereotypical cultural expectations are that you can't really talk to your superiors or don't talk directly, you can set the expectations and teach them in practice and people will adjust. People that I never thought would adjust have adjusted. But, I also think it's important for you as a company to say, 'This is how we want to do things,' and two things happen: one, people can be trained to communicate differently, and two, within every culture there is a subset of people who already speak that way. So, if you can expect that, you can filter for that as you bring new people on." Ohanyan's comments are helpful in viewing how organizations might help employees make adjustments to their communication styles.

On the other hand, **Michael Barenboym**, whose clients have included many CEOs and senior executives, understood the importance of speaking frankly when communicating with people in high-level positions: "What I've seen a lot of times when people get hired and are doing business paid on a contractual basis, they start really being dependent on the people who are paying them. I don't do this, and I always say that I'd rather not do this project if you are not reasonable and if you just want me to do it the way you see things. And that's why I don't hide my opinions from people who are higher than me on a professional level. I'm not afraid of anybody or anything; my style of dealing in business is very open. So I always say whatever I think is right, in my opinion. But I always say that it's my subjective opinion. There's no such thing as right or wrong, it's just a matter of opinions for different people. So, for me, I really like dealing with American companies because I can really teach a lot of people a few things. You always compete with other groups, and cost is not always the decisive factor. It's your credentials, how you present yourself, how charming you are and how you are able to make people believe in you. And psychology is the biggest thing; it's the only thing that drives this world." Both Ohanyan and Barenboym seemed to have overcome any lingering reluctance at managing by speaking directly and clearly to senior executives, in contrast to the typical aversion to doing so in the former USSR.

In contrast to preceding interviewees who preferred maintaining their directness and sometimes bluntness in communication, others modified or softened their styles. **Iryna Everson** explained her style of direct communication, which appeared to be not as harsh as that of some others: "I'm usually not very good at giving hints to people. And I prefer when people speak directly to me, too, because it's easier for me to understand. But I wouldn't say, 'Hey, I want you to send this email this way.' I would say, 'Oh, it would be nice if you would write this email like this,' for example. So, it's definitely polite, but more clear and easy to understand without any opportunities for doubt." Everson also appreciated different cultural attitudes toward feedback: "In Ukraine, people would think of feedback as something like criticizing, usually, whereas here, people think feedback is a gift, for example, to help you be better." Her last comment provides a useful comparison between how feedback might be looked at differently by people in the United States and Ukraine.

Another interviewee who attempted to modify her communication style is **Olga Bazhenova**, E-Commerce Product Manager at Staples, who came from Moscow in 1999 at age sixteen. She explained her direct style of communication and its impact on others she worked with and how she adapted her style over the years: "Russians tend to be very direct. A senior executive had once told me, half jokingly, 'Every time you talk to me, regardless of what you say, I feel like you are scolding me.' You just sort of take it and nod and smile politely. But really, I felt there was nothing I could do about that. But over the years, I've toned it down." Bazhenova also was bemused by feedback she received as a project manager, with some people appreciating her direct style while others did not: 'During one of my yearly reviews, one person wrote that I was too direct and needed to soften it up, while another person said I needed to be more direct and say more clearly what I needed. So I learned to adjust my approach to those specific people."

Another interviewee described how he learned to say negative things without giving offense: "Especially in a large company, there is a more refined way that people communicate failures. For instance, that's where I learned that I could still refine my English in terms of how one can say negative things without offending someone. So I learned to say things like: 'I can understand how you see it this way, but have you ever considered this?' Or, 'Why don't we try to do something differently?'" He felt, however, that polite communication could be taken too far: "There are a good number of people that should and could be more considerate toward others' sensitivities, within reason, but at times I think people tend to smother the meaning by political correctness."

A number of interviewees spoke about the importance of adapting one's communication style to the work context, in terms of both national cultures and individual preferences. **Dmitry Kovalev**, Consulting Systems Engineer at Cisco who came from Moscow in 2008 at age twenty-four, expressed a nuanced view of communication, noting individual preferences within cultures as well as the need for adapting to specific situations: "Communication style might be dependent on the culture, but it's much more dependent on the particular person. Communication is really key, and sometimes it's good when you're over-communicating, and it can be good when you're under-communicating. Every person has to find their balance. It's also specific to the amount of the communication needed by each particular team. That's something that you learn over time. It might be a problem of conflicts sometimes, when you don't say something and you kind of assume that you've said it but the other person didn't hear it, and he's acting differently than he would if he had gotten this information." **Sergey Kononov**, Senior Delivery Manager at EPAM Systems who came from Kharkov, Ukraine, in 2011 at age twenty-eight, explained that he adapts his communication style depending on whether he is working onsite in Cambridge or remotely with teams in Russia: "With teams in Russia, when you are sharp or stern with them, if you've proven your leadership and talent, they understand you have a right to be sharp and you have enough knowledge to guide them. In the US, I have to be much smoother and softer."

One interviewee provided a rather detailed view of how difficult communication can be for any manager in a large and dispersed organization. For **Kira Makagon**, the solution was in embedding an open communication style into the company's culture. Makagon, who as Executive Vice President of RingCentral oversees a large workforce in Silicon Valley as well as remote teams in Russia, Ukraine, and China, explained how she has created an organizational culture based on open communication: "I think you can establish a culture around open communication. People who work for me know that if something is not working, then it's better to give me an early warning sign than later saying, 'Oh, I didn't know it wasn't going to work out,' type of explanation. And that's the culture that I have in the organization regardless of what level people are at, because things don't work out necessarily the way we think they will all the time. There are disasters, you know, there are minor problems, there are major problems. I don't want to know about every minor problem, but I do want to know about anything major as soon as it becomes known. I don't expect to solve it, I expect them to solve it, but I know about it and that's the culture at every level, hopefully."

Still in the realm of verbal communication, two other interviewees provided a clear contrast between communication in the United States and the former USSR. **Sasha Proshina**, Senior Manager at Sanofi Genzyme who came from Tomsk, Siberia, in 2002 at age twenty-two, provided her observations on how Americans relate to one another in contrast to Russians: "What I like about America is that American people show much more gentleness and politeness. Russians don't have that, at least in my family. There is no politeness. They just barrage each other all the time. And if Russians like something, they wouldn't tell you. I'm talking about poor to middle working class people. It's just this vicious cycle of gossiping, criticizing, gossiping, criticizing. After living here in the US for so long, it's like a bitter rude awakening for me when I go back to Russia." Proshina emphasized the adjustment to the American style of communicating as well as the challenge of dealing with the contrast when visiting her homeland.

Andrey Klen, Cofounder and COO of Petcube who was born in Egvekinot, Russia, and came from Ukraine in 2014 at age twenty-six, provided a similar view, but focused on the American style: "Overall, Americans are much nicer in terms of communication and manners, like etiquette, smiling at you, and all that stuff. They are polite on their personal communication level. Formally, everything is done much nicer. And everybody tends to stick to their word, like when it comes to some arrangements and when it comes to promises. Personally, I think that Americans tend to keep a lot of promises and say, 'We can do a bunch of things together.' But sometimes they don't quite follow through on their promises. But when you have a firm arrangement, it is going to work. That's a first impression for me." Like Proshina, he noted that it takes time to adjust to a new cultural style. This is likely due to being imprinted with a markedly different style in earlier years, as were so many interviewees whose observations were included in this chapter.

We end this section on verbal communication with a couple of comments on the importance of humor in communication. Humor is one of the most difficult things to interpret in a different culture because people within a culture are imprinted with a shared experience that may be implicit in a joke or humorous remark. As such, it is difficult for those from different cultures to appreciate or interpret humor from another culture. **Davit Baghdasaryan**, Senior Security Engineer at Twilio, who came from Yerevan, Armenia, in 2008 at age twenty-four explained his experience: "Humor is different in Armenia. Humor in the USSR is different than here in the US. And humor is a very important part of communication, right? So if you make jokes that others don't understand, at least in my world that's probably going to be a show-stopper if you want a better relationship with that person. And it's vice versa, if you don't understand that humor." In the USSR, humor was an especially important means of communication, a means of social bonding, and often served as a safety valve, as a way to ridicule the communist regime among trusted friends. **Dmitry Kerov**, Vice President of Software Engineering at Northern Light, who came from St. Petersburg in 2009 at age thirty-seven, explained that sarcasm and humor helped people cope in the USSR: "We were very limited in the things we could do, and the whole environment was kind of ridiculous, so the only thing you could do was laugh at what was going on. Even in Brezhnev's time, there were a lot of political anecdotes or jokes."

Nonverbal Communication

This concluding part of the section notes the importance of nonverbal communication, so fundamental to every culture and often more difficult to interpret than verbal communication. We focus on the practice of smiling as the form of nonverbal communication described by numerous interviewees and their different interpretations of that gesture in the United States and the former USSR. Some interpreted in a positive way the propensity of Americans to smile frequently with strangers and new acquaintances in addition to family members and close friends, while others saw it negatively. One interviewee initially saw the American custom as being less than sincere: "The biggest issue was expression of emotions. For instance, Americans usually smile, but it means nothing. It's just normal. But Russians' neutral position is not focused on emotions. If you are smiling, it means something. You're happy, you're going to see somebody, like it means something, but in America, it doesn't mean anything, and it was the first issue for me to understand what people mean."

A contrasting view was expressed by **Maxim Matuzov**, Search Program Manager at Apple. Originally from Murmansk, Russia, he came in 2009 at age twenty-one: "So when I say emotional, I mean like someone who shows the emotions on their face, I guess because in Russia we are kind of always serious and don't show anything. Here, it is much easier to tell things from a person's expression. You know, you have a very laid back environment here, and I don't see any point of being super strict, super serious, super smiley or something. There's no point. It's an easy going environment. It's a laid back environment. People like you, people you know treat you well. You don't expect any danger from people, right." Matuzov was one of the few who attributed the difference in the propensity to smile to the environment.

Eugene Trosman, Applications Engineer at Analog Devices who came from Novosibirsk, Siberia, in 1989 at age twenty-four, also commented on the more casual atmosphere the United States: "I think people are more relaxed and more confident here in the US. Although I don't think I smile too much. I never got into the habit of smiling too much. To be honest, the first few years when I came over, and I see people always smiling I thought, well, is it a real smile or is it a made-up smile? How do you know? I couldn't figure it out. And I still sometimes can't figure it out. Russians don't do that, obviously. I think it's a typical American thing. But, yeah, you take the subway in Moscow and among all these people, nobody smiles at all, everybody's tense, everybody's kind of uneasy. Uneasy isn't the right word: I think their expression is tense, it's not relaxed." A similar perspective was provided by **Maria Samarina**, Managing Partner and Cofounder of Yosh Technology who came from Novosibirsk, Siberia, in 2013 at age twenty-seven: "People are more relaxed here and have more money to live and, actually, your cultural difference is that there is not a lot of complaining, like in my country. If we have problems there, we say it directly. Here, when you ask, 'How are you?' you probably don't know how people are really feeling. In Russia when you ask, 'how are you?' people will answer you back with honest answers. If you have a problem here, you just do not speak about it, or you look at the problem from a nice point of view, which I really love. But it's not easy for me to adjust." While noting that smiling was a cultural difference, Samarina explained that it was not easy for her to acclimate to that nonverbal cue in American culture even though she preferred that approach.

A stronger positive view of the American interpersonal environment was provided by serial entrepreneur and investor **Evgeny Medvednikov**, who was originally from Arkhangelsk, Russia and came in 2014 at age thirty-four. Having been in the United States only a few months, he remarked: "People are friendlier here. And they are very happy to connect you with other people, to connect you with partners or investors. And they do it just for free. But, for example, many Russian people know somebody you are looking for, and they may want to get some money from you for this knowledge for an introduction. And it's a serious difference, because Russians still don't understand that you must help somebody first and then expect a return on these investments. First you help and then expect help. But if you sell a service to somebody, it's not a good idea." Medvednikov provided the contrast as well as the probable reason for the difference in approach of Americans and people from the former USSR. And, in an insightful comment about the difference and the reasons for it, **Kira Makagon** said: "Russians don't smile a whole lot, so you've got to break through, you know, break that ice. And it's not that they're that intense. I think it's a culture of introverts, or the behavior is that of an introvert. So you've got to make it a bit more personal, whereas I don't think you actually have to necessarily make it more personal in the US. I think people here just are naturally more relaxed." Her comments emphasize the difference in cultures and inherently the culture of guarded behavior in the former USSR, as opposed to the more open society of the United States.

Getting over the habit of not smiling as well as being less introverted was an important step for **Sasha Proshina**, Senior Manager at Sanofi Genzyme who was originally from Tomsk, Siberia, and came in 2002 at age twenty-two. She learned about smiling from a popular American book mentioned earlier in this chapter by Eugene Baron: "When I was thirteen years old, I was very shy and I hated it, so I read a book by Dale Carnegie, *How to Win Friends and Influence People,*[9] and so I was one of the very few Russians who smiled. I used his lessons a lot and that really helped me to actually make a lot of friends. And when I came to the US, Americans were shocked that I smiled because, you know, Russians never really smile. They have kind of very stern faces." Proshina perhaps unwittingly learned to adapt her nonverbal communication to the American style while still a teenager in the USSR, and it eventually led to a positive turn in her life.

A concluding view of that cultural difference is provided by **Alexandra Johnson**, who grew up in Vladivostok, Russia, came to the United States in 1990 in her twenties, and who travels frequently to the former USSR: "I discovered that when you go to Moscow, people don't smile, and everything you say will be first a negative, then you have to prove that it's a positive, and then you can say it. So that was one thing I noticed; culturally it's still very, very different. At the same time, seeing the geopolitical situation happening now, maybe people have a reason for that. I don't know. Sometimes I wonder when visiting people, because I did teach them to smile and be energetic and whatever, but

---

[9] Dale Carnegie, *How to Win Friends and Influence People*. As noted earlier in this chapter by interviewee Eugene Baron, the book was highly popular in the former USSR. Smiling was one of the book's recommendations.

maybe they're sitting there thinking, 'Sasha is from California and there people have to do all those things to operate.' I don't know, it's complicated." Johnson's reflections provide insights into the difficulties that immigrants from the former USSR face when returning to their homeland, possibly because they might have adjusted so well to their adopted environment in the United States.

Trust

We now provide background on the nature of trust in the former USSR since that concept is fundamental to putting soft skills into context. Without mutual trust, it can be extremely difficult to build the relationships necessary for leaders and managers to effectively utilize soft skills with their employees and vice versa. Trust is also fundamental to success in other business practices since it is critical to external relationships, such as those with investors and customers. Although other formal institutional arrangements are usually in place both internally and externally, the presence of trust adds an important dimension to successful relationships in virtually all areas of business. Two types of trust have been identified: generalized and particularized. High levels of generalized trust among the population overall are typically found in societies with strong legitimate formal institutions including governments, judicial systems, and enforceable contracts. Research has shown that such institutions serve as safeguards against improper actions and in turn facilitate a broad level of general trust within organizations as well as society in general.[10] Generalized trust can produce a more open culture that is accepting of formal and even informal relationships among institutions and between individuals and institutions. In contrast, countries and organizational environments with weak legitimate institutions produce the opposite results, inhibiting trusting relationships beyond one's close network or circle of family, friends, and confidants, creating an environment characterized by particularized trust. The latter is common in many developing and transition economies including those of the former USSR.[11] The institutional environment that citizens experience often creates fear and causes people to be

[10] Angela Ayios, *Trust and Western-Russian Business Relationships* (London: Ashgate, 2004).
[11] Alena V. Ledeneva, *Russia's Economy of Favours: Blat, Networking and Informal Exchange* (Cambridge: Cambridge University Press, 1998).

closed, protective, skeptical, and even suspicious of people outside their close networks.

Andrey Kunov, President of the Silicon Valley Innovation Center who came initially in 1994 at age twenty-two from Zhezkazgan, Kazakhstan, clarified the lack of generalized trust he experienced back in the former USSR, which stood in stark contrast to the abundance that he found in Silicon Valley: "Sometimes you trust people you probably shouldn't trust in the first place, so you're bound to make some mistakes. Yeah, that's where a lot of my failures came from. Sometimes people can exploit your trust or they can manipulate you. Now, when I look back, I feel like I've been manipulated in my life on a recurring basis, I should say. I'm not blaming anybody. It's actually my mistake that I was extending my trust to people who I shouldn't have in the first place. For instance, back in Russia, I was so trusting that my mother told me I was too trusting, and she was right. That's why I would have failed miserably in business in Russia. So I didn't even try to have a business there because the way business was organized there was just so alien to my inner nature. That's probably one of the reasons why I left Russia. Maybe somehow deep inside I was looking for a place where I would not feel so insecure with my trust of other people.

"When I came here, especially in California, I never wanted to leave because there are a lot of people like me who are doing business. I discovered that I am not alone here; there are tons of people like me. And, amazingly, some of them still manage to succeed in building their business. You don't have to be a cutthroat type of a person in order to be successful here. The environment here is much more accepting. The business culture is so accepting here even though there is huge competition. All of those things somehow manage to be combined in Silicon Valley specifically. There are a lot of people who are very trusting, and I love that about this place. I think it's kind of like a self-selection mechanism that is working in Silicon Valley, where there is a microclimate here that supports a lot of trust. I love the fact that we can contact a lot of companies that don't know much about us, but they trust us. We're a tiny company, but interestingly enough, there are huge companies like Microsoft and Intel, and yet they meet with us like equals, which is amazing. And they trust you at your word." Kunov provides a vivid contrast of the difference between particularized trust as found in the former USSR with its low generalized trust, compared to the environment of high generalized trust he found so welcoming in the

United States. And it is clear, even if unspoken in his statement, that strong, legitimate institutions are the foundation of generalized trust in the United States.

Another interviewee who displayed a trusting demeanor was **Roman Kostochka**, Cofounder and CEO of Coursmos. He is originally from Tolyatti, Russia, and came to the United States in 2013 at age thirty-seven: "I don't know, but I always trust people, although sometimes it's not good for me. But if you trust people, they open up to you, and I like these relationships. Whatever places I've been around the world, I've always met very nice people, and if something goes wrong, I just move on, because that's life."

Not all interviewees found their new country as welcoming as Kunov or Kostochka, who evidently did not feel the need to shed imprinting about trust from the former USSR. As Kunov explained, he was already an anomaly in that environment and found comfort being himself in the United States. Others, however, experienced difficulties adjusting because of the imprinting they received in the restrictive institutional environment of their earlier years and that could still inhibit their ability to trust people beyond a close network. One such interviewee illustrates the difficulty of embracing generalized trust, as well as being limited in his ability to receive that level of trust beyond his close network. Serial entrepreneur and investor, **Stas Khirman**, who was originally from Kiev, Ukraine, and who came in 1994 at age twenty-eight, provides clear insight into the complexities of trust: "Again, because I was born in the USSR under the eyes of the KGB, and it's not a joke, you have to carefully consider what you are saying and to whom. I feel that way every day. On a personal level, it took me many years to trust another person. I don't have many very close friends, but those I do have, I have had for five years, ten years, twenty years. It takes a substantial amount of time for me to trust a person to the level I consider necessary for friendship. So about trust: I cannot tell you about the United States generally, because I've only lived in Silicon Valley. Here, it's taken to the extreme with trust and common interests going together. But common interests are long term, throughout your life. It means that someone like me coming from outside doesn't have a credit history, you know no one and no one knows you. And it's hard. Also I have passed through this. Maybe I was successful in my transition from Israel to here, but I lacked a support network, and this can be hard for people. So that's why we founded TEC Club, AmBAR, and

SVOD to help with this situation. And we call these human networks, support networks." Khirman's story seems representative of people who spent much of their earlier years in the former USSR before its breakup. As such, he was likely imprinted deeply with the environment of distrust that existed there. Interviewees who had fewer years in that environment might have been less deeply imprinted, but their own experiences as well as their parents' and even grandparents' stories, almost certainly affected their attitudes toward trust. Yet Khirman managed to help others adjust in important ways that were not available to him as he reached for and achieved his successes.

In a story similar to that of Stas Khirman, **Michail Pankratov**, President and CEO of MMP Medical Associates, who came from Russia in 1974 at age twenty-six, recounted: "It took me twenty-five years on the academic side and twenty years on the business side to develop relationships so that people really trust me. You see what I'm saying? I was always honest, and I never deceived them, and never got involved in anything shady." His story, like that of Khirman's, illustrates how difficult it can be for immigrants to gain the trust of associates and others, even though the United States is usually categorized as a country exhibiting high generalized trust. Both men had to work for years and even decades to persuade others that they were worthy of trust, another indication of the potential difficulties faced by our interviewees as they attempted to integrate into the US innovation economy. **Timur Shtatland**, Development Scientist II at New England Biolabs who came from Kiev, Ukraine, in 1988 at age twenty, echoed: "I was always distrustful of the Soviet Union. I basically hated the Soviets from my formative years." His terse recollection was deeply felt, and his distrust of Soviet institutions reflects the cause of many difficulties that citizens from that country have had in coming to grips with exhibiting and even accepting trust because of the hostile institutional environment of the former USSR that imprinted them so negatively.

Perhaps the most telling story of the difficulties of overcoming such a background was recounted by **Daniel Barenboym**, Chairman and CEO of Collective Learning, who came from St. Petersburg in 1990 at age fifteen. Reflecting back on earlier years, he said: "There's always been the feeling that someone's about to lie to you or cheat you, or do something wrong and harm you in some sort of way. We've gotten away from that feeling, but the DNA in us in reference to that is still there." In contrasting his subsequent experiences in the United States with his Russian experiences, he said: "So, a lot of times, when you call American companies, they will give you the benefit of the doubt to give you the chance to prove that you are good. But the process might be very long and extensive, and it might take a long time to convince them. But they'll give you the benefit of the doubt. With Russians, if you try to approach them and they don't know you, they will instantly assume that you are there to somehow screw them, that you're there to lie to them, to get their money. And they won't give you the benefit of the doubt. However, if you manage to pass that situation, the closing cycle is much faster. Once the Russians trust you, they trust you, and there is an instant rapport. With the Americans, it's not always as easy. Whereas Russians are more like, 'I feel good about it,' Americans are more like, 'This better work for me.' So just different priorities, I guess." Barenboym provides lucid insights into the comparison of gaining trust between Americans and Russians, illustrating both positive and negative aspects of each. It may take Russians longer to establish trust, but once accomplished, they are more likely to continue that relationship, whereas Americans need to know that the relationship will continue to work positively for them. The contrast reflects the difference noted by many interviewees between the more emotional style of relationships on the part of Russians to a more pragmatic one for Americans.

An interviewee who explained the difference between Russians and Americans regarding trust was serial entrepreneur **Sergei Burkov**, who founded Alterra and came to the United States from Moscow in 1989 at age twenty-one: "On the difference in trust between Russia and the US, for Russians, by default, people are bad. You just don't trust people. You trust your childhood friends or your sister's friends, or those with whom you studied together. So, somebody you know, yes, you can trust them, if you know somebody very well. But here in the US, where you might have three people, one from Google, one from Facebook, and one from IBM, who join together to start a company, that would be kind of unusual for Russia." In explaining further how he might accomplish due diligence in order to verify trust, he noted: "Here in Silicon Valley there are reputations. There are people who you could ask for references, people who were with other people in the past – there is a network. So if you want to join forces with somebody, it's very easy to ask people who worked with him in the past, what they think of him. And sometimes they would say, 'No, don't do it.' And sometimes they say, 'Yes, he's good.' But again, you kind of trust, you assume that they are good, that they are honest." In contrast to the more typical approach of those from the former USSR, Burkov began with an acceptance of generalized trust rather than depending only on particularized trust.

We now turn to **Nerses Ohanyan**, Director of Growth Analytics at Viki, who came from Yerevan, Armenia, in 1998 at age fourteen. His perspective has poignant consistency with our opening comments on trust. He clearly articulated the importance of trust in managing remote teams: "Trust is always the number one key when you have remote teams. In fact, it's the number one thing in business and in running a company. But it's especially important when you're in a remote situation. There's nothing more valuable than knowing you can trust a person across the ocean."

Another interviewee who noted the importance of building trust and who learned to do so while working with Western multinational managers back in Russia in the 1990s is serial entrepreneur, **Dimitri Popov**. He was originally from Zelenograd, Russia, and came in 2014 at age forty: "I was trying to be very open about how I made decisions. For instance, even with decisions that everybody expected me to take and nobody actually cared, I would issue a very short paper – written, with my signature on it – and would share it with colleagues. I would say, 'I think this is the right way to do it, and if you don't have any objections, I will do it and later you will have this paper, and if I am mistaken, you will remember that I was that guy who actually recommended it. But if I was right, I will expect the credit for that as well, the credit just to say, 'Dimitri, you were right.' That's it. That's all I wanted. And it kind of worked. It kind of worked because it contrasted with the regular practice of Russian managers who would kind of try to avoid responsibility. In short, commitment, written commitment." Popov's comments and his method of building trust are consistent with a well-accepted definition, the essence of which is that trust involves a person having confidence in what to expect in dealings with another person.

The experience of **Ilya Kabanov**, Head of IT Compliance at Schneider Electric who came from Moscow in 2014 at age thirty-seven, illustrates the fundamental role of trust in business: "Essentially everybody in the world wants to run a business with a company they can trust. It's very natural for a person. You want to buy things from a company you trust, you want to buy products you can trust, and you want to make a deal with a person you can trust. Then you have to think about the vehicles of trust and the instruments of trust. Let's compare the US or European countries, which have mature and enforceable legal instruments, with developing countries that lack those instruments. In business transactions, suppose we have two persons in different countries who want to do business together. And they're trying to assess the instruments of trust that they have. In the US you'd say, 'Okay, I have a contract, I have a court, I have all the instruments to create trust using the standard off the shelf instruments.' And it's fast and convenient because it's a ready to use product. You sign the contract, you know that it's enforceable, it's a level of trust that can be established in a second." He was, of course, describing the environment of generalized trust he found in the United States due to its strong legitimate institutions.

Kabanov then described environments relying on particularized trust: "Then you come to Ethiopia, or Russia, or China, and you try to use the same instruments that you are familiar with. You say, 'Okay, I have a contract. The instrument looks good, very similar to the instrument that I use in my own country.' However, it's not the same tool that can be used to build trust because it's not 100 percent enforceable. I cannot rely on this document to build trust with a partner. Another example can be a judicial system in a developing country. The system looks similar, there are legislations, courts, lawyers, but in the country it's a ridiculous system that can't be used to enforce something transparently, not a system that can be used to build trust. So the next alternative is, 'I have a friend who knows this guy. Probably a solid reference who can become an instrument to build trust.' Because you want to do business. And then you find, 'OK, it's not as good as a contract, it's not as good as being able to go to court, it's not as good as money in a bank account,' but you want to do some business. And you finally use this vehicle as a way to establish trust."

In a twist to his story, Kabanov added: "But what's interesting is I've found so many similarities between Russia and the US. It's the extent of networking here. For instance, I interviewed people in the US, especially people who came from Europe, from Russia, from all over the world, as well as native-born Americans. And all of them said that the value of networks is tremendous in the US and even much stronger than in Europe. Your network is almost the most important vehicle to do business here. In Europe, I would say, 'No, you have to spend more time in advance, and it cannot be so straightforward or direct.' So networks work really well here and for me. It's good, I like the transparency, effectiveness, and efficiency. So is it something new for me? Yes, and it's a thing I absolutely like." Kabanov had analyzed the contrast between gaining trust in a more developed institutional environment and in environments with less legitimate formal institutions and even the importance of informal institutions like networks in the United States

A very different experience and point of view was described by **Dmitry Kuzmenko**, Senior Software Engineer at Google, who came from Ukraine in 2005 at age twenty-three: "So, first of all, in a new environment you just trust people. I mean, you have to trust them. But there are different trust models for different professions, for different psychological types. Some people, if they've read something from a trustworthy source and they trust that source, they would trust what's written. Engineers, however, are usually the kind of people who follow a more mathematical approach. They trust what they can prove to themselves, what they understand and can follow through reasoning." Kuzmenko's story is limited to considering trust in a technological work environment, an important insight given the context of this book with its focus on the US innovation economy.

An even sharper contrast came from several interviewees who expressed resistance to, or at least skepticism in, accepting trust too readily. **Oleg Rogynskyy**, Founder and CEO of Semantria who came from Dnepropetrovsk, Ukraine, in 2003 at age seventeen, recounted: "I learned from my dad's mistakes back in Ukraine. I learned that you can't really rely on partners until the rules of the game are defined and there are checks and balances. I learned that you can't completely trust people until they're proven. So the American mentality is to trust and then pull back when they screw you over. I learned to first make sure that they are trustworthy, and then trust. So I learned to deal with a lot of cultural stuff." A somewhat similar example was recounted by **Alexei Masterov**, Product Manager at Google who came from Penza, Russia, in 1996 at age sixteen: "I had a professor, Daniel Cohen, at Hunter College in New York, and I don't remember the words exactly, but he used to say that the Russian national pastime was complaining and skepticism, or something like that. He had Lenin posters in his office and stuff like that. A very interesting guy." Thinking about himself, Masterov elaborated: "I'm still fighting with my skepticism part, but I hope I don't complain as much as I used to." Regarding his lingering skepticism, he noted: "I just had this conversation with someone recently. I think the skepticism part sort of comes from risk management. It's almost easier to bet on being right if you're skeptical because, you know, good things or risky things or great things don't happen as often as ordinary things and bad things. But if somebody wants to do something crazy here at Google, as many people do with things that might initially sound outrageous, it's almost easier to say that it's not going to work if your goal is to be right. But that's not the attitude that's going to get you anywhere. So I fully realize that.

"But I also realize that sometimes I'm being skeptical just because I came from that culture. So to help overcome this, first of all, I guess I've changed my goals a little bit, like being right is not an important goal anymore. You don't always have to be right to be a good person or to be successful, or whatever. If your goal is to achieve something great, then you have to take risks, and you may have to be wrong every once in awhile. Relationships in Russia were much more informal, and that may be a good thing because they have more friendships, it's all about relationships. But, at the same time, if somebody doesn't like you, they feel entitled to tell you about that and yell at you. It's just part of the Russian culture. And another thing that is very typical of Russian culture is the high level of skepticism. So I mean, historically, I think there was no motivation. In the Soviet era, there was no motivation for people to take risks because the system didn't promote that. It didn't promote risk takers, it didn't promote people who wanted to defy the order. I mentioned that the educational system wasn't designed for that. If you know about the history of innovation in Russia, for the most part, it was done under the point of a gun. Google has this risk-taking culture, and it's actually encouraged. I just used some of the same terminology and the same words that I had seen others use to help inspire them if we were to succeed." Masterov's insights provide additional illumination on the skepticism and risk aversion found in Russian culture, which also applies to that of the broader Soviet culture.

Another interviewee who expressed very clearly that he is highly skeptical – in a business environment at least – is **Eugene Buff**, a technology consultant at Primary Innovation Consulting who came from Moscow in 1994 at age twenty-seven. He said: "Speaking about myself is always hard. I think I have the reputation that I'm a no-BS kind of guy. Some time ago I decided to take an aptitude test that was going around. And, of course, our wives told us that the company was wasting our money. They could have told us the results right away. So it turned out that I had very low numbers on 'nurture,' but I had 99 out of 100 on 'skepticism.' So they kept saying, 'If I'm saying something is OK, it's actually incredible. And if I'm saying something is bad, then it's probably really OK.' So I'm not very good at sugar-coating in that sense. I call things the way I see them. Sometimes I do that with disclaimers that it's just my opinion, and sometimes without that. I think in business people appreciate that. I think my skepticism is a combination of Jewishness and being born in the USSR. Everything you say or hear is probably not true, unless it's proven." As Daniel Barenboym noted earlier in this chapter, Buff seems to indicate that skepticism was in his DNA, likely as a result of imprinting in the former USSR, where, as he noted, there was very little reason to trust people outside of one's personal network, perhaps amplified by further imprinting during his time in Israel.

Another interviewee who might fit into the skeptical group is **Sergei Ivanov**, CEO of Optromix, a company that engages software groups back in Russia in their projects. Originally from Murmansk, Russia, he came in 1992 at age twenty-six. In speaking of trust, he noted: "It's not that easy to earn trust with Russians you don't know. And it was really bad during the 1990s and 2000s when even old friends and classmates betrayed each other like crazy. So, in Russia, you can't really, by default, trust anybody. But once you interact with some-body for a fair amount of time, some level of trust happens. I read somewhere that one difference between Russian and American cultures is that for Russians, a contract is sort of the beginning of negotiations. And that was the case with the company I was working with in Russia." Ivanov had a better experience in a more recent business transaction: "I met some people from Russia in the middle of last year and we started the project at the beginning of this year. One of them was interested in bringing three different products into the market. So far it's been good, and so far the guy hasn't broken his promises." One reason for his reluctance to trust quickly was likely his imprinting experience during the breakup of the USSR, when he noted that even good friends and former classmates turned out to be untrustworthy toward one another, partly due to the chaotic condi-tions of that period.

Other interviewees noted the importance of trust in business, including **Katya Stesin**, Founder of IoT fashion firm, Fit-Any, who came from Moscow in 1991 at age twenty-three: "You can't really expect trust right away. You just have to extend trust before a trusting relationship can develop, but sometimes that doesn't work. You never know even with people you've known for years, because sometimes they change if they get into tough circumstances. I've had that experience. They break their agreements, and almost anybody could do it depending upon the pressure they're under. But you just have to move on, you know. That's life. You have to be positive as an entrepreneur because otherwise you can't do what you need to do."

Another interviewee with an experience similar to Ivanov's back in the USSR is **Anton Voskresenskiy**, CTO of Northern Light, who came from St. Petersburg in 2006 at age thirty: "I think a lot of things changed in Russia in regards to trust. When I was growing up during the breakup of the former USSR, the system fell apart and people were, you know, trying to make a living, stooping to things they would never do otherwise in better times. I think I was experiencing degradation, at least that's my take on it. When I was living there, every year was worse in terms of trust, in terms of people being decent. I think that the pressure of all those things that happened with the chaos in the USSR, and then Russia, ended up being an unbearable burden on the morality of people, at least that's my take, and I was sometimes thinking I would just have to get out. I was not very happy with how things were in Russia in that regard. So when I moved to the US, I was relieved to see that, you know, it's not something that was happening everywhere, but it was something that happened with that particular group of people."

From a more general point of view, Voskresenskiy offered interesting insights in comparing the openness shown by people in the United States, including their readiness to smile: "I'm not sure this is how people do it all over Russia, but in St. Petersburg, at least, you don't see many smiles on people's faces when you walk down the street. And it's not because these people don't know how to have fun. They're not necessarily friendly to you when you approach them, but when you befriend them, they can be very friendly and exhibit sympathy or empathy after you establish a relationship with them. But before that, they envelop themselves in this protective shell, which I kind of grew weary of." Voskresenskiy's time in the former USSR had certainly influenced his thinking about trust. And his dismay at the nonverbal communication of lack of smiling and the withdrawal into one's shell that he witnessed in St. Petersburg was again a barrier to establishing generalized trust. Like others, he noted that even particularized trust with another individual took a great deal of time due at least in part to the low level of generalized trust that imprinted people, but that was susceptible to being overlaid with different imprinting in a new environment.

Shalva Kashmadze, Product Manager at Pocket Gems who came from Tbilisi, Georgia, in 2011 at age twenty-five, had more to say about trust: "In terms of trust, I've never had an issue with that." He then went on to add: "I think the one thing you need to do is to kind of recalibrate how you capture social cues because the way people behave with you, the way people smile at you, and how friendly they are, have different consequences here in America than in the former USSR. People are really friendly here, and, at the same time, they are still distant, and the amount of friendliness, or how friendly they are, does not really reflect how close you are with them. So even someone who you're seeing for a second time might be super friendly, but you should not assume that you're a member of his close circle. This means you just have to be more careful about those cues. But in terms of trust, yeah, I think there are no problems."

Kashmadze added: "Right now I'm feeling much more comfortable than back there in Georgia. I think the environment here, at the end of the day, is more meritocratic. I mean, as long as you're good at what you do and you're getting your stuff done, people are going to trust you and give you more stuff to do. It's not like back home in Georgia, where it's all about navigating who is whose friend and, like, how does the political structure work and who are you going to be friends with." Kashmadze also noted a topic that others mentioned, the importance of cues in both the former USSR and the United States. Smiling, for instance, was noted by several interviewees, either the lack of it in the former USSR and its frequency in the United States. Voskresenskiy noted the lack of smiling back in St. Petersburg and how that bothered him, and how refreshing it was to see people in the United States smiling easily and frequently. Kashmadze, however, emphasized that, on a personal basis, not too much should be read into another person's smile, such as interpreting it as a sign of a close relationship. The stories from various interviewees in this chapter have illustrated different views of openness to trust as well as the differences of being affected by various levels of imprinting that occurred not only in the former USSR, but also in their new environments, that influenced their views and beliefs.

We included several related topics in this chapter that we see as being elements of the soft skills necessary in technology workplaces. We noted that these skills were emphasized by a number of sources as being deficiencies in technological professionals working in the US innovation economy. Interviewees related accounts of probable weaknesses within their soft skills that would likely be transferred to the US business environment, such as in teamwork and communication, as well as in leadership and managerial styles. We first covered teamwork since we wanted to illustrate how well interviewees had adapted to their new workplaces in the US innovation economy, and we illustrated that most were able to do so quite successfully both as leaders or as participants in teams. Most were involved with technical teams and some with remote teams, particularly with their companies' operations back in the former USSR. We believe that they generally illustrated the same success in adapting their leadership and managerial styles, but with notable exceptions. We then covered problems they had with communication, both verbal and nonverbal. The major issue within their soft skill set seemed to be difficulty in developing a more trusting attitude, which a number of interviewees attributed to the skepticism they had developed in the former USSR. That environment was described as fostering skepticism, which led to a lack of generalized trust, characteristics we attributed to the imprinting they had received in environments of weak legitimate institutions.

These conclusions are consistent with literature about life in the former USSR that we described in Chapter 2. We discussed in this chapter how and why trust, especially generalized trust, is a foundation for soft skills. It is important, then, to discuss potential mechanisms for developing and improving one's soft skills to be successful in the US innovation economy, beginning with a willingness to be more open and trusting. Those mechanisms can include the organizations in which the interviewees work, specifically the culture and training that is often provided in various soft skills, as well as the mentoring, networking, and other sources of support noted in Chapter 7 that could accomplish the same objectives. Our interviewees validate these sources for improving an individual's range and depth of the soft skills typically needed to advance careers. In all of these situations that have the potential to help develop soft skills, individuals need to be open and willing to change their attitudes and behaviors and thus receive new imprinting in their adopted environment. In the process, they would receive new layers of imprinting on top of their experiences in the former USSR. This usually disruptive process is fundamental to creating a new identity more consistent with their new environment but also may create potential conflicts with the identity developed during their formative years in the vastly different institutional environment of the former USSR. The next chapter explores the important and complex search for identity that almost uniformly resulted from the migration experience of our interviewees as they experienced a markedly different institutional environment.

9 | Identity
A Constellation of Influences

The previous chapter on workplace adaptation and the development of soft skills ended by noting that individuals need to be open to changing their attitudes and behaviors and to receiving new layers of imprinting on top of the one developed in the former USSR. In Chapter 1, we concluded that this was a disruptive process but usually critical to creating a new or evolving identity for the interviewees, one more consistent with their new environment. By so doing, it could well facilitate their adaptation to, as well as their potential contribution to, the US innovation economy. Furthermore, as we noted earlier, those with bicultural or even multicultural backgrounds could well contribute with a different perspective than those without that background in that they could see situations in a potentially more creative way. Gaining a better understanding and acceptance of their own identities could provide interviewees with insights into ways they might most effectively function and contribute. We have stressed, once again consistent with Chapter 1, that this process could create potential conflicts with the identity developed during those early formative years before coming to the United States. Having the interviewees themselves articulate such complexity could help them understand the continuing process of identity formation they might be experiencing. And such insights could be helpful in dealing with and managing that complexity, moving them toward positive ends in their new institutional environment.

We reach this conclusion because, as noted in Chapter 1, understanding and coming to terms with one's identity is tangential to self-awareness, which is considered a career meta-competency that facilitates the learning of other competencies,[1] and a clear sense of self and identity is essential for making effective career decisions and

---

[1] Douglas T. Hall, *Careers In and Out of Organizations* (London: Sage Publications, 2002).

achieving success.[2] Such introspective insights can be crucial for immigrants like our interviewees not only to advance their careers, but also – in the context of this book – to utilize their abilities to make effective decisions that will contribute to the US innovation economy. It is precisely for this reason that we have emphasized the importance of identity as a cornerstone for analyzing the potential of immigrants from the former USSR to make such a successful transition.

All of these topics will be seen playing out in interviewees' stories. Not surprisingly, many or most considered their identities to be complex and often confusing, reflecting the multiple layers of identity developed in the process of dramatic changes in their environments. Those environments had vastly different institutions, both in broader society and in the workplace. Thus the complexity, whether it adversely affected the interviewees or not, usually involved a number of identity layers that, although not uncommon, are inherently exaggerated in immigrants like our interviewees. These initial insights refer back to Chapter 1, which introduced the theoretical foundations of identity. An individual's identity is largely acquired through imprinting by the environment, and while imprinting can be permanent, it can be altered as a result of a major change in an individual's environment, as will be illustrated by our interviewees. Although such situations can sometimes cause distress and uncertainty, as was expressed by some interviewees, it also raises the possibility, as shown by research, that earlier identities might not be abandoned but remain part of a person's self-concept and continue to influence that self-concept as well as one's professional life. Accordingly, one of the themes from our interviewees revolves around the complexity and confusion many faced in expressing their identities.

In describing their identities, a large group responded that "American" was their primary identity, although many noted that they were "American-Russians," still maintaining a link to their homeland, as did others from Armenia or Ukraine, for instance. Some interviewees described their identities with their homeland first, such as "Russian-American," putting their Russian heritage first because they grew up in the former Soviet Union and/or were steeped in the culture,

---

[2] Brad Harrington and Douglas T. Hall, *Career Management and Work–Life Integration* (Thousand Oaks, CA: Sage Publications, 2007).

and all noted the Russian language as being an anchor to that ethnicity. Some from other former Soviet republics responded similarly about their homelands. Another large group described their identities as being "Russian" for the same reasons and did not include "American." Again focusing on ethnicity, some interviewees responded that they were "Jewish," typically not from a religious perspective but more from an ethnic and cultural point of view. Another group chose to identify themselves by their professions, accomplishments, interests, and passions. A fairly large group noted that the complexity of their ethnicity and experiences prevented them from having a clear view of their identities. However, a rather large group had come to grips with their complex identities by considering themselves to be global citizens, internationalists, or cosmopolitans who were able to adjust effectively to different countries and cultures, usually through travel or friends from different places and an openness to new cultures and environments. We see these responses as being consistent with the research framework presented in Chapter 1.

American or American-Homeland Identity

One of the most frequent responses to the question of their identity was "American" or "American" followed by the ethnicity of their homeland in the former USSR, for instance, "American-Russian" or "American-Ukrainian." We also include a few who might have mentioned their homeland before "American," but whose response seemed to emphasize an American identity. Yet even in this entire group there was often a note of complexity that permeated responses from a large number of the interviewees. As their responses will show, some were unequivocal in their responses, at least initially. Far more prevalent, however, was an element of complexity and even confusion among responses from the various groups that we will describe. In summary, although some interviewees were unequivocal and clear in the descriptions of their identities, we see most responses as containing signs of complexity.

A fitting opening for this section, one that incorporates the elements of complexity and searching, is the story eloquently recounted by **Simon Selitsky**, Cofounder of Coingyft, who came from Moscow in 1991 at age twenty-one: "So I think I'm American. I'm an émigré from Russia who is trying to become American to the extent possible." Yet he began his remarks with: "Well, you know, that's a difficult question, because truth to say, I never had a cultural identity before coming to America. They say that in order to become Russian, you have to come to America, so I was obviously not a Russian there. From a very early age, we Jews proved to be outsiders and discriminated against in our own country. So eventually we were immigrants, and that was the background I grew up in. But I wasn't really Jewish in the traditional sense. I didn't have a Jewish upbringing, and I didn't go to synagogue. I didn't even know what Yom Kippur was until I was over twenty years old, so I never had that culture. The group I could identify with the best was a small subset of intelligentsia in Moscow with Jewish back-grounds who were in the same boat as I was. That was my cultural identity. So they say, you have to come to America to become Russian, because now I'm Russian among my neighbors in Lexington because I have a Russian accent. And even for Jewish people in Lexington, I'm sort of Russian because I'm secular. They call me Russian, so I became a Russian. But before that, I was nobody. I think one of the luxuries of living in America is that you don't need to think about identity unless you absolutely want to, and that's definitely not a focus of my life. So again, I think I'm an American, an émigré from Russia who is trying to become American to the extent possible." Selitsky's experience seems indicative of the confusing journey of many interviewees seeking some satisfying conclusion about their identities.

In a somewhat similar story, **Eugene Boguslavsky,** Release Engineer at Facebook who came at age fifteen in 1989 from Minsk, Belarus, recounted: "I'm a US citizen. As soon as I was eligible to apply for that status, I went for it. You know, I was living in this country, and I lost my Soviet citizenship. They took our passports away, so we were kind of nowhere. We had green cards here as soon as we arrived, but I wanted to feel like a full-fledged citizen, so I got my citizenship right away. I still say I'm kind of Russian-American. I still maintain my Russian identity as much as I can, even though I probably have fewer Russian friends here in California than I do in New York and some other parts of the country. So I'm probably not as close to Russianness because I don't have a lot of Russian friends here." A story similar to those of Selitsky and Boguslavsky on trying to be an American while retaining an appreciation for her ethnicity, was recounted by **Natalie Hill,** a Senior Software Engineering Team Lead at Natera, who also came from Minsk but in 1995 at twenty-one: "I'm an American citizen, but I'm also a working professional mom, and I'm trying to be successful in both of those jobs. Again, it's trying to keep a balance. I would like to see myself as a balanced person who is advancing, and as an American living in this country, to better know its culture. I'm educated enough and want to contribute to the success of this country, but at the same time, I want to keep in touch with my ethnicity and keep up with my traditions, trying to get my children to also have a bit of that in them. In between time, I'm trying to enjoy life."

Another such perspective is that of Sanofi Genzyme Senior Manager **Sasha Proshina**, who came from the Siberian city of Tomsk, Russia, in 2002 at age twenty-two: "I think my identity is American of Russian heritage. Definitely I love America. I love working here. I love the equality between men and women, and although it's not totally real, it's much better than in Russia. And I like the protection you get as a female in the workplace, and in life itself. And I always believed in that. I think there is still a long way to go to be completely equal. I don't think it's going to happen in this generation, but maybe my daughter will be successful in seeing that." **Alex Bushoy**, Founder and CEO of New Concept Group who came from Chisinau, Moldova, in 1991 at age thirty, responded similarly to the four interviewees immediately preceding: "I would say that I still feel part of Russian history and culture, but I'm an American 100 percent, in all aspects of it." Likewise, **Timur Shtatland**, Development Scientist II at New England Biolabs who came from Kiev, Ukraine, in 1988 at age twenty, stated: "I see myself as somebody who is more an American citizen, with a great desire for America to do well. I want to stay here in this country until I die, and for my daughter to stay here, too. Obviously, I have Russian heritage, I mean Russian cultural heritage." Bushoy grew up in Moldova and Shtatland in Ukraine, but both felt an appreciation for Russian history and culture more than Moldovan or Ukrainian. And they felt very American, even while acknowledging their appreciation of Russian history and culture, a common theme among many interviewees.

An interviewee from Russia who identified as Russian American had a party upon receiving American citizenship: "It was a very emotional event in my life because I love America. I can say that with all my heart. I love America for all it is. I think I intellectually learned to love America for the choice, for the freedom, for the safety. But I don't renounce my Russian citizenship. Having spent many years of coming to peace with who I am as a human being, I learned to love my Russian self and to love Russia. I learned to love both of my identities because, without the Russian part of me, I wouldn't see America in its wholeness and as a land of opportunity. I wouldn't be able to appreciate everything America is without being Russian. And vice versa. I understand that my Russian self is probably the driving force behind me being able to fit in here in the US."

Anna Dvornikova, Managing Partner at TEC Ventures who came from Moscow at twenty-one in 1999, explained that, except for the Russian language, she did not see herself as Russian: "The US is the country that gave me everything I have. I think that a big part of it is because of the opportunities that are created here and so, of course, it is a conscious decision for me. I consider my primary identity to be American. Just think about this. I came here when I was twenty-one and I'm now thirty-seven. So I have spent the major part of my adult life here." Like Dvornikova, **Anna Winestein**, Founder and Executive Director of the Ballets Russes Arts Initiative who came from St. Petersburg in 1991 at age nine, noted: "In most environments, I took pains even when going back to Russia to emphasize that I was mostly American. I emphasize to Russians that I may speak the language, and I may be from there, but I'm not one of them. I do this because I didn't want to be treated as a Russian, and as a Russian woman. This is because people there treat foreigners better, especially foreign women. And it's sad, but I just saw that I could do more. I could be freer in America, and being in Russia made me more consciously realize what being in America allowed me to be. And very much through self-invention, I was able to define who I was and what I wanted to do and get a chance to do it in a way that I would not have had in Russia."

Google Senior Software Engineer **Dmitry Kuzmenko**, originally from Ukraine and raised in Latvia, who came in 2005 at twenty-three, had a view similar to those of Dvornikova and Winestein. In explaining why he felt comfortable as an American, he noted: "This feeling of time being valuable was something that I had very early in Latvia and in Ukraine, but most people there don't treat time the same way Americans do. So, yeah, it's probably one of the things that made me closer to America from the beginning, my approach to time. If I want to spend time on something, I want to get the maximum out of it, I want to make it productive, I want to put a lot of energy in it, or I just might not do it at all. I got my green card after three years in the US. It was sponsored by Microsoft and then I got citizenship five years later." **Igor Razboff,** Founder and CEO of Scoros International who came in 1982 at age twenty-eight, had a similar view: "I cannot identify with anything else more than with the US, but I did that from day one, so that's not surprising. I just understand it better now. I understand that there are lots of opportunities. I understand that there's lots of flexibility in this society, and you can find your own niche with whatever you want to come up with. So that feels good to me."

Regarding his identity, **Semyon Kogan,** Founder and CEO of Gen5 Group who came in his late thirties from St. Petersburg in 1989, responded: "Yeah, so it's tough because at the beginning everybody goes through the same phases. For the first couple of years you correct everybody, saying, 'No, I'm not Russian, I'm Jewish,' because back in the Soviet Union, you know, there was a difference. It had nothing to do with your religion but rather with your ethnicity. Culturally, there is a real difference. This is home. Back there, it was not your home, and the more I've lived here, the more I hate them because of how unfairly they've treated Jews. So for me, I always say that America is not a perfect country, but it's the best, especially if you're an immigrant. As to my identity, culturally, of course, it's from a different culture back in the old country. And I would say that there are two things that I don't hate but rather love about it. I take my hat off to education there because to have a good education was always important for my parents, for my friends, for the whole society. First of all, I got exceptionally lucky that in high school I got into a special school that still affects my life. And the second thing is the layer in society where I existed, where it was really important to be smart and professional, and also to be an interesting person. Reading, going to the theater, and learning about different cultures and histories were equally as important as being a great scientist or doctor. I personally divided my entire life between art, the theater, and science, and I used to have my own theater. I miss that here. Work is much more important in America." Kogan appreciated the freedom he found in the United States. He also appreciated the Soviet educational system that gave him the capability to succeed in the United States, a major theme we introduced in Chapter 3 and have woven throughout the book.

Eugene Khazan, Chief Product Officer and Founder and CEO of Braveleaf who came from Riga, Latvia, in 1991 at thirteen, was also

Jewish and responded similarly to Kogan, but with the added complexity of being viewed as Russian while growing up in Latvia: "My identity, that's a tough one because it's still evolving, I think. But definitely, I identify as American. There are a lot of people I know who go back to wherever they are from, Riga or Moscow or whatever, and are very attached to those roots. I never really was. I mean, I have memories and I've cherished those memories for what they are, like when I was growing up. But I never had the pull, and I never went back. And so to me, I definitely don't identify as Russian because I'm not ethnically Russian, but Jewish. My Russianness was a source of problems for me in Latvia, and, in fact, it became a big deal because they became super-nationalistic there after the breakup of the USSR." Khazan, like Kogan, had faced difficulties in his native republic and came to appreciate the greater openness and acceptance in the United States, which, as some interviewees noted, may not be perfect, but it was the best they had encountered.

Practice Manager at LegaSystems, **Valentin Komarovsky**, who came from Moscow in 1995 at fourteen, responded about his identity: "I really don't think about it a lot. I do think I'm more American than the majority of the immigrants here because most of them don't actually know the culture or history of this country, while I've been very interested in the culture and the history. I probably know more about American history going back to revolutionary and colonial days than the average American does. I wasn't born here, but I've been a US citizen for thirteen or fourteen years now. I went to high school and college here. Does that make me American? I'm really not American, but more an Americanized Eastern European." Senior Scientist II at AbbVie, **Anton Manuilov**, who came from St. Petersburg in 1992 at age fourteen, illustrates the inherent complexity recounted by many interviewees, but in his case he had come to reconcile it: "Well, my identity is American. And also, American culture, that is dear to me, as are American values, if there is such a thing. I'm used to this in the US, and I enjoy it. When you go to Russia, they don't have that. I have dual citizenship, but neither of the countries really recognizes the other citizenship. So when I'm on US soil, I'm a US citizen. But if I'm on Russian soil, I would be considered a Russian citizen." Manuilov had not let any complexities of dual citizenship interfere with his clear identity as an American since he identified with and enjoyed American values, values that he felt were lacking in Russia. A somewhat similar situation was described by Science and Technology Associates Management Founder and President **George Gamota** who came from Lvov, Ukraine, in 1949 at age ten. He responded: "Well, as I say, I'm an American, but my origin is Ukrainian. It's very interesting how I perceive myself when I'm in Ukraine. It's totally Ukrainian. But then I slip out of there and I become American, although after working there on and off since 1992 with the science community, I have become more Ukrainian even in the US." Gamota's observation of how location can affect one's identity will be echoed later by others such as Sergei Ivanov.

Several respondents, like some of those presented earlier, identified themselves first as American and then added Jewish to their identity. One such individual was **Michael Fayngersh**, Vice President of Quality and Test at Delta Star, who was originally from Zaporizhia, Ukraine, and came in 1990 at age thirty-six after living in Israel: "I identify myself as American 100 percent, but I also have very strong Jewish identification. For me, what happens in Israel is much more important than what happens in Ukraine. But, of course, the first thing is what happens right here in the US where I live. I really do not identify with Ukraine at all, although I read and follow the events going on there. I would say, in general, for Russia, I really don't talk badly about the people, but I do when I'm talking about the State." Chairman and CEO of Collective Learning **Daniel Barenboym**, who came from Moscow in 1990 at age fifteen, responded somewhat similarly: "What's my identity? I'm a man without a country. No, my identity is I'm American. I'm American, I'm pro-Israel, and I don't care about Russia. I think they had thousands of brilliant people, and they lost it all to Israel and America because they hate Jews. I have no love for Russia whatsoever. I have friends there, and I feel bad for them. Many of them want to leave, but as far as I'm concerned, that country will never change. I don't think anything's going to get better for two, three, or four generations, maybe a couple of hundred years." **Stas Gayshan**, Managing Director and Founder of the Cambridge Innovation Center of Boston, who was born in Tashkent, Uzbekistan, and came from Moscow with his family in 1992 at age ten, noted: "I am very much an American through and through. And I think I also pretty strongly identify with cultural Jewish identity, very much so. I'm not religious in any way, shape, or form, but culturally very much so. And I have been involved with Jewish nonprofits. I don't know the actual demographics of this, but thinking about the Russian-Jewish movement that

happened here starting in the 1970s, the theory I have, and that I want
to actually test, is that most Americans who think they know Russians
don't actually know Russians. They know Russian Jews. And so the
American view of Russian culture is heavily influenced by what they
think Russian culture is based on knowing Russian Jews. But many of
us left Russian culture. And there was a reason. And so there is this very
different view. Ethnic Russians don't think the way Russian Jews think.
But there is a bunch of commonality, obviously, in certain years of
history, and it's very interesting."

Alex Petetsky, Director of Software Engineering at PatientKeeper,
was originally from Berdychiv, Ukraine. He came in 1991 at twenty-
four. "I am an American who is trying to hold very tight onto my
Russian roots. And I also happen to be Jewish. So I guess that would be
a third level of my identity, so to say. One of the reasons I went to
college in Belarus was because I did not want to stay in Ukraine.
I wanted to run away because I had had enough of Ukraine at that
time. I love Ukraine as a country. I love Ukrainians as a people. But
being a Jew in Ukraine is no picnic, not the greatest thing in the world,
and I just wanted to get out of there." Petetsky might not have fully
completed the journey through his complex identity that included
American, Ukrainian, Russian, and Jewish. His situation illustrates
how complexity stems from various layers of imprinting from different
environments, experiences, and institutions.

Medical Device Consultant **Felix Feldchtein**, originally from
Dzerzhinsk, Russia, and who first came in 1990 at age thirty, described
his citizenship: "I would say I'm triple, not dual, triple. So I definitely
identify with the United States, and I love this country. And I identify
with Russian culture. And I also identify with Jewish culture." He, too,
seemed to be dealing with a complex identity resulting from various
layers of imprinting, but, like Petetsky, he emphasized being an
American. **Leah Isakov**, Senior Director of Biostatistics at Pfizer who
came from Moscow in 1992 at age twenty-six, responded more
straightforwardly: "I would say that I am American of Russian descent.
I define myself as American and I really appreciate it. I appreciate this
opportunity that I was able to come to this country and have an
education for my kids. Also, I appreciate the fact that I can feel safe
here."

Sergei Ivanov, who is CEO of Optromix and originally from
Murmansk, Russia, came in 1992 at age twenty-six. He vividly illustrated

how being in a particular context can activate a specific aspect of one's identity and affect behavior – in this case, his choice of something as basic as a beverage: "I think the last time I thought about my identity was flying to Russia. I boarded the plane at Logan going to Frankfurt, and you know, had dinner and a little bit of wine. And I felt quite American. When I changed planes in Frankfurt and got onto the plane going to Russia, it was half-filled with Russians who were drinking vodka for breakfast. At that time, I kind of switched over and I felt myself quite Russian – and poured some vodka for myself. And I don't feel too much of a contradiction between those two parts. I feel like I'm quite American, and I like America. And I wish Russia and America could be allies. In terms of mentality, there is a lot in common between regular Americans and regular Russians. Certainly, the Russian mentality to me is closer to the American rather than the German mentality."

Although his background was also complex, CEO of Grid Dynamics **Leonid Livschitz**, who came from Kharkov, Ukraine, in 1991 at age twenty-four, seemed content with his identity: "What I really say is that I'm a typical American Soviet Jew. In some sense, my family is a typical hybrid like so many Jewish families in Eastern Europe. My mother's side came from Austria-Hungary, my father's side is from Belarus, and I grew up in Kharkov, Ukraine. I say Soviet because there was no Russia or Ukraine as countries at that time. You have to keep looking at history in order to be proud of your roots. So I tell my kids the meaning of the Jewish term for 'never again.' It's an important part of our family and heritage." His daughter, **Louiza Livschitz**, a doctoral student in clinical psychology at Stanford University who was three months old upon emigrating, confirmed how she and her younger sister had come to appreciate their Jewish roots: "My sister and I recently went to Israel on a birthright trip, and we spent a lot of time talking about what it means to be Jewish as part of our culture. We grew up in a very nonreligious family. We were very culturally Jewish. For generations, everyone was Jewish. The values – that's what we spent a lot of time with. We think it's kind of this innate thing where, if you're Jewish, you have these cultural values where it's more than just a religion. So the hard work, the family orientation, all that was very present in our family. We celebrate Hanukkah, but we never went to the synagogue, we never spent time with a rabbi. When we went on this trip, it really did get some thoughts going toward our connection to Judaism, not necessarily in a religious way, but what it means to be Jewish and growing up in society." She described their parents' reactions to their trip plans: "We brought it up over the summer when the war was going on there. I had told my sister, 'There's no way our parents are letting us go. Like, you know how protective dad is.' But when we asked him, he was like, 'Absolutely. Of course. Go. I want you to be there, and I want you to experience that.' So they were 100 percent for it."

Another Ukrainian, Brown University Computer Science Professor **Anna Lysyanskaya,** who came from Kiev in 1993 at seventeen, clearly delineated four important aspects of her identity: "Well, the first thing I say is, 'I'm an American.' The second thing I usually say is that my first language is Russian. The third thing I usually say is, 'I'm from Ukraine.' And the fourth thing I usually say is that I'm Jewish. And sometimes I say the last three in a different order, depending on the context. Now with the developments in Ukraine, I often say that I'm from Ukraine, second." Lysyanskaya included the importance of context, in this case, the political and military struggles in eastern Ukraine.

Homeland-American Identity

Although the responses of this group could be considered similar to some of those from the previous group, the nuance for considering the two groups separately was that the group in this section mentioned their original ethnicity first, followed by "American." In short, America was not the first component of their identity mentioned, but the second. This is in contrast to the preceding group that placed "American" as the person's primary identity or "American" followed by their original ethnicity. We believe this distinction to be important since we saw it as giving clues as to the degree and longevity of earlier imprinting, which for the group in this section was possibly somewhat more lasting than for the preceding group. Yet we realize that, in at least some cases, their responses might have been due to the linguistic practice of putting one's ethnicity before "American," such as "Irish-American" or "Italian-American." Still, many in this section did not just say "Russian-American" or "Ukrainian-American," but gave responses in which they described their ethnicity as being at least as important as any American aspect of their identity.

For instance, **Anastasia Khvorova,** Professor at the RNA Therapeutic Institute at the University of Massachusetts Medical School who came from Moscow in 1995 at twenty-six, stated: "I'm Russian. I was born in

Russia, I was raised in Russia, I speak Russian, I speak Russian in my home, I married a Russian, my parents live in Russia. I'm Russian. But if somebody were to ask me a question, that if I had to give up one of my citizenships since I have dual citizenship, it's going to be Russian. I will keep the American one because my life now is in the United States." Like so many interviewees, the pulls and tugs of different elements of identity caused thoughtful introspection, and Khvorova had resolved her situation by in some ways separating her core identity from her citizenship.

Animatron Cofounder and CEO **Dmitry Skavish**, who came from Yelets, Russia, in 1999 at twenty-nine, responded similarly: "Actually of course, I feel Russian. I came here when I was thirty. So I mean, if you come here when you're in high school or earlier, that's another story, but given my situation, of course I feel Russian. Almost all my friends are Russian, and if you really want to talk about something, like Russian stuff, you can only talk to Russians because you have the same background. And I believe you need to leave earlier than I did so that you can feel that you are really an American. I mean, I feel I am American, but sometimes I don't. I think I belong here. This is my country now. But I care about what's going on in Russia. So I don't know. It's kind of mixed feelings. I feel like I am Russian. So, for example, I have this kind of mixed feeling about taking the oath for US citizenship. You pledge allegiance that you would basically take up arms to defend this country. I discussed it with a Russian friend who obtained US citizenship: 'Yeah, so now you have to choose allegiances.' He also said, 'I guess I had the same allegiance to Russia, but actually it was not really a big deal. But it was kind of a big deal to take the oath in the US, and you're really saying OK, this is your country now.' So that's how I feel." Skavish's Russianness seemed to be the result of a relatively long period of imprinting in the former Soviet Union, as he explained, and that seemed to be responsible for his deep feelings for his Russian identity. Yet he had to deal with taking the oath of citizenship in the United States, which he did not take lightly because of the conflict it created with his Russianness.

Other interviewees described confusion but not discomfort with their identity, referring, like others in the previous group, to their identity changing somewhat depending on where they traveled. **Sergei Sokolov**, Director of Product Management at Altisource who came from Moscow in 1992 at twenty-five, said: "I would call myself Russian-American. When I started working for SmartBear I had to go to Russia on business three or four times a year. So I spent a lot more time there with family and friends, and it was still a small circle. But I got a somewhat different feel for Moscow, seeing it as a glamorous city with all sorts of stuff going on. It's interesting, I thought it was probably more dynamic than New York City, if you will, and of course, that makes an impression. I always have a great time in Moscow, but I still feel I'm a guest." In a similar vein, Founder and President of Auriga, **Alexis Sukharev**, who came in 1990 at age fifty-four, stated: "I think I feel more Russian than American, but I feel American as well. And for me, identification is a pride thing. When I was in science, I lived in Moscow, but I was Russian and American because I was traveling back and forth since very early times which was unusual in the 1970s and even in the 1990s. I spent a year at UC Berkeley as an exchange scholar in 1977–1978. We were always an American company, and that's why I felt American. In business I was more American than Russian because customers were usually American, with a few Russian and European customers. But America was 80 percent of our business." Sukharev recently started a nonprofit foundation to publish documents about the White side of the Russian Civil War right after the 1917 Bolshevik Revolution in honor of his maternal grandfather who was a priest executed by the Bolsheviks, thus illustrating that his Russian roots were clearly still very important.

Alexey Eliseev, Managing Director of Maxwell Biotech Venture Fund, had comfortably sorted out the different influences on his identity. Originally from Moscow, he came in 1992 at age twenty-seven: "I think that at this point I would say Russian-American. Definitely Russian because the ties are strong. Not necessarily to the country but to my friends there, to my relatives, you know, the culture I grew up in, the language, everything. It's strong. But I feel at home here in the US now." **Michail Shipitsin**, Senior Associate Director of R&D at Metamark Genetics who came from Novosibirsk, Siberia, in 1997 at age twenty-two, said: "I guess you could call my identity Russian-American. Russian maybe being in the first place, but certainly a big element of American as well." Northern Light CTO **Anton Voskresenskiy**, who came from St. Petersburg at thirty in 2006, noted the value of biculturalism that was described in Chapter 1: "I'm Russian-American. I definitely feel I'm different because there are certain things I'd like not to have, like my accent. I won't be able to do much with it or be able to pass as a native. But, at the same time,

I think I have some bigger perspective on the world because I've had the chance to live in two cultures that are quite different, and I think that has helped me in a number of ways." Using a slightly different term, eighteen-year-old **Nikita Pashintsev**, son of interviewees Alexander Pashintsev and Tatiana Kvitka, was born in Cupertino, California, and described himself as a Russo-American who speaks and reads Russian fluently and speaks Russian at home.

Sergey Kononov, Senior Delivery Manager at EPAM Systems, was born in a village near Kharkov in southeastern Ukraine. He came in 2011 at age twenty-nine. Sergey added another source of complexity in speaking of himself and his wife: "I was born in Ukraine and so was my wife, but I always called myself Russian. My wife calls herself Ukrainian. I wouldn't say it's because there are some differences between Ukrainians and Russians. She speaks Russian and I speak Russian, but she calls herself Ukrainian. But I grew up in Ukraine and most of my friends are Ukrainians, but then I moved to Russia and got many Russian friends. When I fly to Russia, it's my country, too. So it's hard to say what my ethnicity is. I can say I'm Ukrainian, easily, because I treat Ukrainians the same as Russians and the same as Belarusians. When I go to Belarus, it's like the same culture. There are no big differences – the same smart people." Kononov embraced people in various former Soviet republics in a positive way that helped make him comfortable with the complexity of his multifaceted identity.

Another story of complex identity was recounted by Stanford business school graduate **Shalva Kashmadze**, Product Manager at Pocket Gems who came in his twenties in 2011. He showed his affinity for his native Georgia: "I'm definitely Georgian, but for me my view of Georgian might be different from some others. For me it's really easy to call myself Georgian because my American friends just don't know what that means. So yeah, I'm telling them, 'Hey, I'm Georgian, and I feel like I'm Georgian.' For them, it doesn't mean anything, so I can do whatever I want to do. They probably think, 'Yeah, he's Georgian, and that's what Georgians do.'" In a more serious vein, Kashmadze added: "I do feel more tension when I go home to Georgia because I wonder about how much I actually have in common with my Georgian friends. The nature of what they do is not really something that I find really exciting. They're not working in tech, and now I'm all about tech, so my interests and their interests are diverging and they are really different. There is less and less I can find relatable every single time I go back home.

"Sometimes it's hard for me to understand, so I ask myself a question. I'm curious about what they think is my identity, how my friends are viewing me. Do they think I'm an American? I still feel myself as a Georgian, but for them, I must seem to be a really weird person. So I don't know, it's a question I have to ask myself. There is definitely a certain – I wouldn't say mistrust – but people view Western-educated Georgians when they come back as kind of different. They seem to know that 'He is becoming too American.' That's something my family or friends might be speaking about behind my back." Kashmadze realized, as the writer Thomas Wolfe maintained, "You can't go home again." He realized that his own increasingly complex identity, although clear in his own mind, was an obstacle for others back in his home country of Georgia. Their identities had not changed from decades of imprinting in the same environment, while his had been overlaid strongly with that of his new institutional environment in America. As such, his story is a poignant recollection of the gains and losses experienced by those who struggle to discover their increasingly complex identities.

Another millennial, in this case one who came from Yerevan, Armenia, in 1998 at age fourteen, had similar deep insights about the duality of his identity. **Nerses Ohanyan**, Director of Growth Analytics at Viki, identified with Armenia but also with America and described the conflicting situations he felt in both countries: "I'm Armenian-American. What I love most about being an American is that it takes nothing away from being an Armenian. People in Armenia might say differently, and some do, but I don't care. The ability to say, 'I don't care,' is a very American thing. It really is. It really, really is. Of course, I feel more American when I'm outside America because you identify as American, and it also helps you get your way as an American. When I'm in Armenia, I'm very Armenian, but my Americanness comes out from time to time, especially when certain things aren't up to my standard and I expect certain things because I want Armenia to be better, but I act very American in those instances. I expect it to be this way. And I'm very American when I act like that. When I'm here in the US, I identify with being Armenian more because here, we're all American. It's not as important in the US to be American." Ohanyan epitomizes the duality and complexity of identity experienced by many immigrants. He has clearly settled on his own identity as an Armenian-American, seeing the importance of his American identity, but always as one superimposed on his ethnic Armenian identity. He values both for the contributions each has made to his total identity and to his life, and he is able to decide when either will be more important in any particular environment. In many respects, Ohanyan displays cultural agility by learning to benefit from what for many could be a complex identity.

Homeland-Only Identity

This group includes individuals who basically identified themselves only with their homeland, and, if America was mentioned, it was tangential. In some cases, individuals had not been in the United States long enough to become citizens. This group differs substantially from those discussed earlier, all of whom mentioned American either as their first identity or after their homeland as a concurrent identity. The group discussed here, then, has continued to note their homeland as being the basic source of their identity. One could surmise that the earlier-mentioned groups in this chapter had either not been imprinted as deeply with the institutions and culture of their homelands or were well into a layering process in which the culture and institutions of America had prevailed or at least shared importantly in their identity. We acknowledge that there could be other dynamics involved in the responses of interviewees in all the groups, but we believe that imprinting has been a fundamental cause of interviewees' personal conclusions about their identities. Accordingly, even among the group in this section, interviewees often showed at least some degree of complexity.

Danil Kislinsky, Founder and CEO of IB Consulter who came from Nizhny Tagil, Russia, initially in 2002 at age eighteen, stated clearly: "I am Russian, and if I ever get the nationality of another country, I will still be Russian by nationality and by identity. First of all, I was born there. I respect the culture, and I respect my predecessors. That will never change. It doesn't mean that I do not respect other countries, other cultures, other people. I mean, we are all people of this earth. This may sound very big, but that is the way it is. I am very excited about traveling, and I am very excited about meeting people." Google Software Engineer **Anton Rusanov,** who came from Karaganda, Kazakhstan, in 2012 at age twenty-eight, responded: "I am Russian.

I am trying to learn more about Russian culture to show Russians from a good side, so that people can see more than what the TV shows about them, the other side of Russians."

Iryna Everson (born Yurchak), Manager of Procurement at Pattern Energy Group who came from Kiev in 2008 at age twenty-six, explained her mixed ethnicity: "I'm Ukrainian and maybe there is a little bit of Polish blood, but I consider myself mostly Ukrainian. Both of my parents were born in western Ukraine, so I definitely had a lot of western Ukrainian influence growing up because I would spend summers with my grandparents in the western part of Ukraine. I would also say I'm definitely fluent in Ukrainian, and I'm very proud that I can speak Ukrainian and also read and write it. In western Ukraine you usually speak Ukrainian."

Bella Gorbatcheva, Oncology Translational Project Manager at Novartis Institutes for Biomedical Research who came from Moscow in 1993 at age twenty-six, also had a mixed heritage: "I would say that I'm Armenian. I think it's just because of how much the influence of my dad was in my life. I was born in Moscow and grew up there, but for us, the culture there, and even the food, they were always Armenian and Georgian. My dad was born in Tblisi, the capital of Georgia, but his family came from Iran. They were Iranian Armenians. My mom is Russian, and that's why I am half Russian, half Armenian."

Sergey Markov, Software Engineer for Instagram at Facebook who came from St. Petersburg in 2007 at age twenty-eight, had several ethnic strains but saw himself as Russian: "I see my identity as mostly Russian. I still feel Russian because I grew up in the Russian culture and speak the Russian language. Yeah, that's my sense of identity. But it's a little bit complicated because I think that I am not really Russian. I am one-quarter descended from a small Finnish people in the middle of Russia, one-eighth Jewish, and one-eighth Latvian, so maybe I'm one-eighth Russian. Yet I still feel myself as mostly Russian." **Anna Uvarova**, Founder and CEO of software platform 3DBin who came from Tomsk, Siberia, in 2007 at twenty-three, had a similar response: "I identify as Russian because I was born and raised there. Even though my parents had other ethnicities, they raised me in Russian ways, all the traditions, all the celebrations, even how to treat a grandma. Everything was Russian."

A clear-cut response came from **Polina Raygorodskaya**, Cofounder and CEO of Wanderu who came from St. Petersburg in 1990 at age four: "When people ask me, I say I'm Russian, because that's just the way I feel. I was born in Russia and grew up with Russian parents and Russian culture." **Alexey Bulychev**, Principal Scientist at Moderna Therapeutics who came from Moscow in 1992 at age twenty-four, was also clear on his Russian identity: "Denying who you are is never good. What I mean is, in terms of tastes, habits, and perceptions, I am still very much Russian. I was born in the USSR, and, even though the country is no more, it left an undeniable mark on me. I still watch Russian movies and read Russian literature, even though for the better part of the week I read, write, and think in English. So to maintain balance, when it comes to leisure, I try to do things fifty-fifty, half in English, half in Russian."

An interviewee whose children influenced his linguistic choices, if not his basic identity, was **Rafael Soultanov**. The Founder and CEO of iBuildApp who came from Ufa, Russia, in 2001 at thirty-one and who is a US citizen, quickly responded regarding his identity: "Russian. But I live here in the US and my friends are here and my daughter was born here. I talk to my son in Russian, and I force him to read Russian books. But, my younger daughter, after she went to kindergarten and elementary school here, she speaks English. So I talk in English to her even though she understands Russian. When she comes home, it's hard for her to switch. To me, it's like, 'OK, if it's easier for you in English, we'll do it in English. I'll practice my English with you.'"

Sergei Kovalenko, Software Engineer at Verizon Communications who came in 1995 at age twenty-three, was also clear about his Russian identity, even though he came from Belarus: "I still think of myself as Russian. My friends are almost exclusively Russian. It just happens, we're not filtering out, it just so happens. When I came to work, there were two other Russian guys, and eventually we ended up going to each other's houses to visit. Now we have kids who are best friends, and now the kids go to Russian schools here. That's where you meet other Russian parents. It's a strong Russian community. They're all professionals, they've all got a very high education, they all have very nice jobs. But my daughter has a slightly different cultural identity. In fact, if you asked her, she would say, 'I'm Russian-American.' But when we travel, let's say in Europe, if someone were to ask where she is from, without thinking very often she'd say, 'We're from Russia.' She would not say, 'We're from America.' Not because we tell her for any reason. It's just like once

we're outside of the United States, she feels we're from Russia."
Kovalenko's story shows that his environment, and particularly his
family and friends, reinforce his feeling of Russianness so that he
can quickly respond in an uncomplicated fashion, "I still think of
myself as Russian."

Dmitry Kerov, Vice President of Software Engineering at Northern
Light who came from St. Petersburg in 2009 at age thirty-seven,
uncovered a basic dilemma facing many interviewees, particularly
those who had arrived relatively recently. He is a Russian citizen
contemplating applying for US citizenship. When asked about his
identity, he unhesitatingly responded, "Russian." His complete
response, however, brought out some of the complexity in his situation: "It's actually a really hard question because in the modern world
of globalization it's not clear. So it's hard, and I don't understand
what identity means. If it's citizenship, that raises real questions.
If there were to be a war between Russia and the States, what am
I going to do? I can't answer that question. If I'm going to become
a United States citizen, then I need to answer that I would have to fight
against Russia. I want to be a good citizen of the US because I live
here. I want to participate in voting and in social life, so this is the
citizenship aspect. So in this sense, I'm pretty ready to become
a United States citizen." When asked whether he felt he had figured
out his identity, he replied: "I'm trying my best. I'm OK with being
Russian. It's a great country besides those events in Ukraine and
Crimea, it's a great country. It's a great people."

The next interviewee could well provide a transition to a later
group in this chapter, one that opted for global citizen or cosmopolitan as their identity, as well as to the next section, on those who
chose Jewish as their identity. **Dmitri Krioukov,** Associate Professor
of Physics, Mathematics, and Electrical and Computer Engineering
at Northeastern University who came from St. Petersburg in 1994 at
age twenty-four, responded: "I see myself as Russian, I think. I still
think that. But I'm definitely cosmopolitan, and I don't make any
differences, and I can accept pretty much everything. I think my
mom is one-eighth Jewish, a little bit, and frankly I always somehow liked the Jewish culture, so half of my friends are Jewish and
my wife is Jewish. But I still feel myself as a part of this Russian
culture."

Jewish Identity

Several interviewees in this group also had an element of complexity since they prominently included their Jewish ethnicity as part of their identity. **Anya Kogan,** User Experience Lead and Manager of AdWords Display at Google who came from St. Petersburg in 1990 at the age of eight, noted how growing up in a strong Russian-Jewish community in the United States influenced her identity: "I'm still a Russian-American Jew. It's still a big part of me. It's just the communication piece that becomes kind of hard for me, but it still has shaped who I am so much. Like, when I cook, I like to cook Russian foods. And I'd like to go back to Russia at some point to visit, which I haven't done in a very long time. Also, when my family moved to Fairfield, Connecticut, a bigger Russian community started forming and more Jews were coming over. So we were sort of at the beginning of that wave in that area and it really grew. My parents were able to make a lot of Russian friends, and they have an extremely strong Russian community there. When I was growing up, we never had an American come over to our house. I think maybe once I had a teacher come over. So I grew up in a Russian community even though it was in the US. So it's hard for me not to reference that."

Another interviewee talked of three elements: "Russia is a part of me, but I think I'm more. I really love America, so that's part of it. I don't know if I have just one identity. I don't think so. I'm all of those three things – Russian, American, Jewish, all blended together. I can't just say I'm one thing or the other." A deep insight into the importance and sources of Russian, Jewish, and American identity came from **Kate Torchilin,** CEO of Novaseek Research who came from Moscow in 1993 at age twenty-two: "Yeah, I think probably Russian-American is my identity. Russian, with a strong identity of Jewish as well. My mother is Russian, my father is Jewish and Russian. I mean being Jewish probably has a different connotation in this country because religiously I don't identify as Jewish. But growing up in a country which is very anti-Semitic, and having a big part of my family being Jewish, you identify culturally. And so that's different. In the US, I don't think it resonates the same way. For me, being Jewish is more of a cultural as opposed to a religious way."

Irina Fayngersh, Senior Product Brand Manager at Nacera who came with her family in 1990 from Zaporizhia, Ukraine, at age ten, provided important insights similar to Torchilin's: "I see my identity as a Ukrainian-American Jew. We're not religious but we observe Jewish holidays and the connection to Judaism is very important to us at a very cultural level. So I grew up always being aware that I am Jewish although we weren't practicing, we weren't allowed to practice in the Soviet Union. My parents didn't even know the customs, and my grandparents were not even that aware. They were aware of maybe eating some specific foods connected to holidays, but they didn't know the history behind anything, or the meaning. It's really, really sad that tradition was lost in the Soviet Union, and this was another reason why we left." Raytheon Senior Engineer II **Valerie Gordeski,** who came from Russia in 1995, had a similar experience in Russia: "I identify as a Russian Jew. I'm not really religious because I don't really believe in God or anything like that. It's more of learning about the tradition because I did not grow up with a religion, because religion was banned in the Soviet Union. So Judaism is my ethnicity/nationality. I believe in that, and I know that in the States it's not regarded as such. So people, more often than not, scoff at it and say that it's just your religion, and you must have converted. But coming from a place where it is considered your ethnicity/nationality, my Jewishness is not my religion. My Jewishness is who I am." Gordeski was very clear about Jewish being her identity, specifying that she was a Russian Jew.

Alex Gurevich, Partner at Javelin Venture Partners who, like Irina Fayngersh, came from Odessa, Ukraine, in 1990 as a child, noted that his family background was a major contributor to his Jewish identity. In fact, it is the only identity he mentioned. He had come from Moscow with his parents in 1990 at age seven and settled in a heavily Russian-Jewish community in West Hollywood, California: "Being Jewish is big for me. It's a big part of my identity. It's a big part of my cultural identity. I think of myself as being Jewish first, and all other things second and third. I'm definitely very grateful to organizations like HIAS [Hebrew Immigrant Aid Society] and am definitely very conscious of it. All four of my grandparents are Holocaust survivors, and they went through crazy experiences that I could talk hours about. If I told you all the specific ways their lives unfolded, it could have turned out very bad. It's sad to say, but that's a big part of my identity. It's hard to deny that. I'm not that religious today, but I identify with it strongly." Like Fayngersh, Gurevich's background, and particularly the harsh realities of his grandparents' Holocaust experiences, seemed to have been deeply imprinted into his being and formed his identity.

Michael Schwartzman, President of ValueSearch Capital Management who was from Krasnodar, Russia, and came in 1974 at age twenty-eight, recounted his parents' struggles during World War II that affected his choice of Jewishness as his identity. He explained: "Being a Jew is only one facet of my persona – an important one but not the one that over-whelms other facets. However, it will not diminish to an insignificant one as long as Jews are persecuted, which practically speaking means 'never.' There are many reasons for that, but the simplest and the most powerful one is that it would be cowardly to have this any other way. I am Jewish, but I am also an atheist – and in the past, I also had been for several years a Vice President of a temple. Jewishness is a complex phenomenon."

Another such person was **Kira Makagon,** serial entrepreneur and Executive Vice President of Innovation at RingCentral who came from Odessa, Ukraine, in 1977 at age fourteen. She had attended Hebrew School in San Francisco: "I think more than anything else, it put me in touch with my Jewish identity, which I really didn't have in Ukraine. I knew I was Jewish, but it didn't mean anything except potentially that I might experience persecution. But that's it. It really didn't mean anything else there. So I think that's probably more important than the religious aspects of it. Certainly, we've become traditional Jews and observe certain things, but for the most part, it's not about religion."

Anna Scherer, Science Writer at Custom Learning Designs, came from Sevastopol, Ukraine, at age eleven in 1994. Her original name, Anait Ashovna Tsinberg, reflected her Armenian and Jewish heritage. Of her identity, she replied: "You know, I think the simple answer to that question would be a Russian Jew. That's what I would identify with, but I'm not even from Russia. Maybe Russian-speaking Jew is better, because I always identify with the Jewish culture. I never really learned much about the Armenian culture. That just wasn't part of my daily life, so that is not a part of my identity. And you know, it's funny because in Ukraine I was different because I was Jewish. And then I come here to the US and suddenly I'm different because I'm Russian. So sometimes I struggle with the concept of identity when I fill out the census forms and the only box I can check is Caucasian, and I'm like, 'You don't even know what that means.' To me, Caucasian is someone who is from the Caucasus, but here it means white. And I feel like it's odd to lump me with this category of white Americans. I don't feel like I identify with that culture, even though you look at me and that's what I look like. So yeah, Russian-speaking Jew is probably the identity I would say I identify with most right now." Her response clarified that her Jewish ethnicity was the centrality of her admittedly complex identity, even though she was Russian-speaking, from Ukraine, with a father from Armenia.

Other Jewish interviewees also voiced complexity about their identity. **Nataly Kogan**, Cofounder and CEO of Happier, Inc., who came from St. Petersburg in 1989 at age thirteen, said: "I think my entire identity is informed by my Jewish experience, because I just became a fighter. I mean, a Russian Jew in Russia is a fighter." More broadly, she added: "For me, if I had to define myself in a list of words, immigrant may come before woman or mother, absolutely. I have my Jewish identity and the immigrant identity. I even have two names: Natasha and Nataly, the more Americanized version that I chose to try and fit in better. My immigration experience colors everything I do, the choices I make, how I feel. It wasn't just an event coming here from Russia, but rather, it's an ongoing experience that is part of my daily life and my identity." Like Scherer, Kogan saw her Jewish ethnicity as a foundation for her identity but within a complex set of other characteristics.

MIT Professor **David Gamarnik**, who came from Tblisi, Georgia, in 1990 at age twenty-one, replied: "I'm a Russian Jew, not quite a Georgian. I think because language determines your identity to a large extent, it would probably be Russian. But just saying I'm Russian would not describe me at all, because being Jewish is also very important. Growing up in Georgia, that's very important. Living in the US is very important. Many of my relatives are in Israel, so that aspect is also very important." Gamarnik described his complex identity by noting that being Jewish was one important aspect, but, like Kogan, it was only one of a number of significant influences based in various countries with which he had important connections.

One interviewee saw his Jewish identity somewhat differently from the preceding interviewees. **Gregory Tseytin**, a visiting scholar at Stanford University and a freelance consultant doing R&D in computing, came in 2000 at age sixty-four from St. Petersburg. He summarized his complex identity: "I am an American citizen. I'm not specifically a Jew. I was born a Jew, but I'm not a participant in any Jewish organizations or Jewish campaigns. Of course, I am ex-Russian. But I'm not sure I know what it's like to feel like an American." His short statement summarized rather well the feelings of others who were grappling with establishing a clear identity for themselves. **Yelena Kadeykina**, Cofounder of Startup Access who came from Moscow in 1997 at age eighteen, was another interviewee who did not see her Jewish background as the center of her identity, but who appreciated being Jewish for other reasons: "I don't think I ever really identified myself very much as Jewish. I did become very grateful to the Jewish community in the US, however, because when we came here we went through a bunch of programs that help immigrants, and they were all Jewish. So I definitely developed a great sense of appreciation for the Jewish community."

One interviewee who clearly expressed complexity was **Dennis Bolgov**, Founder and CEO of Tocobox, who came from Astrakhan, Russia, in 1998 at age twenty-five: "I'm half Russian, half Jewish. I have a quarter of Ashkenazi Jewish heritage and a quarter of Bukharan Jewish. I do not have a definitive answer about which group I belong to. My wife and I talk a lot about this because sometimes we want to belong to some group, but we really are like stacks of grain. We are not a standard part of the Russian community, we are also not a part of the American community either. But in some ways we are in between, which does not necessarily mean that we're alone. We have friends with different heritages from all over the world, and we're exposed to many cultures and ideas this way." Bolgov describes well the complexities involved in finding one's identity as a Jewish immigrant from Russia, with feelings of homeland being paramount at times, while at other times feelings about their new country and friends being most important.

Alexey Wolfson, CEO of Advirna who came from Moscow in 1997 at age fourteen, reflected upon the mix of identities that comprise his sense of self: "To me, my identification as a Jew comes probably in the fourth or fifth place, but absolutely not the first. Obviously, all of us identify ourselves in many different ways. For me, well, I'm human, I'm male, I'm a family man, I'm a scientist, I'm Russian. And after that, Russian by culture, and after that, I'm Jewish by origin. So that's like the seventh place. I only learned that I was Jewish when I was thirteen because I had a complete kind of Soviet upbringing. There was some Jewish culture that I realized later, but I did not even feel like I'm Jewish until somebody told me I should. I felt this only two times in my life in the Soviet Union, and both were unpleasant circumstances."

Other interviewees noted that a Jewish or Israeli heritage was important to them but considered it as only part of their complex identities. A senior executive in investment banking responded initially with hesitation regarding her identity: "I don't know, I really don't. You never know. I really feel like I'm bicultural. And when I came here I just remember thinking that I want to keep the best from my culture and take the best from this culture. Somehow I think that's what I was trying to do, but I think people take me for Russian overall." After returning to Russia with her non-Russian husband and their children to work for a few years, she recalled: "I had already thought about the US as the country for our children, so I was comfortable being here. But, for some reason, my husband thought I completely love Russia, which I do. You know, it's somewhere in my heart, but I still think for our children it's a lot better to be here in the US." Regarding her Jewish background, she responded: "While I may have some Jewish background, I definitely don't identify myself with the Jewish religion, but I just happen to have a lot of Jewish friends, and I do get along really well with them." Her story introduces family issues that can affect one's preferences and views and can help resolve an otherwise very complex situation.

In the same vein, **Jane (Evgeniia) Seagal**, Senior Scientist III at AbbVie, originally from Novosibirsk, Siberia, who came in 2004 at age thirty-two, emphasized the role of family as well as the importance of an Israeli and a Jewish heritage in her overall identity and decisions. Seagal had lived in Israel for many years after emigrating from Russia. About her identity, she mused: "I don't know. I say that sometimes it feels like a tree without roots, and it's not a good feeling. It's why I think I'm trying to give something to my kids. They feel at home here, and I like them to feel free here. So I think my big challenge is that I want my kids to have roots. I want them to feel that they have a home and they know where they belong and what they are. That's why we've taught them Russian and Hebrew and about Jewish traditions. We celebrate Russian holidays, Jewish holidays, and American holidays."

We end this section on Jewish ethnicity as a major part of one's identity by emphasizing the complexity that can be involved. We refer to TEC Ventures Managing Partner **Stas Khirman** who was born in

Kiev, Ukraine, and lived in Israel for several years before coming to the United States in 1994 at age twenty-eight. He asserted: "Again, just to clarify, I'm not Russian, I'm not Ukrainian. Today, I would define myself as a Russian-speaking Israeli, born in Ukraine, and living in the United States. So I never feel myself either Ukrainian or Russian, but I was born in the former Soviet Union. I can divide earlier periods in my life into two parts. In the Soviet Union, because I was a Jew, people would say, 'You don't belong here,' and you fight sometimes, but you understand that they're somehow right. I never wanted to stay there. And then you move to Israel and become 'Russian,' and people say again, 'You don't belong here,' and you just laugh because they are so wrong." In reconciling his experiences in three countries, Khirman seems to have resolved the inherent complexity by identifying himself as a Russian-speaking Israeli, born in Ukraine, and living in the United States, the country in which he is a citizen and a highly successful entrepreneur and investor. We again emphasize that not all interviewees have yet reached that successful state where they are fully comfortable with the complexities of their identities. Two of the next three sections show how some interviewees resolved that dilemma, the first group by identifying with their professions, passions, or interests, and the third by identifying themselves as internationalists or citizens of the world. Between those two groups, we felt it was important to present the stories of a substantial number of interviewees whom we believe exemplify a level of complexity that seemed to obscure a path to a clear identity.

Identity from Professions, Passions, or Interests

Many technology professionals, scientists, and entrepreneurs saw their identities as being based heavily or primarily on their professions, passions, or interests, although most also noted that particular identity as being one, but certainly not the only descriptor, of their total identity. One such individual was University Distinguished Professor of Pharmaceutical Sciences at Northeastern University, **Vladimir Torchilin**, who came from Moscow in 1991 at age forty-five. He had received the Lenin Prize, the Soviet Union's highest award, in his case in science and technology: "I see my identity in my credentials as a professional, so I'm pretty satisfied with that. I never asked for anything more than that. I don't believe that it's any special Russianness or Jewishness or anything else." Serial entrepreneur **Yury Lifshits**, Managing Director of Entangled Solutions who came from St. Petersburg in 2007 at age twenty-three, responded: "Entrepreneur. Yes, it comes before Russian or scientist." He made it clear that his complex identity had undergone multilayering since he mentioned "Russian" and "scientist" as other descriptors, but unhesitatingly responded with entrepreneur being his primary identity. In a clarification, Lifshits added: "I don't reduce myself to a single quality. Also, I would think of it differently depending on the context, for instance, casual or professional. There is no competition of identities in me, and I don't really rank them. But if you ask me to name a single one, it's still 'entrepreneur.'"

A related response was given by **Alexei Dunayev**, Cofounder and CEO of TranscribeMe. Originally from Kiev, Ukraine, he came in 2007 at age twenty-six from New Zealand, where he had relocated with his family at age fourteen: "I don't see myself as a Russian-Ukrainian or Ukrainian or Russian. I'm ethnically Russian, so I see myself as ethnic Russian, but I wouldn't necessarily use it as my identity. It's part of my heritage and part of who I am, but it's not necessarily a way that I use to associate my identity. I feel that I'm ethnically Russian, I spent some time in Kiev, and in other places like England and New Zealand, and different parts of Asia, as well as here in the United States. My background does involve a fair amount of traveling and moving to and from different places, and I see all those things contributing to my identity, but not in any overwhelming way. For me, identity questions would really be related to my personal interest and character, not necessarily to my background or cultural places where I've been or grown up. I don't tend to associate myself with places very much. I'm interested in technology startups, and I'm very passionate about building connections between different parts of the world. I'm interested in technology, the latest research findings, and science. I'm also interested in business and how to get ideas and go from an idea to reality. So that would really be more of my identity. I like the Bay Area weather and the New Zealand weather, and they are very similar. So, in a way, that may also be part of identity. But it's really more about my interests and my passions than around geographical places."

David Boinagrov, Research Scientist at Sciton who came from Tbilisi, Georgia, in 2008 at age twenty-two, was a recent Stanford PhD graduate. He, too, mentioned his interests as being the primary aspect of his multifaceted and potentially complex identity, yet he responded without hesitation: "Physicist. Yeah, physicist. I guess a more complete set of identities would be physicist, engineer, Russian, Georgian, American, Armenian, what else – Libertarian." His background growing up in Georgia with some Armenian heritage, moving to Moscow, and being educated and then settling in the United States provided the foundation for his complex identity, one that he seemed to have resolved by identifying primarily with his profession to define his day-to-day being. Yet he understood well the other aspects of his background and experiences that had imprinted various layers to create his overall identity.

In spite of the potential complexity stemming from the various cultural and other influences that might have imprinted their identities, the interviewees in this section exemplify looking beyond ethnicity and other experiences, having emphasized the importance of their professions or other passions as being at the core of their identities.

Complexity Clouding Identity

We now discuss a group who seem to have their identities clouded by greater complexity relative to the previous groups. Such a person is **Andrey Kunov**, Founder and CEO of the Silicon Valley Innovation Center who first came from Zhekazgan, Kazakhstan, in 1994 at age twenty-two: "I guess what I've learned for myself is that I have a complex identity, not only complexity in our business but complexity in my identity. Ethnically, in my blood, and I'm not going to change that, right, I was born with Russian genes in my blood. So that is not going to change what I think of myself. Then, all the other layers on top of that, culturally, are also part of me, but again, it's very hard to change that part of our culture where we came from." Kunov's response could have come from many of our interviewees since it reflects the underlying importance of one's homeland, but demonstrates the complexity added by more layers of imprinting from different environments, particularly that of one's adopted country. His response reflected a fundamental theme of this book: the impact of imprinting on identity formation from various layers of environments and experiences with different institutional characteristics.

The difficulty in understanding one's identity could well be due to an explanation provided by **Shamil Sunyaev**, Professor of Bioinformatics and Medicine at Harvard Medical School who came from Moscow in 2002 at age thirty-one. An American citizen, he said: "I don't know. It's very difficult. This seems to be a very American thing, I guess, that you should have an identity, you should have a certain kind of one box where you put yourself, right? I don't like to think about this in terms of myself, but clearly, by origin, I grew up in the Soviet Union. But that country doesn't exist any longer. So I live here in the US, and I'm interested in this country and in the political process and in the culture. So I think this question about citizenship arises emotionally when we go elsewhere, like the UK, or Europe, or Canada. I think the emotional connection to one's country would probably stay, reading the news, and keeping the idea that you should still vote. I think we've lived here in the United States long enough, and getting citizenship is partly a reflection of that, I guess. And that's interesting. But I still don't want to have this to be the exact structure of my identity, like this is how I see myself." Sunyaev determined that identity cannot necessarily be summed up in one characteristic, including nationality or citizenship, clearly seeing himself as much more than American or Tatar, his fundamental ethnicity. As a scientist, he might even have added, like others earlier in the chapter, that he identified strongly with his professional orientation or with other dimensions.

Grigore Raileanu, Founder and CEO of Noction who came from Chisinau, Moldova, in 2012 at age twenty-nine, provides additional rationale for that complexity of identity, even though he began his response with: "I usually identify myself with Russians. My mother language is Romanian, since my parents are Moldovans of Romanian descent, but I didn't speak it until I was six when my family settled back in Moldova after being in Siberia since my father was a medical doctor with the Soviet army. Before we moved back to Moldova from Siberia, we then lived in Hungary and everyone was speaking Russian around me. I remember that because I learned Romanian when I was six during the summer I went to a camp and I wasn't able to understand Russian, so I had to learn Romanian. I also think my mom has some Ukrainian roots because her father moved from Ukraine to Moldova and then met my grandma and they made this family. My father is Moldovan from the northern part of Moldova. The first language I used was Russian, and still most of the books I read by like age twenty-seven were only in Russian. It was hard for me to read books in my native language in Romanian. Now, I only read in English, and before that I only read in Russian. I still usually identify myself with Russians, but I like the diversity here because there are people from different cultures." As a result of the diverse environments in which he grew up, Raileanu seemed to have settled on Russian as his identity, but understands that his identity is fundamentally far more complex. Like others discussed next, his own complexity has led to an openness in his relationships and his acceptance of other cultures.

Other interviewees also referred to deep complexity and often did not identify strongly with any country at this point in their lives. Financial executive **Aziz Mamatov**, who was originally from Tashkent, Uzbekistan, and came in 2014 at age forty, stated: "You know what's funny, probably the closest depiction that you can get is that I am a Russian-speaking Uzbek. And so this is me. I speak Russian, but I'm not Russian by any means. They're not taking me as one of their own, and I'm not one of their own, because they're different. They're Russians. But I'm not Uzbek because I don't speak the language well, so I'm not one of their own either. So yeah, it's strange." Like others, Mamatov saw language as being fundamental to his identity, and his lack of fluency in the Uzbek language made him feel that his Uzbek ethnicity was not really central to his identity.

The complexity resulting from being an immigrant, along with other causes of complexity often stemming from various layers of imprinting, was voiced by **Ksenia Samokhvalova**, Senior User Experience Specialist at MathWorks who came from Nizhny Novgorod, Russia, in 2003 at age twenty-three. Having obtained US citizenship through her American husband, she responded about her identity: "Is it an immigrant? Is it Russian? Who am I? It's hard to be an immigrant, I'm realizing that now. I feel like part of me didn't think about that very much when I made the decision. Clearly, it's better for me, it's better for my personal life. I could have been a physicist for my whole life back in Russia in the same place where I went to school, and I have plenty of examples of people like that. Maybe not, who knows? It's hard. Sometimes I feel like I don't belong. I don't belong there. I go there and I don't like it. So Russia is no longer home to me. And I look around here and I feel accepted here, but I still don't know half of the cultural references. I didn't grow up playing baseball or softball or whatever it is. My point of reference doesn't apply here because I grew up in a totally different system." Samokhvalova articulated the mixed feelings about identity experienced by many interviewees. Part of her seemed pulled to her homeland, while another part was making adjustments to her new country. Her initial response about her identity, "Is it an immigrant?" is a poignant insight into the root of her confusion and the complexity of her situation.

Marat Alimzhanov, Lead Staff Scientist at Acceleron Pharma who came from Moscow in 2002 at age thirty-one, also noted a disconnect from his homeland. Born in Russia of a Russian mother and Kazakh father, he was raised primarily in Tselinograd, Kazakhstan: "It's a hard question, actually. Culturally, I still feel part of Russian culture. It's something I grew up with. So, even though I grew up in Kazakhstan, I feel culturally Russian. I went to Russian school, Soviet school, but it was all in Russian. But again, I feel Russian, but I don't feel like I belong to Russia. I know it's kind of complicated, but whenever I go to Moscow now, it feels like a foreign city to me. So much has changed there that I feel like I'm a tourist. I don't have any connection any more. People are different, buildings are different, even the language is different. Because language evolves with people, right? So there are a lot of new things that I don't really understand. So I guess I feel like I'm Russian but, you know, mostly when I'm outside of Russia." Even though they grew up in different Soviet republics and felt culturally Russian, Samokhvalova and Alimzhanov sensed they did not really belong to Russia. For both, there was still an identification with their homeland, one likely based in earlier imprinting. But, by the time of our interviews, that identification had been or was being modified or replaced by new layers from the United States, with its markedly different institutions.

Maxim Matuzov, Search Program Manager at Apple who came from Murmansk, Russia, in 2009 at age twenty-one, related his own experience: "My ethnicity is Russian, but I'm more American when I think of Russia right now, in a way that I haven't before this. I can't really say I'm American because, you see, I don't work in an American environment. I work in an international environment here at Apple. I can't say strictly either Russian or American. I think it's lots of the cultures mixed together now, and of course, I'm influenced by them. If I were to go back to Russia right now, there would probably be a huge cultural shock, especially like the homophobic environment right now in Russia." Matuzov's comments about the institutional changes in Russia were important to his seemingly decreasing connection to that country, while his connection to the United States seemed to be increasing.

Another interviewee who referred to the changing institutional arrangements in Russia as affecting his identity was **Nick Bilogorskiy,** Director of Security Research at Cyphort who came in 2006 at age twenty-five via Canada, where he emigrated to as a child: "Canada became a second home for me, and I identified myself very much as a Canadian, or more specifically, a Ukrainian-Canadian living in the US. I was Russian speaking so I identified myself with Russia until recently, until all the things started happening in Ukraine. I got involved with helping Ukrainians by founding Nova Ukraine here in San Francisco, and that's when I really stopped identifying myself as Russian. Before that, you could find me at facebook.com/RussianNick. That was my nickname. Now I can actually make a choice because you can't really be both, with these countries being at war with each other. So at this stage, I'm a Ukrainian-Canadian living in the United States, Russian-speaking, with a little bit of Jewish in me." Bilogorskiy emphasizes how political events and other such institutional upheavals can deeply influence one's identity. In his case, the complexity is exacerbated by his earlier strongly perceived identification as a Russian even though he was born and raised in Ukraine.

Boris Berdnikov, Staff Software Engineer at Google who came from St. Petersburg in 1995 at age nineteen, described the complexity that again involved the changing institutional landscape in Russia. An American citizen, when asked about his identity, he responded: "Confused. So I never became a true US citizen. I never voted here and I'm not planning to, and that's part of my political position. Living here, coming and going, is easy, going to other countries is easy. That's the life here. Most people who live here are citizens. It makes lots of things easier, so why not. But I don't feel like I'm an American, and, by this point, I think it's safe to say that I'll never feel like an American. I identify with Russia much more, except now when I go to Russia, I see how Russia has changed. So my identity is much more Russian, but when I talk to Russians, I understand that I have learned a lot here personally, and I'm a lot closer to the Western or European view of the world.

"So except for those times when I see the American view, my identity is my Russian roots. Russia is my country and is my place. I think this is the way Americans feel here in the US. You fit the environment, it fits like a glove. So I still can't connect to Americans the same way I connect to Russians and will never be able to. And

Russians can't really understand a lot of the things that I'm telling them without me having to explain them from the ground up, from the basics – politics, life, relationships. Politics is the outgrowth of how society works, the social contract, the 300-year-old idea of how people relate to each other and the unwritten rules of contact. They just don't get it. It's not a difference of opinion, but a difference in cultures." Berdnikov had thought a great deal about his own feelings and concluded that he would never really feel like an American and was more comfortable with Russians. Yet he realized that his new layer gained in the United States differentiated him, his thinking, and his viewpoints from other Russians.

Most interviewees described so far in this section illustrate the continuing tension in identity between their homeland and their adopted country. The next three individuals illustrate that this complexity takes many forms, beginning with **Natalia Goncharova**, CEO and Founder of Finance Alpha, who was from Tashkent, Uzbekistan, and came in 2001 at age twenty-five: "Having been in the US for almost 15 years, my relationship with my Russian roots and my relationship with that country has been evolving. And maybe in the beginning, and why I started teaching my son Russian, it was very important. It was like something I was highly identified with. Fifteen years later, it's a part of me, but it's not something that I think defines me. What defines me now is this ability to not just survive but to strive, striving in any environment. I only wish for the world to have less borders. I only wish for us all not to identify with Russian, American, or something, but with humans. How we all are doing with the world we live in. But at the same time, I now think of the US as my homestead. It is my home, and what I wish for the country is to start doing more and more." Goncharova illustrates the transition from a deep affiliation with one's homeland to an evolving identity formed by newer layers from her different environment in the United States.

In a somewhat different vein, **Yulia Witaschek**, Global TAM Customer Engineering Manager at Google who came from Moscow in 1995 at age seventeen, described her identity as being more associated with her own attributes rather than ethnicity. But her remarks clearly showed some confusion regarding her identity: "You know, I don't think I have an identity. I still introduce myself, when people ask me where I'm from, I say I'm Russian. But in reality, I lived in Russia for twelve years and two more in Belgium before coming here. At this point, I think I would describe my identity a bit more with the qualities of who I am and what I stand for, rather than with an actual national boundary. I'm nonviolent, I don't believe in violence. But I do understand the need for it sometimes. I would never attack or be aggressive toward somebody, but I believe in justified force in response to aggression. If somebody hits me, let me hit them back with the same amount of force. I think culturally I'm American; I'm very American with a Russian flavor because I'm fiercely independent." Like the others in this section, Witaschek seemed to see her primary identity as stemming from who she was as a person and what she believed in.

The next three interviewees framed their identities around complexity using the terms "amalgam," "alloy," and "mixture," respectively. The first, who had come from Russia in the mid-1990s as a teenager stated: "I didn't think much about the ethnicity of my friends growing up in the USSR. It didn't matter. I grew up in a country where there was not that much emphasis on religion, actually not at all in the Soviet Union. Everyone was the same, and I didn't care about who I was hanging out with. But when I visited Russia in 2007, I certainly didn't identify myself with Russians. I felt very foreign on every single dimension, in all of my connections with people. So I feel I've been assimilated in the US, but, having said that, my closest friends still happen to be Russian. I am my own person, and you know, I can't really say that I have a very strong identity. Even being half Jewish, in the Jewish tradition you're considered Jewish if your mother is Jewish. So by Jewish standards, I'm not Jewish, but by Russian standards, having some Jewish blood, I'm Jewish. So I was neither Jewish nor Russian at any point in time, but I always had a very strong sense of self and it didn't matter to me. I didn't characterize identity with any particular group because I had so much personal confidence, I guess. So I can't quite say who I am culturally. I'm an amalgam, right?" She was comfortable in seeing her identity through her self-confidence and being her own person, which is the essence of one's identity.

When asked about his identity, **Ilya Yaroslavsky**, Manager of Advanced Product Development at IPG Photonics, who was from Elista, Russia, and came in 1998 at age thirty, responded: "I don't know. It definitely was formed in the Soviet Union, so at the core I am Russian, but it was so heavily influenced by my being American and also having experience living in other countries. It's probably more accurate to describe my identity as an alloy of elements of different cultures." **Tatiana Novobrantseva,** an Immunology Consultant who came from Moscow in 2001 at age twenty-eight, used a term similar to "amalgam" and "alloy." Reflecting for a moment, she found the word "mixture" to describe her identity: "Nonexistent. I certainly do not belong to any of the cultures anymore, not completely. I'm very lucky to have a lot of friends in a lot of places, but I'm certainly not a US-born person, right? I do not share the same cultural references, et cetera. So I feel it in a lot of different things, and it doesn't make me uncomfortable, and it doesn't make anyone else uncomfortable with me. I'm willing to participate in lots of things here in the US. It's just that when you're not originally from here, it certainly shows. I'm not originally from Western Europe where I spent my early twenties. So that was very important, but again, I never felt like a person belonging to that culture. And now, going back to Russia, that country has made such a leap in a different direction that the country I left doesn't exist anymore. It's a different social system, everything is different, so I don't belong to that either. So I'm really a mixture of all of those things, which certainly adds complexity. It certainly gives you a bit more work trying to integrate yourself into different circles of life, right? Professionally, we have it easy because we have this huge international community where everyone brings something different." Novobrantseva's concluding statement provides an important insight into where she felt comfortable with her complex identity – as a biotech professional in a workplace environment that she described as an international community. Although she did not mention her profession as being her identity, it was clearly the place where her complex identity was subsumed by the nature of her professional environment. The previous three interviewees clearly understood the complexity of their identities and, in some ways, the imprinting and layering that had led to the complexity that they described as being an "amalgam," "alloy," and "mixture."

Global Citizens or Internationalists

This next group of interviewees dealt with complexity by responding that, while there were perhaps other identifiers, their primary identity was as a citizen of the world or an internationalist. These individuals did not see themselves identified particularly with an ethnicity or nationality, although a few did add an ethnicity to their broader identity, seeing themselves as global Russians. Still, all seemed to have resolved any confusion or complexity. **Serguei Beloussov,** Founder and CEO of Acronis, who was originally from St. Petersburg and came in 1995 at age twenty-four, explained: "I am a Russian-born, Singaporean Jew, and I do business all over the world. So I am very cynical and pragmatic, and I don't want to be anything. I am technically a citizen of Singapore, but I feel that I am more a citizen of the world. Until I was fifteen years old, I thought I was 100 percent Russian. I have a Russian first name and a Russian last name. The only thing that was suspicious was that my father was apparently Jewish, so more than half of my blood might be Jewish, which is sort of weird. By that I mean I was brought up as Russian. I never realized that I had Jewish blood up until the eighth grade, but from time to time my parents would say, 'Remember, you also have Jewish blood.'" Beloussov's interpretation is similar to that of others who resolved a complex ethnic and national identity by assuming a new and rather creative one.

Maria Adamian, Founding Partner of Ecamb who came in 1994 at age nineteen, was born in Meghri, Armenia, and raised in Tomsk, Russia. She resolved her complex ethnic background and experiences by adopting the identity of an internationalist. She mused: "That's a difficult question. I'm ethnically Armenian, being three-quarters Armenian and one-quarter Russian. Half of my family is from Armenia and the other is from Azerbaijan, and I have Russian and American passports. But I'm not really Armenian or Russian since I don't accept a lot of things in the mentality of either group. I'm definitely not American. I didn't grow up here in the US, and I didn't spend my childhood here, which matters a lot in terms of how your brain works. So, yes, I'm very international. I'm very sensitive to cultural issues, very sensitive to minorities. I can associate myself with a lot of that, even though I'm a white Caucasian here in the States. For instance, in Tomsk I was an Armenian, which is a totally different story there and in Russia in general. I'm also a woman, which is certainly different from being a man, and it was very different there from being a woman here. I'm neither very Russian nor very Armenian nor American. But the opposite side of that is that none of these groups really accept me 100 percent as their own." Like other interviewees in this section, Adamian had resolved her seeming lack of acceptance by or identification with any other ethnic group by choosing to see her primary identity as that of an internationalist.

A similar solution was adopted by **Dmitry Kovalev**, Consulting Systems Engineer at Cisco who came from Moscow in 2008 at age twenty-four: "So in terms of identity, I keep asking myself this question all the time. The answer that I get is that I live on planet earth, and earth is the third planet from the sun. So I really don't like it that people on this planet are being restricted by the borders of their countries. I would rather identify myself as a global person. And what I like about my personal situation right now is that, if I want, I can work in the United States, or, if I want, I can go and work in Russia. So I have this flexibility, and I'm not really restricted by the borders or any rules that are present. Maybe this is a bit of a weird answer, but I would say that I am a Russian who is living in the United States, born in Russia, but I currently live in the United States."

Another interesting response was given by **Sergei Burkov**, Founder and CEO of Alterra who came from Moscow in 1989 at age twenty-one: "Well, I have a Russian word for this, '*bezrodnyi kosmopolit.*' That means I am a cosmopolitan, not belonging to a specific culture or country in that respect, somebody who doesn't have a fatherland or motherland, but is a citizen of the world. '*Bezrodnyi*' is kind of a derogatory term meaning that somebody doesn't have any heritage. Stalin used it in the 1950s to brand enemies of the people, mostly Jewish." Burkov's insights into the historical perspective of his cosmopolitan term did not deter him from adopting it as a description of his own identity. A similar response was provided by **Diana Tkhamadokova**, Founder of I Style Myself and originally from Nalchik, Russia, who came in 2015 at age thirty-two. She began by speaking of her ethnic background: "I come from the Kabardian Carverian republic in southern Russia. Ethnically, it's a mix of Persian and Jewish blood, and the language carries both the Arabic and Jewish sounds." Elaborating beyond ethnicity, she added: "I don't know, an internationalist, I think. Everyone who travels abroad still carries core values in them. But they are also enriched by other nations, cultures, and traditions, and you become much more robust. You become a combination of all that knowledge and skills that are acquired from other people. So hopefully one becomes a lot more nimble and flexible." Having graduated from London Business School and worked in the financial industry in London, she added: "It might seem funny because I lived in the UK for ten years and I could apply to be a British citizen, but I don't think I would want to." Tkhamadokova came to a conclusion very similar to the others in this section in choosing to see herself as an internationalist or a global person rather than from an ethnic or a national point of view.

Michail Pankratov, President and CEO of MMP Medical Associates who came from Moscow in 1974 at age twenty-six responded unhesitatingly: "I am an internationalist. You see, it bothers me when people start beating their chest, saying, 'I am Russian, I am American, I am Jewish.' I feel that I am a human being. I'm a great fan of certain countries, and I'm open. I'm a private person, and I believe in freedoms. I am very liberal in my views, political and otherwise. I am very ultra-liberal, and, in certain years, even radical. My ethnic heritage is such that I'm probably a bag of everything – Asian Bashkiri blood, oriental blood, Russian blood, Jewish blood. You name it, I have everything." Like some others, Pankratov defined himself more by his beliefs and values and expressed his admiration for many countries. That openness appeared to be the foundation for seeing himself as an internationalist.

Isaac Fram, Principal at Biomedical Imaging Solutions who came from Riga, Latvia, in 1969 at age twenty-one, used somewhat different reasoning to arrive at a similar conclusion: "Personal identity, these days I call myself a man of the world. OK, that might be kind of misleading because I have very strong feelings about various ideas and concepts. I am a rigorous believer that there is no easy way to anything. I strictly believe in discipline. And that's kind of why it's like dreams that I had in terms of entrepreneurship kind of connecting with the educational aspects." Fram's response contained elements expressed by some interviewees earlier in this chapter, specifically identifying more with professions, ideas, and concepts as well as their own achievements. Still, Fram's initial response was that he was a man of the world.

Arguably the most complete response about being a global citizen came from **Alexandra Johnson,** Managing Director of the venture capital firm DFJ Aurora, who came in 1990 at age twenty. A US citizen who initially responded that she was a Russian-speaking American, she then elaborated well beyond that description in concluding that she was really a global citizen. She described in detail the path that led her to that conclusion: "I was born in Vladivostok in the Russian Far East, but every summer my dad would take the family to visit his mother in southern Russia, close to Chechnya. When we were living with her, we would travel to Georgia or Armenia, even before I was a year old.

So the story goes that, at some point, my parents had a really tiny apartment, and I was sleeping in a suitcase when I was a baby. The Soviet Union was definitely a melting pot with people from all walks of life and many different regions. I know I was conscious of who I was when I was three or four years old. I didn't consider myself an ordinary girl because I always felt my world was much bigger than that of other children because I had the luxury of seeing the enormous country of the Soviet Union first hand. It was quite unusual. So I remember all those connections I was making as a child, and that I connected with people regardless of where they lived, regardless of what their nationality was. I came and had fun with them, and that is something that I guess was an imprint since the day I was born, and that's what I have cared about throughout my life.

"If you were to ask me again who I consider myself to be, I'm a global citizen. I happen to have been born in Russia, but I feel no particular connection. My friends are from all over the world. Regarding Russia, it's just that I love the country that doesn't exist anymore. In my mind, I had never really left Russia because that's the place where I was born and where people nurtured me and gave me the most fabulous education. And all those people who helped shape my personality in my childhood, I hold dear to me. I came to the US to go to school at UC Berkeley, and I met so many interesting people there, many from countries I had never been to. I realized then that my world, which was already quite big, had become even bigger. So I made the choice that I'm really starting at zero, and I'm starting with my base in the US because it's an opportunity to expand my horizons. So I returned to my old version of the center of the world, which I think is Silicon Valley. That's it. So the globe is my world, but my Russian roots are extremely important to me. I think I am just the lucky one who was given an opportunity to see the globe as your stage. That's how I would say it. It's just that my world became bigger." Johnson recognized her diverse experiences in the USSR as being an early imprint on the identity she formed over time after many other broadening experiences, culminating in seeing herself as a global citizen with the world as her stage, clearly the result of multiple layers of imprinting over the years.

Johnson's comment about Silicon Valley being her center of the world provides a segue to the next three interviewees. Entrepreneur **Tatiana Kvitka**, CEO of Design by Light, who came from Moscow in

1993 at age twenty-five, described her complex identity: "It's a tough one, and right now it's changing. My dad is from Russia and my mom is from Georgia, and I'm of Ukrainian heritage. So you know, my family is a perfect example of the Soviet Union, what it used to be, a melting pot of all sorts of ethnicities and nationalities, republics spread all over. But anti-Semitism was a big issue in the Soviet Union back then, as I know now. Also, you couldn't get some jobs if you were a woman." Turning her remarks to Silicon Valley where she had lived since 1993, she continued: "My home is here, and I am maybe just dual identity, Russian-American, but more a citizen of the world. I would say that because I definitely feel when I'm with Russians these days, I don't feel part of that culture." Regarding the United States, she added: "I don't know if I would stay as long in one place in the United States if I moved somewhere not in California, like in the middle of the States. I think that my identity is of Silicon Valley, I believe more than anything else." Kvitka is an example of someone with a complex ethnic and national identity who has resolved uncertainty by adopting the specific regional culture in which she is now immersed while feeling like a citizen of the world.

Tatul Ajamyan, Cofounder of Wakie who arrived in 2014 from Yerevan, Armenia, at age thirty-two, echoed Kvitka's self-description: "Well, ethnically, I'm Armenian, but we belong to the world, I think. The world is becoming smaller and smaller with the Internet and cheap travel. So costs get lower and the world gets smaller and smaller. Knowing languages makes you convertible, so you can understand other cultures. So I don't think that we should feel ourselves, you know, just limited to some culture. This is best seen from Silicon Valley. When you come here and you see all these people, you know, local Americans and all the immigrants who are building the same things. And sometimes you see people living here for forty years and still speaking with a Chinese accent, but you know, they are part of this culture, and there is not much difference between them and any other American. So it's a mix of cultures, and you can start to understand that the world is a very small place. So I think we identify ourselves as human beings, and we see that in our company, Wakie, with our service that brings people together all around the world. We see that the world is, you know, becoming just one homeland for everyone, and this is a great culture. So I think we all belong to the world." Ajamyan was rather philosophical in that his response exuded the inclusiveness that he has seen in his own company to be a reflection of his observations of the Silicon Valley culture.

Serial entrepreneur, **David Yang**, Founder and Chairman of the Board at ABBYY, was originally from Yerevan, Armenia, and came to Silicon Valley in 2011 at age forty-three: "I have three motherlands – Armenia, Russia, and China. The US is not yet my motherland because I have spent only three years here. But I see myself and my family as belonging to the world. I think that it might happen that someday I would see the USA or Silicon Valley becoming my fourth motherland. But definitely, Silicon Valley is my motherland for the technology world, there's no doubt about that at all. It's a place where all the best minds come together and create something new. This is very important to me." Yang's specific reason for identifying with Silicon Valley being through technology was different from that of Kvitka and Ajamyan who connected more with its broader cultural base. The experiences of all three, however, are a clear reflection of layered identities based on different sources of imprinting. And they, like all the interviewees included in this section, had arrived at "global citizen" or "citizen of the world" as a primary identity.

A number of interviewees did not describe themselves as global citizens or internationalists but emphasized that they felt at home in whatever place they might be at any point in time. In essence, they saw themselves as very adaptable to any new environment they entered and saw no need of any particular identity. One such person was **Roman Kostochka**, CEO and Cofounder of Coursmos who came from Tolyatti, Russia, in 2013 at age thirty-seven and who had traveled and lived around the world: "I don't have an identity because I see many people in many countries, and all these people are very good and friendly, so I don't need an identity. It's really true. I really like different places with different nature and different people, with different culture. It's very interesting and I see this as my own identity." A similar response was given by **Mark Kofman**, CEO and Cofounder of Import2 who came from Tallinn, Estonia, in 2011 at age thirty: "That's a tough question. I don't know. Probably like this: you know the thing we start from, my name and my background. With that common name, I can feel local anywhere, and that's really part of what works for me. I get used to the local environment fast, and sometimes if I move to whatever place in the world, I feel like a local in a couple of months." Technology Executive and Entrepreneur **Igor**

Balk, who came from Moscow in 1994 at age twenty-four, had a similar perspective: "When I go to Israel, I'm a typical Jewish person because I happen to be a Jew as well, and I'm at home there. When I go to Russia, I behave like a typical Russian. Here in the States, I'm like a US businessman. I'm kind of cosmopolitan."

Another person who focused on her ability to adapt seems to summarize well the way in which these interviewees were able to achieve a multilayering of identities, culminating for many in an identity of a world or global citizen. **Umida Stelovska** (born Gaimova), Founder and CEO of parWinr who came in 2009 at age thirty, explained the adaptation process she went through. Born and raised on a cotton-growing collective farm near Samarkand, Uzbekistan, she went to graduate school in Germany, followed by working in the Czech Republic and later the United States. She answered: "I am a world citizen. I feel at home wherever I go. I learn the language, adapt to the local culture and the communities quickly. I always keep the best part of every culture with me and do my best to make new friendships and build my professional network in the community. I have seen people come to a foreign country who are not able to adjust to the local culture, and they became extreme nationalists of their own nation and religion. In my opinion, if you do not adjust and adapt to the new culture where you want to make your home, you will never be happy.

"Sometimes you need to make sacrifices to achieve something better in life even if it means to give up your home country citizenship. I look forward to a future where we have no borders between countries, and no fights and wars because of race, religion, or nationality. I want to see our kids living as world citizens in peace and harmony regardless of where they come from. I also believe that if you can dream and follow your dreams to make it happen, you will overcome any challenges and win big one day. My motivational quote is from Buddha, and I think he made the right point thousands of years ago: 'What you think – you become, what you feel – you attract, what you imagine – you create.'"

A smaller group identified themselves as global citizens but also as Russians, with at least one using the phrase "global Russian." Among them was Muscovite **Ilya Strebulaev,** Professor of Finance at Stanford University who came in 2004 at age twenty-nine. He explained how the different passages and changes in his life led him to an identity of a global citizen with a strong connection to Russian culture: "I think I'm almost a Sovietized Jew because a lot of my Jewish traditions were eradicated in my family. My parents and my grandparents were members of the Russian intelligentsia, which obviously doesn't have the same meaning today. But everything was very cultural where I was coming from. I was lucky that I came from a very cultural family, and for instance, I would go with them at the age of five to the Bolshoi Theater and to museums. My heritage obviously includes a classical Russian education, and I'm still influenced by its traditions. But really, I never think of myself as just Russian. I think of myself as a person of the world. I'm interested in other countries and cultures because I've lived in London, and I'm really interested in British history and British culture, and my favorite authors are British. These days, I'm definitely out of tune with contemporary Russian literature."

Managing Partner at RMI Partners and General Partner at Helix Ventures **Evgeny Zaytsev**, originally from Barnaul, Russia, who came in 2000 at age thirty-three, responded without hesitation that his identity was Russian, ethnic Russian, but that he sees himself as a global Russian. Part of his view could come from his family experiences since he noted that his eighteen-year-old son, Anton, who was born in Russia, sees himself as Russian, while his thirteen-year-old son, Andrei, who was born in the United States, sees himself more as American. His sons spend summers with their grandparents in Russia, and his parents visit the family in Silicon Valley fairly frequently. So family influences, as well as having spent 250 days in 2014 working in Russia, help explain why Zaytsev has no question about his Russianness. But with his younger son identifying more as an American, and family vacations to countries like Cambodia and Peru, Zaytsev may have resolved potential confusion by seeing himself as a global Russian.

Ilya Kabanov, Global Director of Application Security and Compliance at Schneider Electric who came from Moscow in 2014 at age thirty-seven, concurred: "I consider myself as a global person with a Russian background. However, I think that I'm still in transition to becoming truly global, and it will probably take some time. I'm still learning what being global means, and I haven't had enough time yet to reflect on my current experience, what it feels like to work in a global environment and immerse in the global nature of the world. It's a journey, and I'm excited to pursue it." Kabanov notes that he sees globalness in the context of his role in a multinational company, its business, and the world. Balancing such a complicated context is, as he described, a journey that takes time and introspection to comprehend and embrace.

A final interviewee among those who identified themselves as Russian but were greatly influenced by their international experiences was serial entrepreneur **Dimitri Popov**. Originally from Zelenograd, Russia, he came in 2014 at age forty: "I'm clearly Russian, but with significant international exposure and some experience. Of course, I am Russian. I speak Russian, and, as soon as you absorb the language, you absorb every single part of the culture automatically. So it would be completely wrong to say, 'No, I'm not Russian, I am just an international person without any nationality.' It's Russian. You feel it by genetics. When people complain about the cold, you don't. It's not because you're tough, but just because you don't feel it." Popov made it clear that even while embracing his internationalist view, his identity was still primarily Russian and that to deny it would be wrong. Like Kabanov, he may be on his way to developing a more global layer to his identity, but it would not likely replace his basic identity as a Russian. His perspective again illustrates the transitions, within a multilayering identity experience, that will have different results for different individuals. Some, like him, may retain their fundamental national or ethnic identity as paramount, while some will subsume that layer beneath others as they transition to becoming a global citizen or some other related identity, as will be covered next.

Such a person was serial entrepreneur **Vlad Pavlov**, Founder and CEO of rollApp who came from Dnepropetrovsk, Ukraine, in 2010 at age thirty-six: "I consider myself to be a citizen of the world. The world is flat, but as I said, people often think about the world in terms of countries. But the major players in the world are not countries anymore. The most famous example is Apple, which has more cash than the American government. Microsoft has revenue which a few years ago was bigger than the budget of Ukraine. So there are different players now, and it's not about countries anymore. So I see myself as a citizen of the world, a citizen of a high-tech country." Pavlov added the fundamentals to which many interviewees referred, noting that their ethnic identity resulted from their language and being raised and steeped in that culture. He emphasized his deep involvement in technology that helped define him in addition to being an internationalist. In doing so, he articulated an identity layering of one's profession or passion as being fundamental to identity, as did others earlier in this chapter.

Nikolay Vasilyev, Instructor in Surgery at Harvard Medical School who moved from Belarus to Moscow as a child and came to the United States in 2003 at age twenty-nine, described how his identity has evolved: "I consider myself Russian, a Russian living in America, because I think of myself as more Russian than American. I grew up in Russia, my childhood was in Russia, I watched Russian cartoons with Russian characters, and read Russian books. It's very deep in my blood, the Russian heritage and Russian culture. So I still consider myself Russian living in the United States. But, as time goes by, I think more and more about myself as being something like a global Russian. We talk about that with our friends and colleagues, and there are a lot of people who are Russian but who live in other countries in Europe, Asia, and Australia. But they still consider themselves as Russian, and these are the really global Russians. I consider myself like that." Vasilyev's response was very much like others in this group who, while emphasizing their global and international identity, had retained their ethnic identity. The reason given by many was their native language and culture. We believe these global Russians have personally demonstrated the multilayering identity concept introduced earlier. In contrast, the previous group, global citizens or internationalists, had sometimes erased or, more commonly, sublimated their ethnic backgrounds and generally referred to themselves with the singular, primary identity of global citizens or internationalists.

Identity as a Constellation of Influences

As the interviewees in this chapter have shown, identity takes many forms and can consist of many imprinted layers. Regarding the form of their identities, many saw their new country as being a primary influence while others saw their homelands filling that role. But the majority, including these interviewees, expressed complexity and sometimes confusion. In short, their stories acknowledged the importance of the constellation of influences that had and continued to affect their own perceptions of identity. We see this situation as resulting from the accumulation of successive layers of identity, starting with the sometimes indelible layer gained during their formative years in their homelands, overlaid with successive layers from numerous experiences in the former Soviet Union, and certainly from experiences within their new country or countries.

For many, their Jewish ethnicity added another layer, and, for others, their professions, occupations, or work environments added another. And although a relatively large group seemed to experience various levels of confusion from the complexity they saw in their identities, another large group resolved that complexity by being global citizens due to the influence of travel, family, and friends, but primarily by analyzing the complexity inherent in their identities. The various influences all interviewees experienced involved different institutional arrangements that are seen as the primary influences on identity. And as they experienced these different institutional environments, their identities could not help but be modified.

Coming to grips with their identities, then, could be an important facilitator for interviewees' adjustment to the new innovation environment in the United States. Their continuing contributions will likely depend at least somewhat on how well they become comfortable with their evolving identities. The seemingly inevitable confusion that can arise from the inherent complexity involved could become a deterrent for some, while a successful adaptation could facilitate not only their own career development but also their ability to contribute to the success of their organizations and ultimately to the US innovation economy. Because of its vital importance, then, we have placed identity as a prominent focus of this book, one of the five i's that provide its framework as reflected in the subtitle, all concepts that are linked in the experiences of these fascinating and accomplished individuals.

PART III

Conclusion

10 | The Impact of Institutions, Imprinting, and Identity on the Immigration and Innovation Process

In this concluding chapter, we offer our insights on the impact of immigrants from the former Soviet Union on the US innovation economy. Our focus on innovation and immigration rests on an examination of the historical context for three waves of migration from the former Soviet Union and the first-hand experiences of individuals from each of these waves. To interpret the deep and rich information that the interviewees shared with us, we have utilized the social science theories of institutions, imprinting, and identity. We let our interviewees speak for themselves, and we will continue to do so in this concluding chapter. The aggregation of their experiences paints a picture of the diverse backgrounds and experiences these individuals brought from the former Soviet Union as well as the commonalities they shared that imprinted them and shaped their identities prior to coming to the United States. And the patterns that emerged from their subsequent experiences in the United States show the absorption of new layers of imprinting and the impact on their evolving identities.

We started our research by conducting in-depth interviews with 157 entrepreneurs and other professionals from the former USSR who are now living in Silicon Valley or the Boston-Cambridge area and who are active in the high-technology sector. They spoke to us about their lives in the USSR, including their educational backgrounds, their experiences during the migration process and in adapting to American business and social cultures, and the variety of new identities that they continue to form in their adopted country. They discussed their contributions in organizations large and small, from startups to multinational corporations, as well as in universities, hospitals, and other research institutions. We hope that this book helps expand knowledge and understanding of high-technology entrepreneurs and professionals from the former USSR by presenting their real-life experiences in their own voices.

In writing the book and documenting the significance of this select group of people, we realized that it was essential to put their experiences in the context of a number of complex issues of history, immigration, geopolitics, and innovation. We begin with our conclusions on the background issues and then present our main conclusions in the context of institutional influences, imprinting, and identity, relying again on the voices of our interviewees.

Immigration, Innovation, and the Soviet Diaspora

The topic of innovation is timely as well as fundamental and has fostered a wide range of analysis, including business-oriented how-to and descriptive books and articles. What we have observed is that the topic of the relationship of immigration and innovation is underrepresented and underrecognized. As evident from the immigration legislation presented in Chapter 4, many of the leading entrepreneurs and researchers associated with America's innovation economy would have been excluded from the United States before 1965. An essential element in the innovation economy is human resources in the form of highly intelligent, well-trained, and creative people. The leading innovation centers in the United States – Silicon Valley and the Boston-Cambridge area – have been able to attract such people in large part through their concentration of universities and to hold them through the vibrant ecosystem of finance, government support, serial entrepreneurs, and established, innovation-oriented companies. The role of immigrants has been an understudied aspect in understanding the flow of human resources helping to fuel the innovation economy.

The US government made a dramatic change in immigration policy with the Immigration Act of 1965, which opened the doors for talented people from around the globe to come to the United States and to participate in the high-technology sector. As described in Chapter 4, among other provisions, the exclusion of Asians was finally ended and a new system of visa and immigration categories replaced the previous national origins criteria. These developments brought rapid change and internationalization to the high-technology sector, with well-known contributions from Indians, Chinese, Taiwanese, Israelis, and many other nationalities that could not previously work or live legally in the United States. While a body of literature has accumulated about such groups, the contribution of the diaspora of people originally from the Soviet Union or its successor states is much less known. Our interviewees are a narrow slice of a much broader movement of brain power and technological skills from around the world into the United States.

Immigration: Labor Market Pull and Political Regulation

We review US immigration policy, which was covered in Chapter 4, to put our interviewees' experiences in that important institutional context. Immigration policy is determined by the interaction of labor market needs bounded by political considerations. It was not until the 1880s that there was any regulation at all since a rapidly expanding country needed vast human resources. The demands of the labor market, particularly for Asian railway laborers, ran into political resistance that led to a ban on Chinese immigration through the Chinese Exclusion Act of 1882 and the Japanese Exclusion Act of 1924. The growing demand for industrial labor led to huge waves of immigration from Italy, Austria-Hungary, Russia, and Eastern Europe, which was essentially halted by the Immigration Act of 1924 that set nationality quotas based on the national composition of the US population in the 1890s before the eastern and southern European wave of immigrants. The ongoing need for agricultural workers was filled by Mexicans, who were not subject to the 1924 Act, and the need for industrial manufacturing workers was met by the massive migration of African-Americans moving from the agricultural south to the industrializing north and later to the expanding west. This system was in place until the great reform of immigration law in 1965, which opened the United States to immigrants from around the world under more equal criteria.

Consciously or unconsciously, this new framework fit the demands of a new labor market in the United States for the knowledge-based economy developing in key innovation hubs. The new quotas and priorities for admittance set the regulatory framework for the normal bureaucratic operation of the immigration system that allowed for the diversity now observed in universities and innovation hubs. Within this framework, the United States, as the leading center of the new knowledge-based economy, was able to attract the best, brightest, and most entrepreneurial individuals from around the world. Alongside the standard immigration process, the United States instituted a parallel system of enactments allowing for immigration in response to political or geopolitical situations. These enactments permit immigration out-side of the regular quotas, mainly as refugee exceptions. These enact-ments include the Hungarian Refugee Act of 1958, Azores and Netherlands Refugee Act of 1958, Cuban Adjustment Act of 1966, Indochinese Refugee Act of 1977, and the Refugee Act of 1980.

Many interviewees in this book came to the United States as refugees, as defined under the terms of the Refugee Act of 1980. They came primarily through the enactment route, outside the standard system. In the beginning, they were pawns in the political contest between the United States and the Soviet Union, and their status as refugees had a political meaning and purpose. With the collapse of the Soviet Union, the exodus of the intelligentsia became a valuable human resource for the United States. By that time, both refugee status and other forms of legal residence allowed relatively large numbers of immigrants from the former Soviet Union.

Three Waves of Immigration

We have presented the movement of immigrants from the former USSR to the United States in three general waves spanning the later twentieth and early twenty-first centuries. Within each wave, some portion ended up in the US technology sector. By the mid-1970s, a trickle of Jews with world-class math and science training began to emigrate from the Soviet Union, pushed out by anti-Semitism and the desire for better opportunities. With the collapse of the Soviet Union, this trickle turned into a deluge caused partially by economic hardship along with ongoing ethnic and religious discrimination. As American universities and high-tech companies absorbed part of this deluge, it became clear that the collapse of the Soviet Union offered an opportunity to access world-class scientific talent. For the countries of the former Soviet Union, this was a period of a tremendous brain drain. This altered the character of immigrants active in high technology as well, as they increasingly came to the United States for economic and academic opportunities, rather than being pushed out of their homelands by anti-Semitism and economic hardship.

We used the term "immigrants from the former Soviet Union" as the general organizing criterion for the interviewees because this cohort of people is very diverse although still unified in some basic core aspects, such as the Russian language and the direct or indirect influence of the Soviet Union and its culture. In some sense, they are similar to the people interviewed by Svetlana Alexievich,[1] whom she refers to as *Homo sovieticus*. While she interviewed people who did not emigrate, our interviews were with people who came from the same cultural milieu but left it behind and had to adapt to an entirely new way of life. As individuals, they reflect the broad experience, institutions, imprinting, and identity of the Soviet Union, even though many of them have very negative feelings toward the Soviet Union and some bristle at being linked together as Soviet citizens or identified as Russians. As a result of these many unresolved grievances, there is no Soviet or even Russian diaspora that operates on the level of the Indian, Chinese, Taiwanese, or Israeli diasporas. Nonetheless, the educational, political, and social institutions of the Soviet Union have had a huge impact on them as individuals and continue to influence the institutions of all the successor states. Over time, national and cultural differences between the successor countries of the Soviet Union will likely grow and be reflected in new institutions, imprinting, and identity formation.

Natalia Goncharova, originally from Tashkent, Uzbekistan, and now a certified professional accountant and business advisor, spoke to this categorization based on her own life experience while living through the breakup of the Soviet Union: "In comparison to the Indian and Chinese immigrants, they still have India and China. But for us, the parent country, the Soviet Union, is dead. I wasn't a Russian, I guess I was more like a Soviet. I think that dissolution of the country is not understood enough in the world, and I think everyone was celebrating something that shouldn't have been celebrated. Twenty-four years after, there are huge dislocations of people. Families have been pulled apart. My dad is still in Uzbekistan, and I'm thousands of miles away. And you know what's happening in Ukraine. We're not any more representing Russia or Uzbekistan or something, we don't even call ourselves Russians anymore. We're called 'Russian-speaking,' because we do share that common ground, but the former Soviet Union is fifteen different countries at this time. We say that the Indians in the US do this or that, or that the Chinese in the US build this or that. But then when we say 'Russians,' people who are not Russian but speak Russian feel offended. So there is this kind of a void. No one has been able to come

---

[1] Svetlana Alexievich, *Secondhand Time: The Last of the Soviets* (New York: Random House, 2016).

up with a new story, with a new name to it, with a new category that
appeals to everyone involved and still refers back to those times.
Uzbeks feel Russians invaded us, and Russians say, 'No, Uzbeks
opened the door and then we built a lot for you.' So there's still,
maybe not an actively boiling pot, but a pot that had all these kinds
of pieces to it that unfortunately we were not able to transform into
something completely peaceful and positive."

Many of our interviewees became research scientists in leading
US universities or important contributors to major US companies.
Others became founders of new software companies like Evernote,
Abbyy, EPAM, Auriga, and Acronis. Still others formed science-
based companies, like industry-leading laser manufacturer IPG
Photonics, and a host of other companies in biotechnology, medical
devices, and other industrial applications. Beyond this entrepreneurial
activity, others have made significant contributions at Apple, Google,
Microsoft, Facebook, Verizon, and other large US companies. Given
their pervasiveness and integration into the fabric of the innovation
economy, it would be impossible to present a comprehensive account
of the value of their contributions, but those who work in the high-
technology and science sectors in the United States are aware of them in
their midst.

Institutions: Catalyst for Emigration

The principal institutional impetus for emigration to the United States
changed over the different periods considered in our study. The push to
emigrate during Wave One was the failure of the Soviet Union to deal
with deep-seated anti-Semitism and the unwritten but understood rules
limiting Jews from reaching their potential, combined with the economic
hardships of the post–World War II Soviet Union. As geopolitical conditions opened the opportunity to emigrate, Jews left the Soviet Union
for Israel and the United States, continuing the emigration of European
Jewry rooted in World War II and its aftermath. In the Second Wave, the
number of emigrants dramatically increased, with the principal cause
being the collapse of the Soviet Union and the near collapse of most of its
institutions.

During Wave Three, the causes for emigration became more complex as the former Soviet republics became independent countries, each
with its own legislation and economic and cultural adjustments. After

2000, institutions in Russia strengthened based on rising world energy prices, and, for a period of time, Russia was integrated into the global technology sector. Ukraine did not experience the same economic and institutional revival, but many Ukrainians did become part of the global technology sector based on their strong educational foundation from the Soviet period. Meanwhile, the technical intelligentsia in countries like Georgia and Armenia continued to look outward for economic opportunities because their countries also lacked a robust economic recovery. As a result, those who emigrated in this period came to the United States for opportunity, and many remained more connected to their home countries. All of these varying institutional factors that acted as catalysts for emigration are reflected in our interviews.

The political, career, and social bias against Jews was a motivating factor for many interviewees, particularly in Wave One. **Leonid Raiz,** who became a significant contributor to Boston technology companies, recounted how anti-Semitism ultimately caused him to emigrate. He was initially denied admission to the university of his choice because he was Jewish. He managed to get into night classes and worked his way into day classes, ultimately graduating first in his class. He went on to say: "After graduating, there was my second lesson in anti-Semitism. I couldn't find a job. I graduated magna cum laude, couldn't find a job for various reasons, so I ended up going to work in a place where they designed turbines, hydraulic turbines. That was in 1973. Eventually it turned out to be blessing that they did not hire me in any of the prestigious places. If they had hired me in one of those places, I would never have been allowed to leave Russia. In 1979, my wife's family decided to emigrate, and I was faced with a decision myself because that year they were leaving. We had to decide what we wanted to do, and my wife convinced me that it was time for us to go."

A Wave Two interviewee who emigrated as a teenager recounted how anti-Semitism impacted her family's life in the Soviet Union and ultimately their decision to leave: "The way the system worked in Russia was that, after you graduated, you were placed into research universities. In some fields, Jews were essentially prohibited or there were quotas for certain research universities that they could enter. So my dad, being Jewish, couldn't choose his position, and he was basically sent off to a certain research university. On the other hand, my mom, who was Russian, had a choice of where she could go." What we heard most often were patterns of restrictions on professional assignments or achievements based on being Jewish. These patterns of discrimination were the basis of the refugee status awarded to these individuals.

The decision to emigrate for other Wave Two interviewees was a matter of survival amid the institutional collapse and economic hardship resulting from the dissolution of the Soviet Union. While many in the United States viewed this dramatic political event as the collapse of America's main adversary on the world stage and the end of an authoritarian, repressive state, this was also a period of extreme hardship for the technical and scientific intelligentsia, as well as for the general population. Once it became legally possible to emigrate, many did so, joining the continuing stream of Jewish emigration. In total, this period of emigration constituted a huge brain drain of Soviet-trained mathematicians, scientists, and researchers.

Vadim Gladyshev, who came in 1992 during Wave Two under a US National Institutes of Health grant and who is a Professor of Medicine at Harvard Medical School and Director of the Center for Redox Medicine, described the choices available during that period: "At the time when I was finishing graduate school in Russia, society was rapidly changing. It was the time when the Soviet Union had just collapsed and there were, I guess, three options for young scientists like me. One option was to stay in science in Russia, but then you could not do much because there was no support for it from the state. Another option was to leave science and move to business, and the third option was to leave the country. I just loved science so much, there was not really a choice for me among these options." Beyond the love of science, there was also the harsh reality of finding a way to support oneself that pushed people in science to look abroad.

Patent attorney **Maria Eliseeva**, who came in Wave Two as a physics PhD candidate, described her application to US universities as "a way to enhance my education and educational opportunities." She applied to dozens of universities, and four sent her admission letters, but three declined her admission because she did not have money for the application fee: "At that time in Russia, I had a teaching assistantship and my salary at the Academy of Sciences was maybe twenty bucks a month, and there was absolutely no way to pay for anything. So I took the highest scholarship offer and that was from Ohio State University, and I'm very grateful." Along with the immediate economic pressure and hardship were considerations of the future that also pushed many others to emigrate. In Eliseeva's case, she came to the conclusion from history books she had read that democracy and peace would not come to Russia soon. She reflected: "You can go back in history to maybe the Spanish Inquisition, and it was about, essentially, killing the prideful and the successful. It doesn't matter what the ideology was, it can be communism, it can be Christianity, it doesn't matter. They scoop away this fertile sector of the society – people who are educated, who are smart, who want to do something, who are entrepreneurial. It takes a long time to develop fertile soil and grow new grass and trees, and it's the same with the Soviet Union. I grew up in that society that essentially eliminated the best and the brightest, you know, either physically or by making them emigrate. Then you had World War II, which not just decimated, but eliminated, so many of the bright and young. And this fertile layer of the soil was completely eliminated. It can't regenerate just because we want it. It takes time. Even under the best of circumstances, it takes time. You first get grass, then bushes, then trees, and it takes probably a generation at the very, very least, or maybe more. Not that I'm a pessimist, but I have only one life."

While the crisis of the time was devastating for many, it also opened opportunities for people who emigrated at a young age with their parents. **Andrey Kunov,** CEO of the Silicon Valley Innovation Center, came during Wave Two to Stanford University in 1994 at the age of twenty-two from Kazakhstan. He described the contradictions of the times for his parents and for him personally: "Of course, the collapse of the Soviet Union was a pretty dramatic event for them. It was a huge opportunity for someone like me, their son. But for them it was like a crumbling, something that basically devastated their lives. They were like many, many other Soviet citizens who had invested the better part of their life into building that country, and all of a sudden it just fell apart right in front of their eyes. So they were looking at that differently than I did. I mean, I didn't really look at it in any way at that point, ideologically, I would say. Well, I guess I did actually. I had my opinions about that. But they, my parents, of course, were much more – how should I say that – were much more practical. It was sort of like an investment, right? They were not ideologically motivated to uphold the communist ideology in Soviet times, though my father was a member of the Communist party. It was a habitual thing for many people to have to do if you wanted to advance your career. But for them it was more, you know, like a house that you built for the first part of your life so that you could enjoy the second half of your life. And then, all of a sudden, you end up having no roof and no walls. And they lost all of their savings, of course, during the early Gaidar, or even pre-Gaidar times."[2]

The economic pressure was even worse in other newly independent republics. **David Gukasian**, who came during Wave Two from Armenia in 1995 at age seventeen and is Senior Systems Administrator at Battery Ventures, described those times. He referred to how his life in urban Yerevan took a terrible turn: "Things were not good at all, and when I say not good, let's say we had electricity maybe one hour a day. I'm talking about the winter. My friends and I, we actually had to go to the forest to chop some wood to bring it back to burn in the oven to make some food. It was just really bad. That was because of the collapse of the Soviet Union, the economic collapse, and the huge inflation that happened in 1992. At that time, you know, the entire ruble currency just basically crumbled because of the war and because of all kinds of conflicts. Just before that, in 1988, Armenia suffered a pretty big earthquake which damaged the majority of the infrastructure. Armenia is a pretty small republic so such a blow was devastating."

As political institutions strengthened in the Russian Federation, supported by rising world oil prices, the mindset of many emigrants in high technology also changed. In many ways, this period began to conform much more closely to the pattern described by Saxenian[3] regarding Indian, Chinese, and Taiwanese technical professionals in Silicon Valley. Of the former Soviet republics, there was a strong impetus in Russia to integrate into the global technology and innovation sectors. The notion of a "brain circulation" replacing the "brain drain" of the 1990s became a popular idea in technology circles. This concept was supported by corresponding institutional exchanges with counterparts in the United States and other countries. Still, Wave Three continued to see emigrants from the former USSR. This change is exemplified by the experience of **Nikolay Vasilyev**, pediatric cardiac surgeon and medical device entrepreneur at Children's Hospital in

---

[2] Yegor Gaidar was a Russian economist who served as Acting Prime Minister of Russia from June to December 1992 and was instrumental in implementing "shock therapy" by removing price controls and state subsidies.

[3] AnnaLee Saxenian, *The New Argonauts: Regional Advantage in a Global Economy* (Cambridge, MA: Harvard University Press, 2006).

Boston: "I finished my residency and then started working at the Bakulev Center in Moscow. I was looking to see where I could learn more about cardiovascular surgery. I was looking all over the world, at the US in particular. I applied to several places and I got a reply from the Cleveland Clinic from the Chief of Pediatric Cardiovascular Surgery there at that time. I accepted and went there. It was 2003. As a medical doctor, as a practicing physician trained in Russia and as a researcher trained in the US, I have had a multicultural, and a big-picture view. I've seen how clinical and basic science research is done in Russia, I've seen it a little bit in Europe, and I've seen it in the premier institutions in the US. I was also lucky to have great teachers who taught me how to look into a problem, how to create a hypothesis and how to run an experimental study. So I think this multi-angular perspective, if you can say it like that, was quite unique."

Institutions: Capability to Add Value

We have examined the institutional backgrounds that our interviewees brought with them that motivated them to leave but that also facilitated their ability to make contributions in the United States. We acknowledge that the people we interviewed are unique, highly intelligent people whose personal attributes account for much of their success. It is important to also acknowledge certain institutional factors in the former USSR that have enabled their drive and capabilities to be effective. The primary background institutional support for their capability to add value to the US innovation economy is the Soviet educational system.

The rapid, forced industrialization of the Soviet Union required large numbers of engineers and a robust system of education in mathematics and science. In the Golden Age of Soviet Mathematics in the 1960s, some Soviet universities became world leaders in those fields. Many of the brilliant mathematicians trained within this system emigrated under the pressure of anti-Semitism and then the collapse of the Soviet Union, making way for them to contribute to the US technology sector. Many non-Jewish scientists and researchers later followed in their wake. Ultimately, the United States benefited from the excellence of the educational institutions of the Soviet Union.

Leonid Raiz spoke about how his mathematics training in the USSR led to his ability to contribute to the technology sector in the United States: "None of my experiences in the style of working were applicable. But my understanding of mathematics and understanding of software development, that was all applicable. As a result, my strength always was having the general idea of how software should be organized and work. When given a task, I didn't just do the task, I tried to understand the context. Typically, I ended up kind of figuring out a bigger system, how my portion fits into it, and quite often adjusting the bigger system to make it more logical and coherent. Back in Russia, after university my work was very close to what here is called computer-aided design. So, when I came here, my first job was with a company called Computervision, not exactly but very close to what I had been doing in Russia. With Computervision, I kind of pioneered one of the modern approaches for what they were trying to do. Shortly after that, I started moving between various companies. I spent two, three years at Computervision, then there was a company called Prime Computer, and from Prime Computer I joined the group that was starting the company which became PTC."

Harvard Medical School Professor **Vadim Gladyshev** reflected on the positive aspects of Soviet education that contributed to his later success: "Often, the Soviet Union is demonized. But when I grew up, I really benefited from the Soviet system. In school, there were all sorts of support and activities. You could study any subject or take after-class activities for free. For example, I studied chess, and we played in competitions all across the country. We stayed in hotels, and the government paid for all our accommodations, meals, and travel, everything. So, in that sense, there was significant support from the government. The university was also free, students received stipends, which were higher if you studied well, and the dorms were almost free. So I would say that I really benefited from that support for education."

Imprinting: Impact of Institutions Old and New

The imprint of prominent features of an institutional environment has a deep influence on formation of an individual's attitudes, values, standards, and forms of behavior, as described in Chapter 1. Usually, this basic imprinting takes place from ages six to twenty-five. Our interviewees began this process of imprinting in one environment and then emigrated into another, and, for some, their experiences were also affected by the breakup of the USSR before they emigrated. As emigrants, many had staged some kind of internal rebellion against the dominant institutional environment prior to leaving. Research, again as noted in Chapter 1, has identified conditions under which individuals might replace all or part of their earlier imprinting. Clearly, a number of the people interviewed for this book did make fundamental changes in their worldview and patterns of behavior in the process of moving from the former Soviet Union into the US technology sector. Most presented a combination of influences imprinted from their pasts layered with the influences of their present environment. Thus, as in many circumstances, generalizations about individual personalities based on place or culture of origin is inappropriate. At the same time, we were able to capture significant impacts of imprinting that provide some general understanding of the three waves of immigrants interviewed.

Imprinting: Soviet Influences

Some areas of influence imprinted from the Soviet institutions were a significant backdrop, including attitudes toward business, truth, science, trust, and creativity. A number of interviewees spoke about the fact that not only was private business illegal in the USSR, but it was also culturally denigrated. Third Wave entrepreneur **Mikita Mikado Teploukho** noted that the same attitude prevailed in Belarus even during his childhood in the 1980s, when opportunities for entrepreneurial activity began to be allowed: "As a kid, I perceived selling as something ugly, dirty, and totally not cool. Coincidentally, those are the same 'fine' words society attributed to salesmen in the Soviet Union." Teploukho learned this firsthand, having sold his family's produce and his grandfather's home brew at markets as a child during hard times.

The divergence between official ideology and actual practice in Soviet society also led to a constant struggle for individuals to find the boundaries between truth and fiction. **Eugene Shablygin**, a successful entrepreneur in both post-Soviet Russia and in the United States, discussed the issue of truth in general and the impetus toward the study of science as an alternative to politically dominated professions: "Russian society is very interesting because, in many respects, it's based on constant lying. Yeah, you constantly need to lie, all the time. You need to think one thing, do another thing, and say the third thing, and they're totally different. So the only way you can do something, and think something in the same area is when you do something in natural science."

Thus the institutional environment, dominated by communist ideology, hypocritical avoidance of that ideology, and authoritarian institutions, led to a complicated imprinting in the area of trust. In Soviet society, reliance on family, school friends, and trusted acquaintances was crucial, while suspicion of strangers was the norm. Wave Two interviewee **Eugene Shablygin** stated directly: "But the biggest, biggest challenge is that the Russians, unfortunately in many cases, do not trust themselves. And especially they don't trust each other." At the same time, these immigrants have created new Russian-speaking cultural, civic, and business organizations in the United States, similar to patterns of other immigrants in professional, cultural, and educational institutions.

The imprint from the trials and tribulations of navigating complicated bureaucracies and hypocritical institutional structures is also seen in a certain kind of creativity that is often mentioned regarding Russian programmers, researchers, and scientists. Eugene Shablygin expressed the opinion that "Because Russians have an enormous amount of creativity that has been suppressed by the thousand years of horrible traditions, the only place where they can really apply this creativity is outside of Russia."

Nikolay Vasilyev, pediatric surgeon and entrepreneur, elaborated on this theme, illustrating the impact of institutional imprinting on the development of creativity: "Again, it's very subjective because it would be easier probably to ask people who work with me. But I think that through our Russian history and through my personal history going through the *perestroika* time and some other difficult times of my country, I think what I bring and the Russian-speaking scientists and physicians also bring, is we always try to think about the problem outside the box. We also are trying to find solutions that are not typical because we are used to that in our personal lives, in our country's history. In our lives, we had scenarios in which we could solve the problems in the regular way. But we didn't have enough resources, we didn't have enough manpower or financial resources. Yet we still solved the problems successfully. So I think that is one of the key points: we always think broadly – at least I can speak for myself – and try to solve the problems not like head on, but rather assess the big picture, assess the problem from various angles, and then sometimes find a solution that is not obvious. That is one of the biggest advantages."

Imprinting: Transition to Entrepreneurship and the US Innovation Economy

The strengths and weaknesses brought to the United States from the institutions and imprinting of their homelands were predominantly modified and adapted to life in the United States generally and specifically to the practice of entrepreneurship as well as working in technology. **Leonid Raiz** spoke about the fundamental adjustment he had to make in the United States: "One of the first things which I kind of figured out for myself, and that was one of the big shocks for me, was that back in Russia everything is predetermined. You are born, you go to school, you work, and so on. You don't really have to think much. One of the big shocks for me initially was how much I had to think for myself, how many decisions I had to make here. I had trouble with even the simplest decision, buying my first car, which was a used car. Which one to buy?"

The entire meaning of business in the United States was vastly different for most. Wave Two Armenian immigrant, **Nerses Ohanyan**, Director of Analytics at Viki, spoke about this paradigm shift: "Yeah, I never thought about business because the word for business in Armenian means to buy and sell. So for me, business was really about this idea of arbitrage, right? Where you buy cheap, sell expensive, and you're essentially trying to cheat people out of money. That's the mentality that I had, where you're essentially trying to extract value out of something that isn't there to just buy and sell. This was kind of – I don't want to say – it may sound weird in English – but more of a dirty thing for me. It wasn't my thing. I wasn't interested in it all. That's not something that ever attracted me for a long, long time."

In contrast, the institutional imprinting in the United States promotes the idea of entrepreneurship as a social good, quite the opposite of the Soviet imprinting as expressed by Ohanyan as well as by Wave Three interviewee **Davit Baghdasaryan**, Senior Security Engineer at Twilio, also formerly from Armenia. He remarked: "I think it came from here, from this environment after moving to the US, perhaps. I haven't thought about that question. I'm pretty sure that probably something existed in me because not everyone has that spirit, right? But

I think it's been accelerated being here because, you know, with the media helping entrepreneurship, presenting it in a way that people are getting excited about it, for good or bad reasons. I can't really judge, but it makes people think like that. The media is playing a big role in that."

Another Wave Three interviewee, **Shalva Kashmadze**, who is Product Manager at game developer Pocket Gems, spoke about the impact of university and specifically Stanford's imprinting on him the ideas of entrepreneurship: "I think it's probably Stanford. It's mostly Stanford because, before I came to the United States, I think growing up in Georgia in a family that had to deal with all sorts of financial hardships, all these things made me kind of risk averse. All these things made me really oriented toward getting my predictable salary in my pocket and to contribute to my family because these are the things I had to do. I had to help my family, I have to help my parents, I have to be this financially responsible guy that kind of builds the family future. So, before I came to the United States, I never would have begun a startup in Georgia because it's just a really hard environment to do it, but also because it would have been a real risky move for me. I would never have done it, but Stanford taught me. I think it gave me a bigger sense of confidence. It taught me a totally different attitude and totally different approach toward failure, what failure is and what it is not. It gave me more sense of financial security and gave me those basics that right now help me to take the next risky move."

Alexander Vybornov, a Wave Two immigrant and Product Line Manager of Medical Products at IPG Photonics, discussed the dynamics of the different institutional imprinting and the role of money as an incentive in the United States: "You kind of know that in US society, it offers a much higher potential. In Russia at that time, most of the work in academia was driven by state and government contracts, which tended to be very long and very solitary, individual work. You had people working for ten, twenty, thirty years on the same project. And at the end of the day, these projects either succeeded or they failed. And I think there's also the inherent desire to see the results of your work measured not only by recognition or publications, but also money. Money, not in the sense of things you can buy, but as a very nonsubjective indicator of success. So I think it's just that fundamental difference between the noncapitalistic socialists and the capitalist structures. It's very enticing."

Identity: Soviet, National, or Ethnic

We presented in Chapter 9 the complicated question of identity and how that concept has developed among the people interviewed. The Soviet Union as the successor state to the Russian Empire contained within it the complexity of numerous nationalities and religions. There was a long history of conquest, antagonism, and politically instigated violence from czarist times that presented a set of historical and practical problems for the new Soviet state. The Bolsheviks tried to manage these inherited problems by creating territorially based political subdivisions of the state, primarily the fifteen national republics of the Soviet Union; suppressing organized religion; and creating an overarching Soviet culture encompassing all citizens.

One of the principal failures of this approach was in the treatment of the so-called "Jewish Question." The Bolsheviks treated Jews as a nationality, and they were identified as such in their passports. The formal resolution of the "Jewish Question" was to designate a province in the Russian Far East as the Jewish Autonomous Region. This was an attempt to consider Jews as a nationality with their own homeland, the same as the homelands established in the Soviet Union for other nationalities. Even superficially, this solution did not correspond to the reality of Jewish life, history, and places of residence in the Russian Empire. Very few Jews actually lived in their designated "homeland." The majority lived in European Russia or Ukraine. However, based on the Soviet system for designating nationality, Jews living in Ukraine, for instance, did not have Ukrainian as their nationality. Their nationality was Jewish, within the general Soviet State.

The official ideology of the Soviet State was based on equal treatment for all nationalities. However, this was true for only limited times and limited purposes. The designation of nationality on passports was more often a means of identifying Jews for discrimination rather than for ensuring equal treatment. This disparity between official policy and actual practice had a great impact on many interviewees, particularly those from the First Wave and early Second Wave who were primarily Jews.

Having been identified as Jewish by nationality, but having practically no religious education or traditions due to the secular and anti-religious policies of the Soviet Union, many interviewees have a very different concept of identity as Jews when compared to Jews living outside the Soviet experience. Once living in the United States, they were free to define this identity in any way they chose, a freedom that led to many variations from the same starting point, as described in Chapter 9.

Maria Eliseeva, a Boston patent attorney, discussed the complications and political consequences of designated nationality in the Soviet Union: "I have a half-Jewish, half-Russian mother who designated herself as Russian even though she had a Jewish last name. I was designated as Russian because of my grandmother and her mother, despite having a Bulgarian father and a half-Jewish, half-Russian mother. For political reasons, everyone who could be designated as Russian did so. On the other hand, as we say in Russia, there is your motherland and there is the government, and these are different things. So, I think we will feel good feelings about the motherland, but not necessarily feel the same about the government."

The devastation and suffering endured by the Soviet people as a result of World War II, or the Great Patriotic War as it was referred to in the USSR, was a shared experience that the Soviet government drew on to help create a common Soviet identity. A Wave Three executive gave a view of its pervasiveness and impact: "As a Soviet child you grow up with the Great Patriotic War being part of our national identity. You grow up honoring that event in Russian history. May 9 is a hugely celebrated day for any Russian person or Soviet person, where you honor the veterans who freed the country. The thing is, everybody in the Soviet Union had someone who died during the Great Patriotic War. I believe 20 million people died. Growing up as a child this was a fact, that we had had such a war. You always have to be alert; you have to understand that things can change to hard times immediately. And, as a child, we knew about what war is because throughout school you learned about the Great Patriotic War, about how we suffered, and about how Soviet soldiers died."

We have sought to develop an appreciation for the concept of identity and its role in immigration and social interaction. Our interviewees have shown a complex interplay of factors shaping identity that include ethnic or national background, family values and beliefs, stated societal values, observed societal values, and expressed attitudes toward the political environments in which they reside or have resided. The identity that they brought with them led to many successes, but inevitably involved adjustments in self-concept and behavior in both work and social environments. Based on their statements, we can see a variety of approaches to this adjustment. These in turn relate to social adjustment as detailed in Chapter 7, soft skills in business relationships as described in Chapter 8, and, ultimately, to one's personal identity, the subject of Chapter 9.

Identity: Transition and Change in the United States

The identities that our immigrants brought with them became subject to challenge, change, and choice in the United States. What we saw from our interviews was a diverse set of choices of identity that likely came from their common experience but reflect widely different responses. For instance, we found Jews who remain identified with being Jewish as a nationality, some of whom continue to see themselves as part of a broader Russian-speaking cultural community. Some Jewish immigrants have become more religious as they embrace the tenets of traditional Jewish worship and spirituality, while most identify in a nonreligious way. Some Jewish immigrants have chosen integration into broader American society, acknowledging their Russian and Jewish roots but not necessarily as primary considerations in their life. Some non-Jewish immigrants have worked very hard to protect what they value in Soviet or Russian culture and impart that value to their children. Others have seen their children become increasingly American. And some of our interviewees have identified themselves as internationalists or global citizens.

Wave One immigrant and venture capitalist **Michael Schwartzman** described how, in going through the emigration process open primarily to Jews like himself, he had been perceived as an example to his colleagues back in the USSR: "In 1973, when I was applying for the exit visa, I was the first academic to do so in the massive 35,000-people scientific establishment of Akademgorodok ["Academic City," near Novosibirsk, Siberia]. The eyes and ears of hundreds of scientists in Akademgorodok were firmly centered on the travails of my exit process that took about three intense months and five years of meticulous preparation and planning. Dozens of would-be emigrants, some of whom would follow me, observed my exit process and its results, deciding whether they wished to be the next ones, meaning the next ones to go to prison, if I went there; the next ones to get fired, if I got fired; the next ones to be lucky and get a visa, if I got one, and so on. I needed to plan and foresee the reactions of many institutions and entities: the KGB, the regional and local Communist Party ruling committees, the scientists I worked with and for, the administrators of the research facilities I worked at, the university – my alma mater – where I was teaching, and, last but not least, the reaction of the Academy of Sciences in Moscow, including the decisions of the head-quarters of the KGB in Moscow, and the overall management of the emigration issue by the Communist Party. It was a full plate for me."

Other interviewees described how they came to embrace their heritage and pass it on to their children. **Tatiana Kvitka** of Wave Two who is CEO of Design by Light, provided an example of Jewish self-identification in her family: "My children, surprisingly enough, want to associate themselves as Jews. Whenever it comes up, they say, 'We're more Jewish than anyone.' So, I don't know where that comes from. But they kind of heard that if your mother is Jewish then you can become Jewish, and so they kind of take it as 'If we were to choose, we would probably be Jewish.' But, at this point, none of them is religious."

Others focused on instilling a Russian identity in their children. **Alexey Wolfson**, a biotech entrepreneur also from Wave Two, described how he and his wife provide opportunities for their children: "So once a week they have Russian school, and it's very long and intensive. The mental exercise the kids are getting at this school is absolutely wonderful. The teacher there is building the Russian community there. It's not so much writing in Russian; it's about Russian culture. As a result, the kids are engaged in the community in a way. They also know Russian literature. They are reading the Russian classics of the eighteenth and nineteenth centuries. They are performing classical Chekhov plays, which gives them exposure to the deep roots of Russian culture."

Alexander Vybornov, Product Manager of Medical Products at IPG Photonics, expressed a different view in choosing to embrace broader assimilation into US society: "For me, and I would say it's probably true for most immigrants I've observed, normally in the Russian enclave, like in others such as Asian or Indian communities, they often tend to stay together. So these communities frequently become enclosed. Reflecting backward, I think that one of the keys to success is breaking out of that little group and integrating into the broader society. Those people who can make the transition, I think they ultimately become a success. And those who stay in that little enclave, well it depends on how you judge success, but professionally I think it's a very important factor that I learned. I think it was more of what my father taught me from the very beginning, and actually something that you learned as a Jew in Russia, that the bar is always set a little higher. So, in some ways, not to get off on a tangent, but it's somewhat similar to how you see some of the minorities in this country, for example if you're African-American. The official storyline is that everyone is equal, but the bar is always set a little higher. It's a lesson that I learned and have carried with me from the very beginning of my time in this country, that I always knew I had to reach a little higher. This translates in all of the basic rules of engagement, whether it's your hygiene, how you always smile in a conversation, how you exhibit high work ethics, and how you try to underpromise and overdeliver."

Presenting an even broader view of identity evolution is **David Gukasian**, Senior Systems Administrator at Battery Ventures: "I would probably say I'm a citizen of the world. I don't really claim any specific nationality or say that I belong to this political party or to that belief. I just see advantages and disadvantages from everywhere. Everything is good and bad. It's all about yin and yang. It's finding the right balance among things. Would I go back to Armenia? I don't think so. I was in Armenia in 2011, and I found my mentality is just very different from theirs. We speak the same language, but I felt like I came from Mars."

We close this section by emphasizing that the identity complexity experienced by virtually all of the interviewees is not necessarily negative, as evidenced by the insight expressed by MIT professor **David Gamarnik**: "I'm Georgian, I'm Jewish, I'm Russian-speaking, and I'm an American citizen. So my identity is complex. And I'm proud of it."

Some Conclusions: Innovation and Immigration in the Transition from the Industrial Era to the Knowledge-Based Economy

The waves of immigration from the Soviet Union and successor states from the 1970s through 2015 have played an important role in the broad movement of brainpower and technological skills from around the world into the United States. This movement began with the passage of the Immigration Act of 1965, which opened the doors to recruitment of talented people from all over the globe. The removal of the old national quotas and the institution of work-related, family reunification, and refugee visas made this process possible. This new wave of immigrants helped fuel a period of booming innovation and technological change in the United States, led by innovation clusters in Silicon Valley, the Boston-Cambridge area, and other regions of the United States. The policies and practices of this era of openness are now under challenge in the late 2010s, making an examination of their impact all the more important.

The United States has been a magnet for aspiring technology entrepreneurs, world-class researchers, and top scientists. Whether the country continues to attract the same volume and level of technology professionals is an open question for a variety of reasons, not the least of which is the recent anti-immigrant tenor of the country in general. Yet there is a growing recognition around the world that the transition from primarily industrial production to knowledge-based economies depends on the availability of human resources. This has been leading to increased competition among countries to attract or retain talented individuals. For America, the competition from countries like China and Russia is being financed through generous funding by their governments for improving and modernizing universities, as well as by grants for talented researchers and programs to develop ties with diasporas of their citizens living in other countries. With regard to Russia and other countries of the former USSR, it is unlikely that we will see such a large-scale transfer of brain power to the United States any time soon, absent a general political collapse of the Russian state or other countries of the former USSR. Yet the ongoing economic and political crisis in Ukraine will provide continuing impetus for potential emigration from that country.

At the same time, we expect that the contributions of the immigrants we interviewed, as well as those of their counterparts, will continue to grow over the coming years. In this respect, the contribution to the United States from the brain drain associated with the collapse of the Soviet Union will continue to grow in importance. The example of this community of immigrants and their undeniable contributions to innovation, technology, and job creation is powerful support for continuing an enlightened approach to US immigration policy that would encourage others like them from countries of the former USSR, as well as from other parts of the world, to come to America.

This scenario raises the main question of the willingness of the US government to continue to accept and absorb immigrants who historically have contributed so much to American dominance in the global technology and science sectors. We hope that this book and the stories related by our interviewees will provide support for immigration policy driven by deeper insights into the competitive considerations so important to sustaining and strengthening the US innovation economy.

Index

396